Energy, Physics and the Environment

Third Edition

Ernie L. McFarland
James L. Hunt
John L. Campbell

University of Guelph

CENGAGE Learning™

Australia • Brazil • Japan • Korea • Mexico • Singapore • Spain • United Kingdom • United States

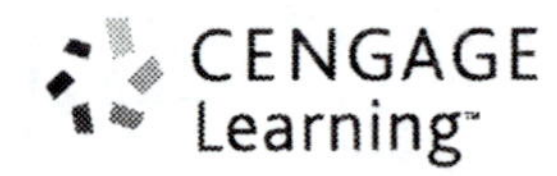

Energy, Physics and the Environment
Third Edition

Ernie L. McFarland
James L. Hunt
John L. Campbell

University of Guelph

Executive Editors:
Michele Baird
Maureen Staudt
Michael Stranz

Project Development Manager:
Linda deStefano

Senior Marketing Coordinators:
Sara Mercurio
Lindsay Shapiro

Production/Manufacturing Manager:
Donna M. Brown

PreMedia Services Supervisor:
Rebecca A. Walker

Rights & Permissions Specialist:
Kalina Hintz

Cover Image:
Getty Images*

* Unless otherwise noted, all cover images used by Custom Solutions, a part of Cengage Learning, have been supplied courtesy of Getty Images with the exception of the Earthview cover image, which has been supplied by the National Aeronautics and Space Administration (NASA).

ISBN-13: 978-1-4266-2433-9

ISBN-10: 1-4266-2433-6

Cengage Learning
5191 Natorp Boulevard
Mason, Ohio 45040
USA

Cengage Learning is a leading provider of customized learning solutions with office locations around the globe, including Singapore, the United Kingdom, Australia, Mexico, Brazil, and Japan. Locate your local office at: **international.cengage.com/region**

Cengage Learning products are represented in Canada by Nelson Education, Ltd.

For your lifelong learning solutions, visit **custom.cengage.com**

Visit our corporate website at **cengage.com**

Printed in the United States of America

PREFACE TO THE SECOND EDITION (2001)

The last quarter of the twentieth century has seen growing concern among the public, in the media, and in the scientific community about humankind's dependence upon ever-increasing use of energy.

Concerns about energy supply were focused sharply by the "energy crisis" of 1974, when the suppliers of much of the world's oil increased prices by a factor of four. While that event caused significant economic dislocation, it provided impetus to the conservation of energy in industry, commerce, transport and the home. More efficient processes of production, better building insulation, and automobiles with improved fuel consumption all appeared rapidly, as the realization that energy supplies were finite took hold. Alternative energy sources have been developed, although at a slower rate than might have been expected. As this new edition goes to press, the term "energy crisis" has re-appeared, and the issue of alternative sources again becomes large.

A second area of concern was the impact upon the environment both by the extraction of fuels and by their subsequent consumption. Mine tailings and wastes were now seen as more than unsightly, and the issue of visible degradation in formerly pristine areas of natural beauty arose. The ozone layer was observed to be shrinking, thus increasing the exposure of people, plants and animals to hazardous ultra-violet radiation. The debate over the global warming, predicted to result from increasing carbon dioxide in the atmosphere, is a highly charged one in both the scientific and political arenas. In an increasingly global economy, it is now realized that the effects of energy consumption are also global.

The need for a quantitative approach to these matters is obvious. The great physicist Lord Kelvin wrote: "I often say that when you can measure what you are talking about, and express it in numbers, you know something about it". Citizens and decision-makers must be provided with more than half-researched media articles if reasoned decisions are to be made.

We introduced the first edition of this book in 1994 to provide a quantitative account of energy and related environmental issues for university students in science who have a first-year preparation in Physics. Many alternative texts then available provided excellent qualitative approaches, directed mainly to students in the humanities and social sciences. Our objective was to deal with the numbers involved, in a manner that would help the reader develop a quantitative grasp of the various dimensions of the overall energy issue.
....

At our university, a large number of students majoring in Biological and in Physical Science have taken the course and it is a part of our recently introduced degree program in Environmental Science. We sought and received valuable feedback from many students in order to assist us in developing the third edition, and we express our thanks to them.

As with the first edition, it is not our aim to cover the subject exhaustively; that would not be feasible in the standard North American one-semester course. The book is a foundation. We hope that its readers will be stimulated by it to continue their reading, and to expand their knowledge of the issues. With such a foundation, the reader will be able to profit from the excellent in-depth articles upon energy and environment that appear from time to time in widely available journals such as *Scientific American, Science, The New Scientist,* etc.

We certainly do not claim that energy issues can be dealt with solely through a quantitative understanding of the basic science involved. Decisions upon energy matters will be made, as they are now, on the basis of a complex set of considerations, including aesthetic, geographic, economic, political, health, and social factors. However, we continue to assert that the quantitative dimension, too often lost in emotional public debate, is indispensable as background knowledge.

Guelph, Ontario
June 2001
E.L. McFarland, J.L. Hunt and J.L. Campbell

PREFACE TO THE THIRD EDITION (2006)

The years since the Second Edition of 2001 have been particularly turbulent in the energy sector which even in "normal" times is rapidly changing and developing. After the new developments, and sometimes wild speculations, spawned by the "energy crisis" of 1974, interest in alternative and renewable energy sources waned in the wake of the oil glut of the 1980s. Several things of global consequence have changed this apathy in the last decade. Increasing political crises in the Middle East impinging on the global oil market, the recognition by almost every one of the fragility and possible permanent damage to the planet's atmosphere, and the obvious rapid growth in the use of private automobiles, being just three.

This almost revolutionary change in the energy sector has meant that, for this new edition, there must be substantial change. Some possibilities that looked promising in 2001 are now off the table and some technologies that were barely noticed have come to the fore. Some energy sources that looked utopian in 2001 have since been seen to have their "dark side". All of this has meant that certain sections of the book have had to be substantially rewritten particularly the material on fossil fuels and on renewable resources.

We have also taken the opportunity to re-arrange the material in what we think is a more rational order.

1. The material on Heat Transfer seemed not to be terribly relevant until the discussion of Conservation so it now appears as Chapter 16 rather than 5.
2. A short discussion of 3-phase power has been added to the chapter on AC electricity.
3. The Chapter on the automobile, which is a "use" not a "source" of energy, has been moved much later in the book and the section on conservation in the transport sector added.

4. The former two chapters on radiation and its biological effects have been combined.
5. The chapter on Energy Conservation has been distributed among the various relevant technologies.

We have taken the opportunity to correct a few errors and add a few problems but the major objective has been to update the material, certainly qualitatively but also quantitatively in keeping with our credo that the subject is nothing if it is not understood with numbers.
ELM, JLH, JLC
August 2006.

TABLE OF CONTENTS

CHAPTER 1

ENERGY AND POWER IN TODAY'S WORLD

An enormous amount of energy is used by humanity on a daily basis around the world, with unfortunate environmental consequences such as smog, acid rain, and global warming. This first chapter explores the energy sources that are employed at present, discusses the ultimate fate of all the energy that is used on Earth, and provides quantitative measures of how much energy is used from a variety of sources.

1.1 SOURCES OF ENERGY

Various sources of energy are available to the human race for day-to-day functions such as cooking, transportation, and heating. It might seem at first glance that there are many energy sources. For instance, buildings are often heated with oil, gas, wood, solar energy, or electricity generated from coal, gas, falling water, nuclear energy, or wind. Nonetheless, we will see that at the most fundamental level, there are only three energy sources available on Earth. In the next few pages, we give a brief introductory discussion of energy sources, all of which will be described in more detail in later chapters.

Fossil Fuels

About 90% of the energy used in the world today is provided by fossil fuels: oil, natural gas, and coal. The energy stored in these fuels — often referred to as chemical energy — is a combination of electric potential energy of the electrons and nuclei that constitute atoms and molecules, and the kinetic energy (energy of motion) of these electrons.

Where did this fossil-fuel energy come from? As you probably know, fossil fuels were created by the action of pressure and heat on the remains of plants and animals that lived hundreds of millions of years ago. The fundamental source of energy for these plants and animals was the Sun; plants use the energy of sunlight to convert carbon dioxide and water to carbohydrates (sugars and starches), some animals eat plants, other animals eat these animals, etc. So when we use gasoline to power our automobiles, natural gas to heat our buildings, and coal to generate electricity, we are actually using solar energy that has been stored for a long time.

But what provides the Sun's energy? In the Sun, hydrogen nuclei are fused together in a multi-step process, creating larger nuclei of helium and releasing energy. This process of combining small nuclei into larger ones — *nuclear fusion* — is the subject of much research in the hope of developing commercial fusion-energy sources here on Earth, but this prospect appears to be decades in the future.

Wind and Wave Energy

Although people are now starting to use the wind to generate electricity, it has been used for hundreds of years for sailing-ships and windmills (Fig. 1-1). Like fossil fuels, the wind is a form of stored solar energy: the Sun heats Earth's surface unevenly, thus developing regions of high and low air pressure, and the resulting air movements often cause high winds at ground level. Of course, solar energy is stored in the wind for a much shorter time than in fossil fuels.

(a) (b)

Figure 1-1 (a) In the Netherlands wind has been used for centuries as a source of energy for small industries such as mills. (b) Modern electric-generating windmills.

Another source of energy, derived primarily from the wind, is wave energy. Except for wave-powered navigation buoys, wave energy is not yet economically competitive with other energy sources (especially inexpensive fossil fuels), and small pilot projects are progressing slowly.

Wind and wave energy are both examples of *kinetic energy*, or energy of motion. Mathematically, the kinetic energy (*KE*) of a moving object is defined as:

$$KE = \frac{1}{2}mv^2 \quad \textbf{[1-1]}$$

where m and v represent the mass and speed, respectively, of the object.

Biomass Energy

There are many different energy sources that are categorized as biomass. Biomass refers to carbon-based material produced by plants and animals in recent years (up to decades). The largest contributor at present is wood, which is primarily burned for heating or cooking. Also included in this category are:

- crops such as sugarcane whose sugars can be fermented to produce ethanol, which can be used as a fuel;

- manure or garbage which produce methane gas through anaerobic digestion by bacteria;
- woody biomass which can produce either combustible gases (methane, hydrogen, and carbon monoxide) through a gasification process, or methanol, which is a liquid fuel.

Since all biomass results directly or indirectly from photosynthesis, biomass energy is yet another form of stored solar energy. Biomass provides medium-term storage of solar energy, whereas wind gives short-term storage, and fossil fuels represent long-term storage.

Hydro Energy

Another example of short-term stored solar energy is falling water, often called hydro energy. The Sun evaporates water from bodies of water and soil, the water forms clouds and eventually falls as rain, snow, etc. Where local geography happens to provide a large drop in elevation, then there is the possibility of using falling water to provide energy, usually for the generation of electricity.

The energy of the water (before it falls) is an example of *gravitational potential energy* (PE_{grav}). As the water falls, this gravitational PE is converted to KE of the water, and in the case of a hydroelectric station, this KE is used to turn a turbine generator, which produces electricity.

The gravitational PE of an object of mass m at an elevation h is defined as:

$$PE_{\text{grav}} = mgh \qquad \textbf{[1-2]}$$

where g is the magnitude of the gravitational acceleration (or gravitational field). Near the surface of Earth, the value of g is: $g = 9.80\ \text{m/s}^2$.

(Notice that when we say that an object has an elevation h, we mean that the *centre of mass* of the object has this elevation; for a symmetrical object such as a sphere or a rectangular box, the centre of mass is at the geometric centre.)

Direct Solar Energy

So far we have been discussing various types of stored solar energy, but of course solar energy itself can be used directly to provide heat or electricity. Solar energy is in the form of electromagnetic radiation, emitted from the Sun in many regions of the electromagnetic spectrum: ultraviolet, visible, infrared, microwave, and radio. The energy of this radiation is quantized in packets called photons.

Nuclear Energy

In common parlance, nuclear energy refers to the energy released in the fission, or breaking apart, of a large nucleus (typically uranium) to produce two mid-sized nuclei. This *nuclear fission*, which occurs in a controlled way in a nuclear reactor, generates large amounts of heat that can be used to make steam and turn a turbine for the generation of electricity or for motive power, as in a nuclear-powered submarine.

Uranium is a naturally-occurring element in Earth, and its formation was not due to the Sun; thus, nuclear energy is the first energy source discussed here that is not derived from solar energy.

Geothermal Energy

Most people are aware that Earth's temperature increases with depth below the surface, and that there are regions where Earth's surface is extremely hot, producing hot springs and geysers. This surface heat can be exploited as an energy source. But what generates this heat? Some elements in Earth — uranium, thorium, and radium, for example — are naturally radioactive and decay by emitting high-energy particles that heat the surrounding material. And so geothermal energy is actually a derivative of natural nuclear energy.

Tidal Energy

The final energy source in this discussion is the tides. In a few locations around the world, the tides are high enough on a regular basis that it is feasible to consider trapping the water with a dam at high tide, and allowing the water to run out through turbines at low tide to generate electricity.

Tidal energy is neither solar nor nuclear in origin, but comes from the gravitational *PE* and the *KE* of the Earth-Moon-Sun system. The gravitational *PE* is associated with the separation of the three bodies, and the *KE* is associated with Earth's rotation and the orbital motion of the bodies about each other.

Electricity

You might be wondering where electricity fits into the energy picture. Electricity itself is not a source of energy — it must be generated by a source such as a nuclear reactor, falling water, or fossil-fuel burning. Electricity is sometimes referred to as an *energy currency*, which (like monetary currency) serves as a medium of exchange between a raw source of energy such as coal and a convenient end-use such as lighting. Another term used to describe electricity is *energy carrier,* that is, something that carries energy from a producer to a user. Another potential energy currency or carrier is hydrogen, which is discussed in a later chapter.

Fundamental Sources of Energy

You can now understand why it was stated at the beginning of this Section that there are only three fundamental sources of energy that are available to us. All the sources that we have mentioned — fossil fuels, geothermal energy, tides, etc. — are derived either from solar energy, or nuclear energy, or (in the case of the tides) the gravitational *PE* and the *KE* of the Moon, Earth, and Sun. Table 1-1 summarizes the connection between the various derived sources and their fundamental sources. Of course, since solar energy results from nuclear fusion, we could classify solar energy as nuclear energy, and list only two fundamental sources. However, because of the importance of solar energy to life on Earth, it has been considered separately as a fundamental source.

Table 1-1
Fundamental Energy Sources Corresponding to Various Derived Energy Sources

DERIVED ENERGY SOURCE	FUNDAMENTAL ENERGY SOURCE
Fossil fuels	Solar energy (nuclear fusion)
Wind	"
Waves	"
Biomass	"
Hydro	"
Direct solar energy	"
Nuclear energy (reactors)	Nuclear energy (fission)
Geothermal energy	Nuclear energy (radioactive decay)
Tides	*KE* and grav. *PE* of Earth-Moon-Sun

1.2 ULTIMATE FATE OF EARTH'S ENERGY

Section 1.1 briefly listed several energy sources. It is important to consider what eventually happens to all the energy that people consume, whether for transportation, heating, communications, lighting, or whatever. The electricity generated in a coal-fired power plant is a representative example.

In the electric power plant, coal is burned in a furnace, thereby producing heat to convert water to steam; some of the *heat is lost* up the chimney. The hot high-pressure steam is used to turn turbine generators to produce electricity, and is then condensed back to liquid water (to begin the cycle again) by being cooled by water from, say, a lake or river. This cooling water is then returned to the lake or river, which *becomes hotter* as a result.

The electricity that is generated is transmitted by power lines to the customer; some of the energy is lost as *heat* in the transmission lines themselves. This lost energy is typically about 10% of the total energy transmitted.

The consumer now uses the electricity in a light bulb, for example. For an incandescent light bulb, only about 5% of the electrical energy goes into light; the rest appears as *heat*. (Put your hand near a light bulb if you don't believe this.) The light is radiated outward from the bulb, and is absorbed by surrounding objects: walls, furniture, people, etc. As a result of this absorption, the surrounding objects *become slightly hotter*.

By considering together all the statements that appear in italics in the preceding paragraphs, you can see that *all* the energy in the coal is eventually converted to heat. This complete conversion to heat is true for all energy processes here on Earth. (As another example, consider the kinetic energy of an automobile travelling down a highway: when the brakes are applied, all the kinetic energy is converted to heat in the brakes.)

And what happens to this heat? It gets radiated away into space as electromagnetic radiation. Any object at a temperature above absolute zero emits such radiation; for objects at typical temperatures on Earth (about 20°C), this radiation is primarily in the infrared region of the spectrum, although there is also some in the microwave and radio regions. This invisible radiation is sent out into space and lost to us forever. In other words, there is only one chance to use the energy available to us, and it is then irretrievably gone.

1.3 UNITS OF ENERGY

One of the unfortunate facts of life in the energy field is the plethora of units in common use. Energy values can be quoted in calories, barrels of oil, electron-volts, British thermal units, etc., and in order to make sense of all these units, you will need to be adept in converting one unit to another. See Appendix I for a link to a Web-based tutorial on this subject if you need help.

In this book as throughout the scientific and technological community, the units used are generally those of the SI (Système International d'Unités). The SI unit of energy may be determined by using the definition of, say, kinetic energy given by Eq. [1-1]:

$$KE = \frac{1}{2}mv^2$$

On the right-hand side of this definition, the SI unit for mass m is kg, and the unit for speed v is m/s. (The "½" is unitless.) Thus, the unit for $\frac{1}{2}mv^2$ is $\text{kg}\cdot\text{m}^2/\text{s}^2$. By convention, this unit is called the *joule* (J), named after James Prescott Joule (1818-1889), an outstanding British experimental physicist who was one of the leaders in the development of the principle of conservation of energy.

$$\text{kg}\cdot\text{m}^2/\text{s}^2 = \text{joule (J)}$$

Although this unit was developed using kinetic energy, the same unit arises regardless of the type of energy considered. You might wish to use the definition of gravitational potential energy given in Section 1.1 to show that its SI unit also works out to be $\text{kg}\cdot\text{m}^2/\text{s}^2$.

Another energy unit is the calorie (cal), originally defined as the heat required to raise the temperature of one gram of water from 14.5°C to 15.5°C. The calorie was introduced as a unit of heat about two centuries ago, when it was thought that heat and energy were different quantities. Today the US National Bureau of Standards defines the calorie in terms of joules, with no reference to the heating of water:

$$1 \text{ cal} = 4.184 \text{ J (exact)}$$

You have undoubtedly encountered another type of calorie in connection with food consumption: the food calorie, or Calorie (Cal). (Notice the upper case "C.") This Calorie is 1000 cal, or 1 kcal (kilocalorie), which is equivalent to 4184 J, or 4.184 kJ.[1]

$$1 \text{ Cal} = 1 \text{ food calorie} = 1 \text{ kcal} = 4184 \text{ J} = 4.184 \text{ kJ}$$

In North America, food energy is still commonly measured in the archaic Calories, but civilization is proceeding at a more rapid pace in some other countries, where joules are being used (Fig. 1-2).

Figure 1-2 SI units are taken seriously in Australia.

Another unit of energy frequently used in North America is the British thermal unit (Btu); furnaces and air conditioners are often rated in terms of how many Btu of heat energy they can supply or remove per hour. The definition of the Btu is similar to the original definition of the calorie: 1 Btu is the amount of heat energy required to raise the temperature of one pound of water from 63°F to 64°F. In joules, this is [2]:

$$1 \text{ Btu} = 1055 \text{ J}$$

A huge quantity of energy is consumed in the world each year, and to express this amount of energy handily, we need correspondingly large units. One such unit is the quad:

$$1 \text{ quad} = 1 \text{ quadrillion Btu} = 10^{15} \text{ Btu}$$

Although calories, Btu's, and quads are still in use, they are gradually being replaced by SI units. For example, a quad and an exajoule ($1 \text{ EJ} = 10^{18} \text{ J}$) are roughly equal (as shown below), and exajoules are now being used more and more frequently instead of quads.

$$1 \text{ quad} = 10^{15} \text{ Btu} = 1.055 \times 10^{18} \text{ J (using 1 Btu} = 1055 \text{ J)} = 1.055 \text{ EJ}$$

World energy consumption in 2005 was 445 EJ, of which 99 EJ was consumed in the USA. The US, which has about 5% of the world's population, is responsible for approximately 22% of the world's energy consumption. Canada consumed 13 EJ of

[1] A table of SI prefixes, such as "k" for kilo or 10^3, appears in Appendix IV.
[2] There are also other definitions of the Btu, which result in values from 1054 J to 1060 J.

energy in 2005, which although obviously less than the US total, is actually somewhat higher than US consumption on a per-capita basis. The per-capita energy consumption in the USA and Canada (as well as the United Arab Emirates and Kuwait) is considerably higher than that of other countries.

Oil is a dominant energy source today, whether measured in calories of energy, in dollars of value, or in importance to international development. The price of crude oil in $US per barrel is an important general indicator of world energy prices, and is widely published in daily newspapers. As a result, consumption of energy resources other than oil is often quoted in terms of the amount of oil that would be required to provide the same amount of energy. The quantity of oil is usually given in barrels or in tonnes. (1 tonne = 1 metric tonne = 1000 kg ≈ 7.3 barrels.) For example, North American coal consumption in 2005 is tabulated in a variety of publications as "614 million tonnes of oil equivalent." Conversion to more conventional energy units is straightforward using the following (approximate) conversion factor:

$$\text{1 barrel of crude oil is approx. equal to } 5.8 \times 10^{9}\ \text{J}$$
$$\text{1 tonne of crude oil is approx. equal to } 4.2 \times 10^{10}\ \text{J}$$

A final energy unit, which is important in the study of subatomic particles, and which will be useful in our later discussion of nuclear reactors, is the electron-volt (eV). This unit is defined as the energy gained by an electron in passing through a potential difference of one volt. In joules:

$$1\ \text{eV} = 1.60 \times 10^{-19}\ \text{J}$$

Energies in the keV to MeV range are common for single subatomic particles.

1.4 ENERGY CONSUMPTION AND SOURCES

In Section 1.3 world energy consumption in 2005 was stated as 445 EJ. Figure 1-3 shows a breakdown of this energy according to source.[3] You can see that fossil fuels (oil, gas, and coal) account for 92% of world energy. This heavy dependence on fossil fuels is rather disturbing — the reserves of these fuels are finite, and there are serious environmental problems (climate change, air pollution, etc.) associated with using them.

How is all this energy used? About 38% is used by industry, another 38% by the residential and commercial sectors (including public buildings such as schools and government offices), and 24% in transportation. Oil is the dominant source for transportation, of course, and road vehicles account for half of the annual consumption of oil.

The hydroelectric and nuclear energy components in Fig. 1-3 together provide 5% of world energy as electricity. However, do not be misled into thinking that electricity

[3] Sources: BP Amoco Statistical Review of World Energy 2006; U.S. Energy Information Administration

constitutes only 5% of world energy — fossil fuels are also used to produce electricity. Approximately 50% of annual coal consumption generates electricity, along with smaller percentages of gas and oil. Combining this fossil-fuel use with hydro and nuclear energy, a total of about 30% of world energy goes into the production of electricity.

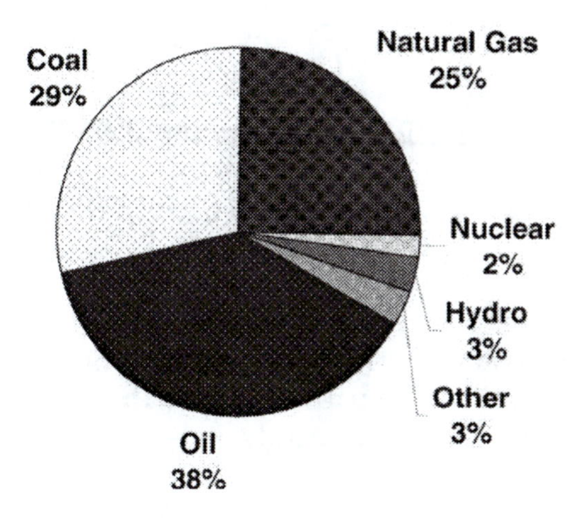

Figure 1-3 World Energy Consumption by Source in 2005

In some publications (including the previous edition of this book), the nuclear and hydro components of world energy consumption are stated as about 2.5 times larger than the quantities shown in Fig. 1-3. This is because nuclear and hydro electricity production is sometimes quoted in terms of the quantity of oil that would be needed to produce the electricity. When oil is used to generate electricity, about 60% of the energy goes into waste heat and only 40% into electrical energy. Hence, to produce a given amount of hydroelectric energy, for example, about 2.5 times that amount of energy would be needed if oil were used.

US and Canadian energy consumption by source are illustrated in Figures 1-4 and 1-5, respectively. Although the US source percentages are very close to the world values in Fig. 1-3, this is not the case for Canada. Fossil fuels account for 82% of Canadian energy, compared with 92% for the world, largely because of a smaller dependence on coal (13% for Canada versus 29% for the world). Hydroelectricity plays a larger role in Canada, supplying 12% of the energy, in contrast to only 3% in the world.

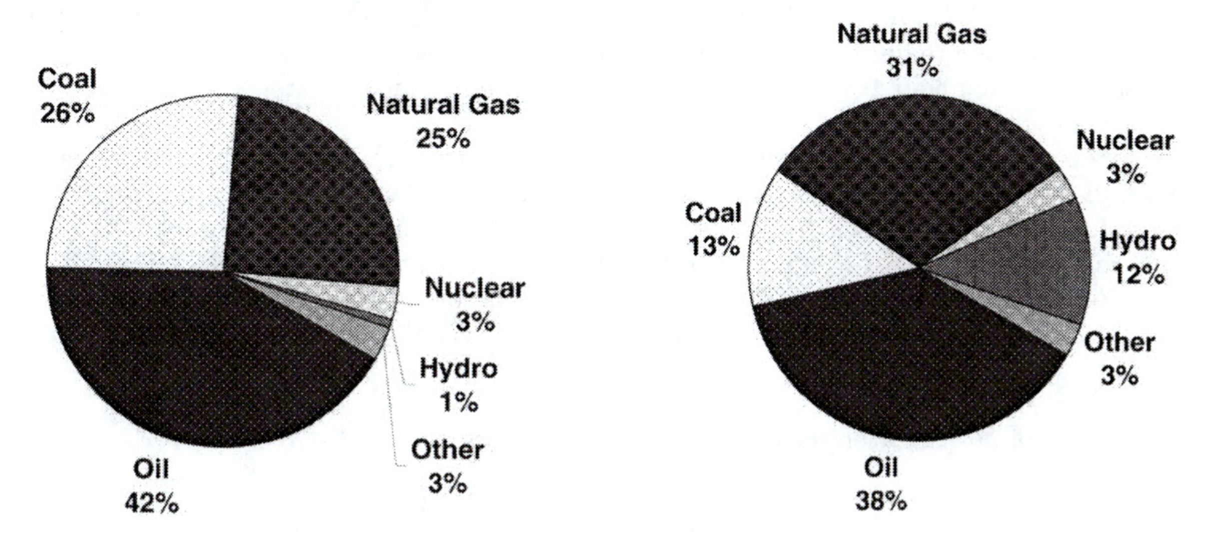

Figure 1-4 US Energy Consumption by Source in 2005

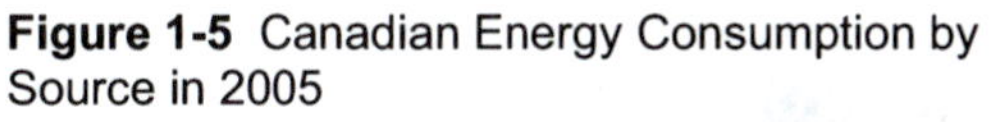

Figure 1-5 Canadian Energy Consumption by Source in 2005

1.5 POWER

In everyday speech, the words "energy" and "power" are often used interchangeably, but the scientific meanings of these words are different. Energy has already been discussed above, and now our focus turns to power. The SI unit of power, the watt, is one that you have undoubtedly heard of; for example, hair dryers and light bulbs are labelled according to their "wattage."

Power is the rate at which energy is used or provided, that is, power P is simply energy E divided by the time t during which the energy was used. For example, if a car's *KE* increases, the power input to the car is the increase in the KE divided by the time taken. Mathematically, power can be written as:

$$\text{Power} \quad P = \frac{E}{t} \qquad \textbf{[1-3]}$$

With energy in joules and time in seconds, the unit of power is joule per second (J/s), which is conventionally referred to as a *watt* (W):

$$\text{W (watt)} = \frac{\text{J}}{\text{s}}$$

Watts and kilowatts (kW) are often used to indicate the power consumption of home appliances such as light bulbs and hair dryers, and megawatts (MW) are used in describing power output of electrical plants. A 60-W lightbulb uses 60 J of electrical energy per second (but as stated in Section 1.2, only 5% of this energy is converted to light). As you sit reading this book, you use about 100 W of power, that is, each second you convert 100 J of food energy into other forms of energy (mainly thermal energy, which is then radiated away). A typical Canadian nuclear power reactor produces about 600 to 800 MW of electrical power, enough for 300000 to 400000 homes. At Niagara Falls, Ontario Power Generation produces a total of 2300 MW of power at five hydroelectric stations. The electrical power demand in the province of Ontario has a peak of about 26000 MW or 26 GW, which occurs in midsummer. The solar power striking Earth's surface is 178000 TW, or 178 PW.

Did You Know? The watt is named after James Watt (1736-1819) who did not, contrary to common opinion, invent the steam engine. Using scientific principles he improved the efficiency of the engine, invented by Thomas Newcomen (1663-1729), making it a commercial success.

Another unit of power, often used in the automotive industry, is the horsepower:

$$1 \text{ horsepower} = 1 \text{ hp} = 746 \text{ W}$$

EXAMPLE 1-1

A small hydroelectric plant produces 250 kW of electrical power from water falling through a vertical height of 18 m. What volume of water passes through the turbine each second?

(Assume complete conversion of the water's gravitational potential energy to electrical energy.)

SOLUTION

In a hydro plant, the gravitational potential energy of the water is converted to kinetic energy as it falls, and then to electrical energy. Hence, the electrical power can be expressed as the gravitational potential energy lost by the water divided by the time:

$$P = \frac{mgh}{t}$$

where m is the mass of water falling through a vertical height h in time t, and g is the magnitude of the gravitational acceleration. Since we are asked for volume of water, the mass m is replaced with the product of density ρ and volume V:

$$P = \frac{\rho V g h}{t}$$

Re-arranging to solve for volume, and substituting numerical values:

$$V = \frac{Pt}{\rho g h} = \frac{(250 \times 10^3\ \text{W})\,(1.0\ \text{s})}{(1.00 \times 10^3\ \text{kg/m}^3)(9.80\ \text{m/s}^2)(18\ \text{m})} = 1.4\ \text{m}^3$$

Hence, a volume of 1.4 m^3 of water passes through the generator each second.

Kilowatt·hour — An Energy Unit

One of the most confusing units for the general public (and even for physics students) is the kilowatt·hour (kW·h). Because part of the unit is kilowatt, which is a unit of power, many people believe that the kilowatt·hour is a power unit. However, since the kilowatt is multiplied by a time unit (hour), the kilowatt·hour is actually a unit of energy, as shown below.

Since power is energy divided by time ($P = E/t$), then energy can be written as $E = Pt$. This relationship states that the product of power and time is energy (regardless of the particular units used). If SI units are used, power in watts multiplied by time in seconds gives energy in joules, that is, joule = watt·second. But power and time can be expressed in other units; if power has units of kilowatts and time has units of hours, then the product — still an energy — has units of kilowatt·hour (kW·h). An energy of 1 kW·h can be expressed in joules through a unit conversion, using 1 kW = 10^3 W, 1 W = 1 J/s, and 1 h = 3600 s:

$$1\ \text{kW}\cdot\text{h} \times \frac{10^3\ \text{W}}{1\ \text{kW}} \times \frac{1\ \text{J/s}}{1\ \text{W}} \times \frac{3600\ \text{s}}{1\ \text{h}} = 3.6 \times 10^6\ \text{J} = 3.6\ \text{MJ}$$

Hence, 1 kW·h is equivalent to 3.6 MJ.

The kilowatt·hour is a very handy unit of electrical energy consumption, since the total power requirement of a house is often in the kilowatt range, and time can easily be measured in hours. Using joules for electrical energy would result in extremely large numbers. For electrical energy consumption at the national and world levels, gigawatt·hours (GW·h) and terawatt·hours (TW·h) are often used.

Electrical energy used in homes and businesses is normally priced by the kilowatt·hour. For example, in Guelph, Canada as of May 1, 2006, the residential price is 10¢ per kilowatt·hour for the first 600 kW·h used per month, and 11¢ per kilowatt·hour for all remaining kilowatt·hours in the month. In most other countries around the world, the residential price lies between 15¢ and 30¢ per kilowatt·hour.

EXAMPLE 1-2

What is the cost of the electrical energy consumed by a 100-W lightbulb that is turned on for 12 h? Assume a price of 11¢ per kilowatt·hour.

SOLUTION

To determine the energy in kilowatt·hours consumed by the lightbulb, the power in kilowatts is multiplied by the time in hours. Since the power is given in watts, this must first be converted to kilowatts:

$$100\ \text{W} \times \frac{1\ \text{kW}}{10^3\ \text{W}} = 0.10\ \text{kW} \quad \text{(assuming two significant digits)}$$

$$\text{Then, } E = Pt = (0.10\ \text{kW})(12\ \text{h}) = 1.2\ \text{kW·h}$$

$$\text{The cost of 1.2 kW} \cdot \text{h is: } 1.2\ \text{kW} \cdot \text{h} \times \frac{11¢}{\text{kW} \cdot \text{h}} = 13¢ \quad \text{(to the nearest cent)}$$

1.6 ELECTRICAL ENERGY

Now that the kilowatt·hour has been introduced, we turn our attention to electrical energy around the world. A brief summary of important events[4] in the development of commercial electrical energy is presented in Table 1-2.

Table 1-3 provides an international comparison of electricity generation (in terawatt·hours) by fuel type for the world's 10 largest electrical energy-producing countries in 2003. Notice that over 65% of the world's electricity is generated by burning fossil fuels, primarily coal. This is followed by hydroelectric and nuclear generation, both approximately at 16%. The category "other," accounting for only 2% of world electricity, includes wind, solar, geothermal, and biomass. The USA leads all countries in total electricity generation, and Canada is the leader in hydroelectric production. Almost 60% of Canada's electricity is generated from hydro, compared to

[4] Sources: *Electric Power in Canada 1990*, Energy, Mines and Resources Canada; A. Hellemans and B. Bunch, *The Timetables of Science*, Simon and Schuster, New York, (1988).

Table 1-2
Significant Developments in Electricity Production

Year	Event
1876	Alexander Graham Bell made first long-distance telephone call (218 km, in Ontario, Canada)
1879	Thomas Edison (U.S.A.) and Joseph Swan (England) each produce carbon-thread electric lamps that can be used for practical lengths of time
1880	Thomas Edison's first electric generation station opened in London
1881	First practical electric generator and electric distribution system were built
1881	First electric streetcar introduced in Berlin
1882	First hydroelectric power plant went into operation in Appleton, Wisconsin
1883	Canada's first street lights installed in Hamilton, Ontario
1884	Charles Algernon Parsons designed and installed the first steam turbine generator for electric power
1888	Nikolai Tesla developed an alternating current motor
1897	Canada's first long-distance high-voltage transmission line (11 kV) carried power 29 km to Trois-Rivières, Quebec
1921	Sir Adam Beck No. 1 generating plant, then the largest in the world, opened in Niagara Falls, Ontario
1956	First large-scale nuclear power plant designed for peaceful purposes opened in England
1967	Canada's first commercial-scale (220 MW) CANDU nuclear generating station entered service at Douglas Point, Ontario
2005	The contribution of wind energy to electricity in Denmark increased to 20%

only 16% in the world overall. Notice the large nuclear production in France: 78% of the electricity in this country, which has no oil or gas and only a little coal, comes from nuclear power.

As seen in Table 1-3 the mix of energy sources used for electricity generation varies greatly from country to country. As well, it varies on a local basis within a country, depending on resources available. For example, in the mountainous province of British Columbia, Canada, 90% of the electricity is generated from hydro resources, whereas in the relatively flat prairie province of Saskatchewan, burning fossil fuels provides 78% of the electricity.

Table 1-3
Electricity Generation in Terawatt·hours by Fuel Type, 2003
For the World's 10 Largest Electricity-producing Countries[5]

Country	Conventional Thermal (primarily coal)	Hydro	Nuclear	Other	Total
USA	2891	306	788	97	4082
China	1578	284	43	2	1907
Japan	681	104	240	22	1047
Russian Fdn.	606	158	150	2	916
Germany	377	24	165	33	599
Canada	164	338	75	10	587
France	56	64	441	6	567
India	535	75	18	5	633
U.K.	296	6	89	8	399
Brazil	33	306	13	13	365
World Total	11057 (66.0%)	2726 (16.3%)	2635 (15.7%)	324 (1.9%)	16742 (100.0%)

Installation Cost for Electricity Generation

There are many energy sources that can be used to generate electricity. However, there is wide variation in the cost to install the power-generating facilities. Figure 1-6 shows this cost in US dollars per kilowatt of power generation.[6] The technologies associated with generating electricity from each of these sources are discussed in later chapters.

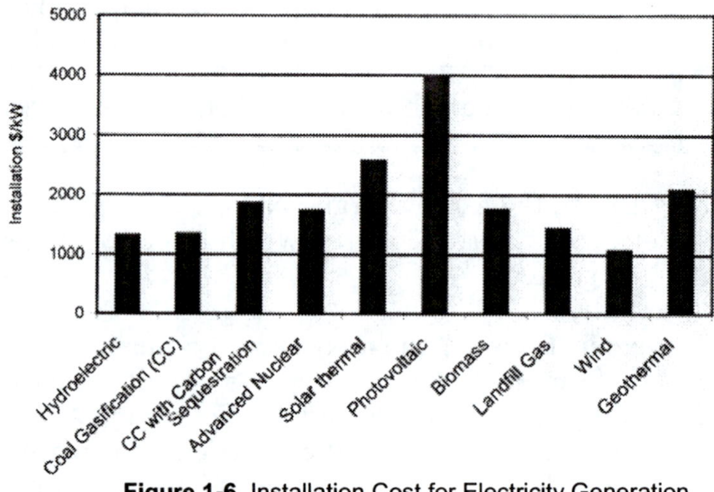

Figure 1-6 Installation Cost for Electricity Generation

[5] International Energy Agency
[6] *Assumptions to the Annual Energy Outlook 2006*, US Department of Energy Website

EXERCISES

Please refer to the Appendices for useful information concerning unit conversions, significant digits, numerical constants, etc.

1-1. A truck of mass 4500 kg brakes to a halt from a speed of 82 km/h in a distance of 77 m.
(a) How much energy (in megajoules) must be dissipated?
(b) What happens to this energy?

1-2. By using conversion factors provided in Section 1.3 or Appendix IV, perform the following unit conversions.
(a) 5.82 Btu to Calories
(b) 22 tonnes of crude oil to quads
(c) 52 billion barrels of crude oil to quads

1-3. A sprinter, starting from rest, has a power output of 5.1×10^2 W for a time of 7.2 s. Neglecting losses due to production of heat, etc., determine his final speed. The sprinter's mass is 67 kg.

1-4. (a) By using the basic definition of a watt (and the meaning of kilo), determine how many joules there are in one kilowatt·hour.
(b) Convert 581 kW·h to joules.

1-5. The solar energy striking Earth's surface every year is 178000 TW·yr. Convert this to joules.

1-6. (a) Given below are the electrical power requirements for five household appliances. Determine the number of kilowatt·hours of electrical energy consumed if all these appliances are running simultaneously for two hours in a house.

solid state colour TV: 145 W
automatic washing machine: 512 W
furnace fan: 500 W
clock: 2 W
humidifier: 177 W

(b) If the average residential cost for electricity is 12¢ per kW·h, what would be the cost (to the nearest cent) for the energy calculated in (a)?

1-7. The residential rates for electricity in Guelph, Ontario (as of May 2006) are 10¢ per kilowatt·hour for the first 600 kW·h used per month, and 11¢ per kilowatt·hour for all remaining kilowatt·hours in the month. If a family has a consumption of 1475 kW·h in a month, what is its average cost per kW·h?

PROBLEMS

Please refer to the Appendices for useful information concerning unit conversions, significant digits, numerical constants, etc.

1-8. In a 770-kW hydrolectric plant, 300 m^3 of water pass through the turbine each minute. Assuming complete conversion of the water's initial gravitational potential energy to electrical energy, through what distance does the water fall? Assume two significant digits.

1-9. Calculate the power in megawatts available from a tidal power scheme where the incoming tide fills up a catchment area enclosed by concrete walls, and then at low tide this water is allowed to fall through openings at the bottom of one of the walls to spin turbines. The square catchment area has 1.2-km sides, and the tide rises by 3.7 m. Assume that the process of emptying takes 1.0 h, and that all the energy of the water is converted to energy of the turbines. Hint: when using the expression mgh for gravitational potential energy, remember that h represents the height of the centre of mass.

1-10. (a) Suppose that an automobile is travelling at a constant speed on a horizontal road; the engine is running and obviously producing energy. Where does this energy go? (It does not go into kinetic energy of the automobile, because the kinetic energy does not change if the speed is constant.)
(b) An automobile moving at 90 km/h ascends a hill of gradient 1 in 25 (i.e., a vertical rise of 1 unit for a horizontal distance of 25). Its mass is 1300 kg. What power (in kilowatts) is needed for the climb up the hill over and above the normal power used when moving horizontally? Assume two significant digits.

1-11. The intensity of radiation from a distant source like the Sun varies inversely with the square of the distance from the source. At the top of Earth's atmosphere the solar intensity (power per unit area) is 1.35 kW/m^2. If a space ship with a solar panel of area 125 m^2 were halfway between the Sun and Earth, determine the total solar energy that the panel would receive in a day. Express your answer in kilowatt·hours and in joules.

ANSWERS

1-1. (a) 1.2 MJ (b) given in Section 1.2
1-2. (a) 1.47 Cal (b) 8.8×10^{-7} quads (c) 2.9×10^{2} quads
1-3. 1.0×10^{1} m/s
1-4. (a) 3.6×10^{6} J (b) 2.09×10^{9} J
1-5. 5.61×10^{24} J
1-6. (a) 2.67 kW·h (b) 32¢
1-7. 10.59¢ per kW·h
1-8. 16 m
1-9. 27 MW
1-10. (b) 13 kW
1-11. 1.62×10^{4} kW·h, 5.83×10^{10} J

CHAPTER 2

ENERGY CONSUMPTION: PAST, PRESENT, AND FUTURE

The data shown in Chapter 1 demonstrated that about 90% of world energy consumption comes from fossil fuels. A question of tremendous importance is: How long will our fossil-fuel resources last? An accurate answer to this question requires an understanding of trends in energy consumption, and some mathematical "crystal-ball-gazing" as projections are made for the future.

2.1 HISTORICAL ENERGY CONSUMPTION

First, consider the question, "How much energy do you need for daily survival?" Ignore, for a moment, the energy consumed by people in driving automobiles, in manufacturing and using all sorts of commercial products, in heating, cooling, and lighting, etc.; our most basic need is food energy. A typical person uses about 2000 to 3000 food calories (or roughly 10 MJ) per day. This was the per capita energy consumption of primitive humans.

Throughout history and even in pre-history, humans have been cleverly inventive in tapping energy from many sources and for many ends. Primitive people learned to use fire in about 100,000 BC, they domesticated animals for agriculture and transportation around 5000 BC, and began to exploit the wind about 1000 BC. Roughly 2000 years ago, the Romans developed the waterwheel, and a thousand years later, coal started as an energy source. The Industrial Revolution of the late 18th century, with steam engines driving machinery, greatly increased the consumption of energy. Nowadays, there is scarcely a single facet of our daily activities that does not use some external source of energy. Oil is used for transportation; gas for heating and cooking; and coal, hydro, and nuclear energy to generate electricity for lighting, heating, cooking, cooling, entertainment, computing, etc. In addition, fossil fuels and other energy sources are used in a variety of ways in manufacturing and in resource-based industries such as forestry and mining.

As a result, the 10 MJ daily per capita energy consumption of our primitive ancestors has now increased to a world average of about 160 MJ. Figure 2-1 shows the per capita energy consumption[1] for a number of countries and the world. The United Arab Emirates has the largest per capita consumption by a wide margin. People in Kuwait, Canada, and the USA use considerably more energy per capita than people in other industrialized countries, who in turn use considerably more than people in developing countries.

---

[1] Sources: BP Statistical Review of World Energy 2006; U.S. Census Bureau

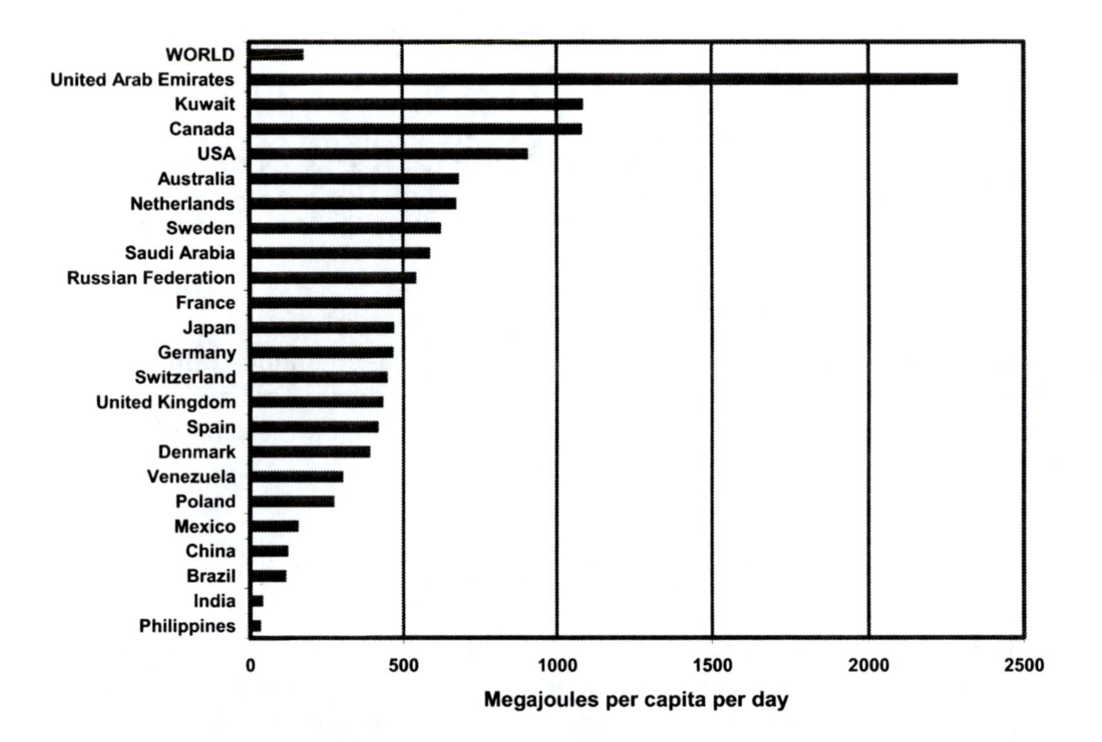

Figure 2-1 Daily Energy Consumption Per Capita in 2005

Total world energy consumption depends not only on per capita consumption, but also on world population, and both of these have been increasing. From 1995 to 2005, world energy consumption increased by an average of 2.1% annually — 1.2% due to population and 0.9% due to per capita consumption. Figure 2-2 shows world energy consumption[2] from 1860 to 2005. Although this graph shows a couple of downward segments, the general trend is steadily upward. The slight dips are themselves interesting: notice the drop due to the economic depression in the early 1930s, and the decrease in the early 1980s because of rising oil prices and economic recession.

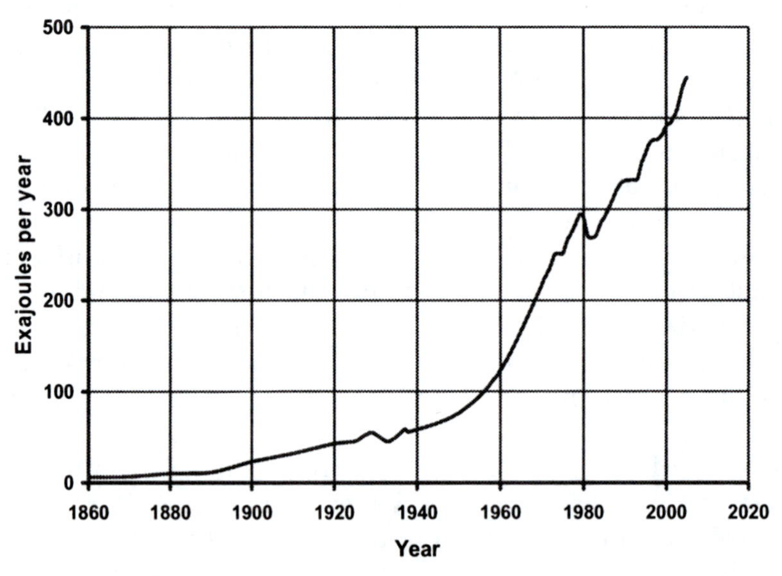

Figure 2-2 World Energy Consumption 1860-2005

[2] Sources: BP Statistical Review of World Energy (various years); J. Fowler, *Energy and the Environment,* McGraw-Hill, New York, 1984, p. 76

2.2 EXPONENTIAL GROWTH

Population and energy consumption have both been increasing. In order to set the stage for making predictions of future population and energy consumption, and for calculating how long fossil fuels might last, the mathematics of *constant-percentage growth*, or *exponential growth* needs to be considered.

As an example of exponential growth, suppose that you invest $5.00 in a bank account, which pays 10% interest per year, compounded annually. After one year, you have the original $5.00, plus $0.50 interest, for a total of $5.50. In essence, the original $5.00 has been multiplied by a factor of 1.10, where the 0.10 represents the 10% interest. The $5.50 is then invested for a second year, and is multiplied by another factor of 1.10, giving a total of $6.05 after two years. The original $5.00 has now been multiplied twice by 1.10, i.e., by $(1.10)^2$. Table 2-2 shows how the original deposit of $5.00 grows with time, with each year contributing a multiplication by 1.10. After *n* years, the total is $5.00 times $(1.10)^n$. Notice that the *exponent* of the 1.10 is growing — after 2 years we have $(1.10)^2$, after 3 years $(1.10)^3$, etc. — hence the name *exponential growth.* Exponential growth results whenever a quantity grows by a certain constant percentage (such as 10%) per unit time (such as a year).

Table 2-2
Investing $5.00 at 10% interest

Time (yr.)	Amount ($)
0	5.00
1	5.00 (1.10) = 5.50
2	$5.00\ (1.10)^2 = 6.05$
3	$5.00\ (1.10)^3 = 6.66$
n	$5.00\ (1.10)^n$

In order to write a mathematical equation for exponential growth, define the following:

N = quantity that is growing (such as money in a bank account, energy consumption, etc.)

t = time

N_0 = quantity at time $t = 0$ (beginning of time interval of interest)

C = constant multiplying factor per unit time (such as the 1.10 used above)

Following the example of the money invested in the bank account, it is straightforward to write an expression for the quantity N at any time t:

$$N = N_0 C^t \qquad \textbf{[2-1]}$$

Although Eq. [2-1] is a perfectly good equation for exponential growth, it is not in the form normally used. Instead of using the base C, it is much more common to write the equation with the base "e," where e has the usual value of 2.718 . . . We can change to base e by first writing C as $C = e^{\ell n\, C}$, where $\ell n\, C$ is the natural logarithm of C. Now, since C is a constant, then $\ell n\, C$ is just another constant, which can be written as k, that is, $\ell n\, C = k$. Thus, C can be expressed as: $C = e^{\ell n\, C} = e^k$. Substitution of $C = e^k$ into Eq. [2-1] gives $N = N_0(e^k)^t$, or more simply:

$$N = N_0 e^{kt} \qquad \textbf{[2-2]}$$

Equation [2-2] is the usual way that exponential growth is expressed mathematically. The constant k is referred to as the *growth constant*, and it is useful to relate k to the percent growth rate, symbolized by R. (If the growth rate is 10% per year, then R is 10.) C is related to R by $C = 1 + R/100$ (for $R = 10$, then $C = 1.10$), and since $k = \ell n\, C$,

$$k = \ell n(1 + R/100) \qquad \textbf{[2-3]}$$

Figure 2-3 shows a graph of $N = N_0 e^{kt}$. Notice that as t increases, the curve rises ever more steeply, that is, the rate of change of N increases with time. In other words, with exponential growth, the growth gets bigger and bigger as time progresses. For a larger value of k, the curve would rise even more steeply. Notice the similarity between Fig. 2-3 and Fig. 2-2, which shows total world energy consumption as a function of time.

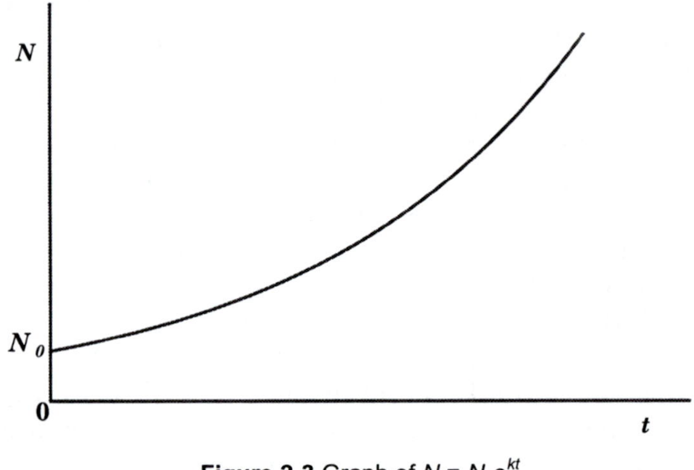

Figure 2-3 Graph of $N = N_0 e^{kt}$

The equation $N = N_0 e^{kt}$ can be written in an alternative way involving logarithms. Start by re-arranging $N = N_0 e^{kt}$ as $N/N_0 = e^{kt}$, and then take natural logarithms of both sides, giving

$$\ell n \frac{N}{N_0} = kt \qquad \textbf{[2-4]}$$

This logarithmic form of the equation for exponential growth is very handy when dealing with situations in which N and N_0 are known (or the ratio N/N_0 is known), and one of k or t is known and the other is to be determined.

EXAMPLE 2-1

Between the years 1950 and 1975, world energy consumption grew from 76 EJ/yr to 252 EJ/yr, with a growth that was approximately exponential.

(a) Determine the growth constant.
(b) What was the average annual percent growth rate?
(c) What would the consumption be in the year 2020, assuming continued exponential growth at the same rate?

SOLUTION

(a) The given quantities are $N = 252$ EJ/yr, $N_0 = 76$ EJ/yr, and $t = 25$ yr. To determine k, start with Eq. [2-4]: $\ell n(N/N_0) = kt$. Re-arranging and substituting:

$$k = \frac{1}{t}\ell n\frac{N}{N_0} = \frac{1}{25\text{ yr}}\ell n\left(\frac{252\text{ EJ/yr}}{76\text{ EJ/yr}}\right) = 0.048\text{ yr}^{-1}$$

Note that the units of k are yr^{-1}, if time t is in years. This means that the exponent kt in $N = N_0e^{kt}$ is dimensionless, as all exponents must be.

(b) To calculate the average annual percent growth rate, R, use Eq. [2-3], $k = \ell n\,(1 + R/100)$, since k is known. To remove R from inside the logarithm, perform exponentiation using the base e:

$$e^k = e^{\ell n(1 + R/100)} = 1 + R/100$$

Solving for R: $R = 100\,(e^k - 1) = 100\,(e^{0.048} - 1) = 4.9$

Hence, the average annual growth rate is 4.9% per year.

(Perhaps you are wondering about units for e^k, since it appears that the exponent k is not dimensionless. Implicitly multiplying the k in this exponent is a time of one year, since R is the growth rate for one year, and hence the complete exponent is the dimensionless product kt.)

(c) To determine the projected energy consumption in the year 2020, use Eq. [2-2], $N = N_0e^{kt}$. To find N, either

- set $t = 0$ in the year 1950 and use $N_0 = 76$ EJ/yr; then $t = 70$ yr for the year 2020

or
- set $t = 0$ in the year 1975 and use $N_0 = 252$ EJ/yr; $t = 45$ yr.

Either way, the same answer for N will be obtained. Using the latter set of values:

$$N = (252\text{ EJ/yr})\,e^{(0.048\text{ yr}^{-1})(45\text{ yr})} = 2.2\times10^3\text{ EJ/yr}$$

Thus, if exponential growth at the 1950–1975 rate had continued until the year 2020, the annual consumption in that year would be 2.2×10^3 EJ/yr.

2.3 DOUBLING TIME

One significant feature of exponential growth is that for a given growth constant, there is a constant time required for the growing quantity to double. This time, called the *doubling time*, is closely related to the growth constant.

To show this relationship, start with Eq. [2-4]: $\ell n(N/N_0) = kt$. For a doubling, $N/N_0 = 2$, and t is the doubling time, represented as T_2. Hence, $\ell n\,2 = kT_2$. Solving for T_2:

$$T_2 = \frac{\ell n\,2}{k} = \frac{0.693}{k} \qquad \textbf{[2-5]}$$

Notice that if k has units of, say, $year^{-1}$, then the units of T_2 are years.

EXAMPLE 2-2

In Example 2-1, the growth constant for world energy consumption between 1950 and 1975 was determined to be 0.048 yr^{-1}. What is the doubling time associated with this growth?

SOLUTION

Use Eq. [2-5]:

$$T_2 = \frac{\ell n\, 2}{k} = \frac{0.693}{0.048\ yr^{-1}} = 14\ yr$$

Thus, the doubling time is 14 yr.

It is important to recognize that the doubling time applies to all doublings of the growing quantity. In Example 2-2, energy consumption doubles *every* 14 yr. This means that in 14 years energy consumption doubles, after another 14 years it doubles again (so that it has quadrupled relative to the original value), then after another 14 years it doubles so that it is 8 times the original value, etc. To put this another way: when growth is exponential, the amount of the increase is itself a rapidly growing quantity.

2.4 HANDY APPROXIMATIONS

Thus far, the equations used for exponential growth have all been mathematically exact. However, it is often convenient to use approximate relationships for exponential growth that permit some calculations to be done in your head. Begin with Eq. [2-3], which relates the growth constant k and the percent growth rate R:

$$k = \ell n\,(1 + R/100)$$

The right-hand side of this equation has the form $\ell n\,(1 + x)$, where $x = R/100$. Now, if $|x| < 1$, $\ell n\,(1 + x)$ can be written in a Taylor series expansion:

$$\ell n(1 + x) = x - \frac{x^2}{2} + \frac{x^3}{3} - \frac{x^4}{4} + \ldots \quad \text{for } |x| < 1$$

This expansion can be simplified if $|x| << 1$; in this case, the first term, x, is much greater than any of the subsequent terms involving x^2, x^3, etc. Thus, the expansion can be approximated by x alone:

$$\ell n\,(1 + x) \approx x \quad \text{for } |x| << 1$$

Returning to $k = \ell n\,(1 + R/100)$, the above approximation can be used to write the right-hand side as

$$\ell n\,(1 + R/100) \approx R/100 \quad \text{for } R/100 << 1$$

Hence,

$$k \approx R/100 \quad \text{for } R/100 << 1 \qquad \textbf{[2-6]}$$

Approximation [2-6] is a very handy way to relate k and R provided that the growth rate is relatively small, that is, *less than about 10% per unit time.* This approximation can also be used to develop a relation between doubling time and percent growth rate. Start with Eq. [2-5]: $T_2 = 0.693/k$, and then use approximation [2-6] to replace k with $R/100$:

$$T_2 \approx \frac{0.693}{R/100} \approx \frac{69.3}{R}$$

Since this is just an approximation, the value 69.3 can be replaced with 70:

$$T_2 \approx \frac{70}{R} \quad \text{for } R/100 \ll 1 \qquad \textbf{[2-7]}$$

EXAMPLE 2-3

World energy consumption between 1950 and 1975 grew at a rate of 4.9% per year. Determine approximate values for the growth constant and doubling time.

SOLUTION

Since the growth rate is less than 10% per year, approximations [2-6] and [2-7] can be used. The growth rate of 4.9% is very close to 5%, and since we are dealing with approximations, it is legitimate to use 5% — that is, $R \approx 5$. Approximation [2-6] gives

$$k \approx R/100 \approx 5/100 \approx 0.05 \text{ yr}^{-1}$$

Using [2-7]:

$$T_2 \approx \frac{70}{R} \approx \frac{70}{5} \approx 14 \text{ yr}$$

Thus, the growth constant is approximately 0.05 yr^{-1} and the doubling time is approximately 14 yr. Compare these values with the more accurate numbers determined in Examples 2-1 and 2-2: 0.048 yr^{-1} and 14 yr. The approximations are obviously very good.

2.5 SEMI-LOGARITHMIC GRAPHS

It is often useful to graph data in such a way that the graph appears as a straight line. With exponential growth, a linear graph is obtained if the data are plotted on a semi-logarithmic graph, as explained below.

Start with Eq. [2-4], the logarithmic form of the exponential growth equation: $\ln(N/N_0) = kt$. Rewriting the left-hand side as $\ln N - \ln N_0$, and then re-arranging to solve for $\ln N$ gives:

$$\ln N = kt + \ln N_0$$

This equation has the familiar form of the equation of a straight line: $y = mx + b$, where y is $\ln N$, the slope m is the growth constant k, and the y-intercept b is $\ln N_0$. In other words, under conditions of exponential growth, a plot of $\ln N$ vs. t will yield a straight line with slope k and y-intercept $\ell n\, N_0$ (Fig. 2-4). Such a graph is referred to as a *semi-logarithmic (or semilog) graph*, since one of the two axes involves a logarithm.

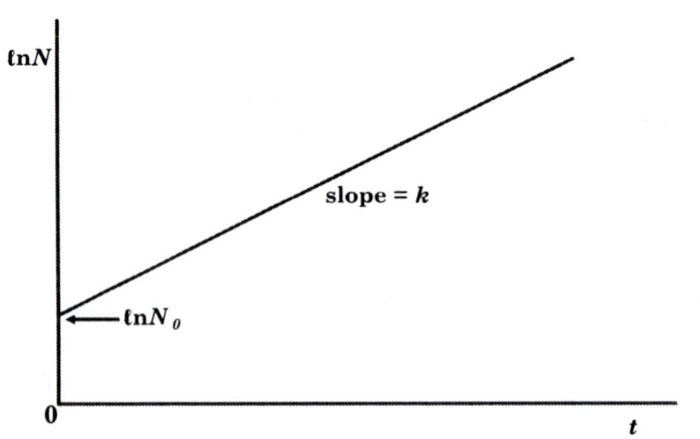

Figure 2-4 Graph of ℓn N vs. t for exponential growth.

When a semi-logarithmic graph is being plotted, it is customary to use semi-logarithmic graph paper, which has spacings along the y-axis that automatically space the data points as if logarithms had been taken. Figure 2-5 shows the data of Fig. 2-2 for world energy consumption replotted on a semilog graph.[3] Although this graph has some bumps and wiggles, it is nonetheless roughly linear, indicating that world energy consumption has indeed been approximately exponential from 1860 to the present day. Therefore, it is quite reasonable to model world energy consumption as an exponential function.

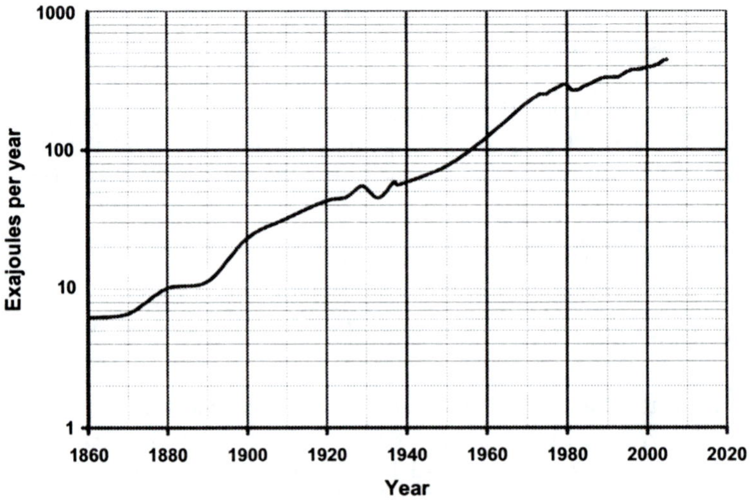

Figure 2-5 Semilog graph of world energy consumption.

EXAMPLE 2-4

The data shown in the semilog graph of Fig. 2-5 can be approximated quite well as a straight line joining two points: 10 EJ/yr in the year 1880, and (approximately) 250 EJ/yr in the year 1980. (Use a straight edge to confirm this.) Determine the growth constant associated with this line.

SOLUTION

The growth constant k is just the slope of the line; normally the slope is calculated simply as rise/run, but since this is a semilog graph, it is necessary to take natural logarithms of the y–variable (consumption N):

$$k = \frac{\ell n N_2 - \ell n N_1}{t_2 - t_1} = \frac{\ell n(N_2 / N_1)}{t_2 - t_1}$$

Thus, $$k = \frac{\ell n(250/10)}{100 \text{ yr}} = 0.032 \text{ yr}^{-1}$$

[3] Semilog graph paper with a more detailed *y*-axis is shown in Appendix II.

Notice that the calculation of this slope is equivalent to solving for k in the equation $\ell n(N/N_0) = kt$, using $N = 250$ EJ/yr, $N_0 = 10$ EJ/yr, and $t = 100$ yr.

2.6 LIFETIME OF FOSSIL FUELS (CONSTANT PRODUCTION MODEL)

It is now possible to provide one answer to the question: How long will fossil fuels last? Since energy consumption has been increasing exponentially, it seems appropriate to use exponential growth as a model for fossil-fuel consumption, and the next section deals with this topic. In the present section, as a first step, consider fossil fuel lifetimes if consumption is constant.

Production and Consumption

Although the previous paragraph refers to "consumption," the actual quantity that is important in determining the lifetime of a resource *in a given country* is "production," which refers to the quantity of fossil fuel removed from the ground, say in a particular year. Considering the *world as a whole*, the annual production of a fuel is virtually identical to its annual consumption (the amount of fuel actually used), and hence the distinction between consumption and production is not especially important globally. However, for any one country, production and consumption can differ markedly because of imports and exports. For example, US oil production in 2005 was 13 EJ, whereas US oil consumption was 40 EJ. Obviously, 27 EJ of oil was imported. In contrast, oil production in Norway in 2005 was 5.8 EJ, but its consumption was only 0.4 EJ; exports accounted for the difference of 5.4 EJ.

Reserves and Resources

The lifetime of fossil fuels depends on two factors: the production rate and the amount of fossil fuels remaining. This amount remaining is often quoted in two categories:

Reserves (or proven reserves) refer to fossil fuels that have already been discovered, their quantity measured, and are known to be extractable at a competitive price.

Resources (or recoverable resources) include the reserves and, in addition, fossil fuel deposits that are inferred or expected, for which recovery is anticipated to be technically and economically feasible. For the purposes of estimating fossil fuel lifetime, resources are the important quantity.

Table 2-3 shows fossil fuel reserves and resources (in exajoules) for the world, Canada, and USA at the end of 2005. The data are averages from a number of sources. The resources data have large uncertainties, and should be taken only as "ball-park" figures. The reserves are known more accurately, but of course even they change in time as fuels are consumed, new discoveries are made, new extraction technologies are developed, and the prices of fuels change. Notice the large resource of coal, compared to oil and gas. It should be mentioned that roughly 60% of world oil reserves are located in the Middle East, and only about 16% in North America.

Table 2-3
Fossil Fuel Reserves and Resources (exajoules)

Region	Oil Reserves	Oil Resources	Natural Gas Reserves	Natural Gas Resources	Coal Reserves	Coal Resources
World	7500	12000	6000	8000	20000	150000
Canada	1000	1200	60	400	140	2700
USA	170	600	180	700	5000	36000

The world oil reserve base has actually increased in the last decade, partly as a result of new technologies. Starting around 1990, experiments were begun with horizontal drilling whereby channels are drilled sideways from the standard vertical shaft so that more of the oil deposit can be accessed. As well, oil exploration firms now use computer-based three-dimensional seismic technology to produce a 3-D multicoloured image of the various rock layers, making it easier to determine where to drill.

Another major reason for the increase in the oil reserves (and resources) is that the price of oil has risen high enough that the Athabasca tar sands in Alberta have become economical to exploit, and the extractable energy there is now included in tabulations of reserves. A discussion of these tar sands is included in this chapter in a following subsection "Unconventional Sources of Fossil Fuels."

Fossil Fuel Production and Lifetimes

The production of fossil fuels[4] in 2005 is presented in Table 2-4. Note that the production of oil is larger than that of gas or coal. Since the resources and present production rate are known for these fossil fuels, it is possible to determine how long the resources would last *if* production remained constant at its 2005 value by simply dividing each of the resources in Table 2-3 by the appropriate production rate in Table 2-4. The results (rounded-off to one or two significant digits) are shown in Table 2-5.

Table 2-4
Fossil Fuel Production in 2005
(exajoules/yr)

Region	Oil	Gas	Coal
World	164	105	122
Canada	6.1	7.0	1.5
USA	13	20	24

[4]Source: BP Statistical Review of World Energy 2006

As can be seen from Table 2-5, there is not much oil left in the world — about 50 years' worth. The situation in the USA and Canada is even worse, with only about 30 years of oil left in domestic supplies. The situation for gas is somewhat better, with almost 100 years of resource remaining. For coal, there still remains almost about 1700 years' supply, but of all the fossil fuels, coal is the one that has the most damaging environmental effects.

Table 2-5
Lifetime (yr) of Fossil Fuel Resources
Assuming Constant Production (2005 Rates)

Region	Oil	Gas	Coal
World	75	75	1200
Canada	200	55	1800
USA	45	35	1500

It is important to remember that the lifetimes presented in Table 2-5 are calculated *with the assumption that production remains constant.* It is obvious that if production increases, as it has for many decades in the past, then the lifetimes will be smaller. Lifetimes for exponentially increasing production are calculated in the next section.

Tar Sands

The data for oil resources in Table 2-3 include the Alberta tar sands, as well as conventional deposits of oil. The tar sands, now commonly called "oil" sands although the deposits are a thick hydrocarbon (bitumen) mixed with sand, have reserves second only to those of Saudi Arabia. Tar sands now account for approximately half of Canadian oil production annually. About 20% of the deposits are close to Earth's surface, and are mined by huge power-shovels that load the tarry sands into dump trucks, each of which can hold 380 tonnes. To remove the bitumen from the sand, hot water is mixed with the sands, the mixture is aerated, and the tar and sand separate out.

For the 80% of deposits that are too deep to mine, steam is injected into the deposits, and the hot tar seeps downward through the sand. A pipe is drilled to this level and the hot tar is pumped to the surface. Tar-sands production uses large amounts of natural gas to heat water, both for separation of the tar from the surface deposits and for generation of steam for the underground in-situ recovery. The tar is sent by pipeline to refineries, but because of its high viscosity, it must first be upgraded using natural gas or other petroleum products. Once at the refinery, it is processed further to produce synthetic crude oil.

As well as using large amounts of natural gas, tar-sands production requires copious amounts of water, and there is increasing concern about this water usage. In a newspaper article[5] in May 2006, it was reported that "in situ projects, which use steam to melt bitumen before it is pumped to the surface, used almost three times as much water in 2004 as originally projected," and one observer commented that "I fear we're going to run out of water before we run out of bitumen in northern Alberta." Improved

[5] Globe and Mail, "Alberta's Thirst for Oil Leaves a Dry Taste," May 2, 2006, pg. B1 and B6

methods to extract and process the tar using less energy and less water are being explored, both by petroleum companies and by researchers at institutions such as the University of Calgary's Institute for Sustainable Energy, Environment, and Economy.

In addition to the tar sands in Alberta, there are significant tar sands deposits in Venezuela, roughly comparable in size to those in Alberta. However, tar-sands development in Venezuela has been much slower, and the political situation in that country might result in further delays. As of January 2006, the tar sands deposits in Venezuela have not been included in oil-industry tabulations of reserves and resources.

Other Unconventional Sources of Fossil Fuels

Another unconventional source of oil is oil shale, in which a heavy hydrocarbon (kerogen) is deposited throughout rock. In a few places in the world, the kerogen content is large enough that the rocks can actually be burned, usually mixed with coal. In other cases, the oil shale can be processed by pyrolysis (heating without oxygen) to produce a liquid oil, along with some gas and residual carbon. At the present time, this oil from shale cannot compete economically with other sources of oil.

Natural gas (primarily methane, CH_4) is normally found in association with oil deposits, and in the early days of the petroleum industry, it was considered a nuisance that was simply burned off at the well. It was not until storage and pipeline facilities became available that gas became commonly accessible as a fuel. There are a number of unconventional sources of gas:

- Gas in non-porous rocks, that require fracturing to allow the gas to flow freely to a well.
- Coal-seam methane, i.e., gas found in coal deposits.
- Geo-pressured methane, which is gas found in subterranean deposits of hot water beneath an overlying layer of rock.
- Gas hydrates, also known as *clathrates*, which are frozen mixtures of water and natural gas. These crystalline solids are found buried on the ocean floor around the world, at depths greater than 400 m and temperatures below 1-2°C. Most of these deposits are at depths of several kilometres. Gas hydrates are also found in polar regions beneath the frozen permafrost. Although there is an enormous amount of natural gas frozen in clathrates, an easy and economical way to extract the gas has yet to be found. A concern about the gas hydrates is that if global warming (Chapter 6) continues for another 100-200 years, the methane might be released as a result of melting of the permafrost or warming of the ocean depths. Since methane is a strong greenhouse gas, global temperatures would increase even more, releasing more methane, and a dangerous positive feedback loop would have been established.

There are also deposits of coal not included in the category of recoverable resources because the coal seams are too thin, too deep, or too inaccessible (under a town or lake, for example).

2.7 LIFETIME OF FOSSIL FUELS (EXPONENTIAL GROWTH MODEL)

When determining fossil fuel lifetimes in Section 2.6, constant production at 2005 levels was assumed. However, since energy production has historically risen exponentially, it is natural to wonder how long fossil fuels would last if exponential growth were to continue in the future. The mathematics of exponential growth provide a quick analysis.

Suppose that the annual production N of a resource is expressed as an exponential function of time t: $N = N_0e^{kt}$. To emphasize that N is a function of time, write this as $N(t) = N_0e^{kt}$. At some arbitrary time t, consider a small time interval of duration Δt (Fig. 2-6). The production at this time is $N(t)$, which is the "height" of the graph above the origin at time t. In Fig. 2-6, a narrow rectangle has been drawn with a base of width Δt and a height of $N(t)$. The *area* of this rectangle is $N(t)\Delta t$, and represents the *quantity of resource produced* in the time interval Δt. Note the units: $N(t)$ has units such as exajoules/year and Δt would be in years, and thus their product is in exajoules.

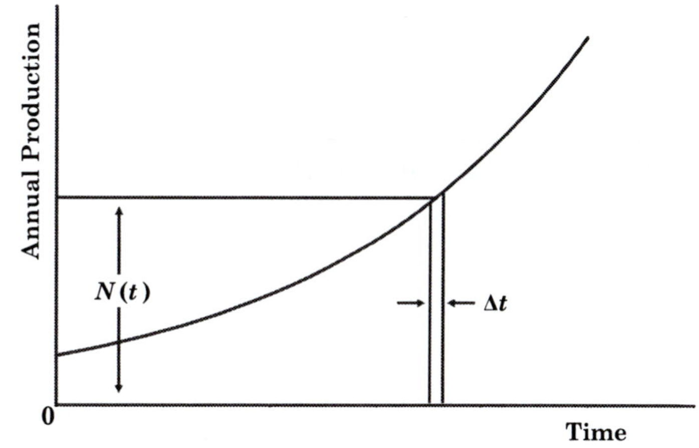

Figure 2-6 The area of the narrow rectangle is $N(t)\Delta t$, representing production during time interval Δt.

In Fig. 2-7, consider the total area under the exponential curve from the present ($t = 0$) until some time T in the future, when all the resource will be just depleted. T is the *resource lifetime*, that is, the time required for production of the resource that remains now (at $t = 0$). The area under the curve is essentially just the sum of the areas of a series of rectangles such as the one in Fig. 2-6, and since each of these areas represents a quantity of resource produced, the total area is the total quantity of resource produced from now until time T. This is just the quantity of resource now remaining, which is represented as Q_T. This area is also the integral of the function from $t = 0$ to $t = T$. Hence,

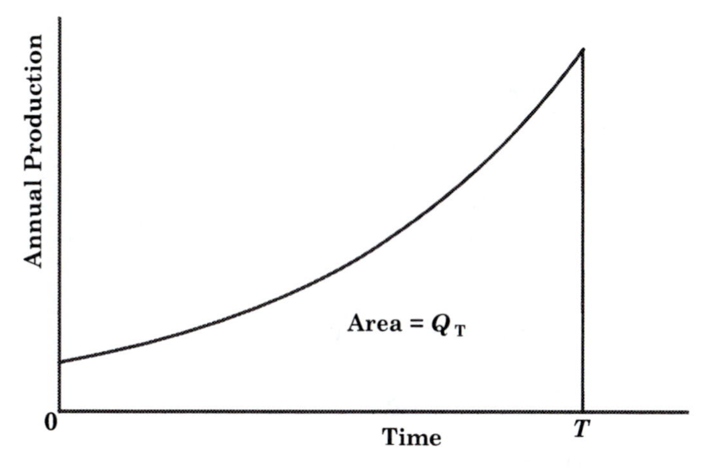

Figure 2-7 The area under the curve from time $t = 0$ to T (the resource lifetime) equals the resource Q_T produced during this time.

Q_T = resource remaining

= quantity produced from the present (t = 0) to time T

= area under curve

$= \int_0^T N(t)\,dt$

Recalling that $N(t) = N_0e^{kt}$, and integrating:

$$Q_T = \int_0^T N_0 e^{kt}\,dt = \frac{N_0}{k} e^{kt}\Big|_0^T = \frac{N_0}{k}\left(e^{kT} - 1\right)$$

This equation must be solved for the resource lifetime T. First re-arrange to isolate e^{kT}, and then take natural logarithms:

$$e^{kT} = \frac{kQ_T}{N_0} + 1 \quad \text{and thus,} \quad kT = \ell n\left(\frac{kQ_T}{N_0} + 1\right)$$

$$\text{Therefore,} \quad T = \frac{1}{k}\ell n\left(\frac{kQ_T}{N_0} + 1\right) \qquad \textbf{[2-8]}$$

Equation [2-8] is extremely useful — given the resource remaining Q_T at a particular time, the production rate N_0 at that time, and the growth constant k, it is possible to determine how long the resource will last.

EXAMPLE 2-5

In the two decades prior to the "oil crisis" of 1973, world oil production had been growing at about 7% per year. (From 1995 to 2005, the average growth rate was much smaller, only 1.7% per year.)

(a) If production were now to return to 7% annual growth, how long would world oil resources last?

(b If the actual resource remaining were discovered to be twice the current estimate, how long would oil last, assuming 7% annual growth?

SOLUTION

(a) Use Eq. [2-8]: $T = \frac{1}{k}\ell n\left(\frac{kQ_T}{N_0} + 1\right)$

Since the growth rate is 7% per year, $k \approx 0.07$ yr^{-1} from approximation [2-6]. Use $Q_T \approx 12000$ EJ from Table 2-3, and N_0 = 164 EJ/yr from Table 2-4. Substituting values:

$$T \approx \frac{1}{0.07\ \text{yr}^{-1}} \ell n\left(\frac{(0.07\ \text{yr}^{-1})(12000\ \text{EJ})}{164\ \text{EJ/yr}} + 1\right) \approx 26\ \text{yr}$$

Hence, if world oil production were to grow at 7% per year, the world as a whole would run out of oil in about 26 yr. Contrast this value with the 75 yr calculated in Section 2.6 assuming no growth.

(b) If the resource is doubled, $Q_T \approx 24000$ EJ. Re-doing the calculation with this value gives $T \approx 35$ yr. Doubling the resource extends the lifetime by less than 10 yr!! It is apparent that exponential growth has an insidious effect on resource lifetime.

Table 2-6
Approximate Lifetime (in yr) of Fossil Fuel Resources
Assuming Exponential Growth With Various Annual % Growth Rates

Region & Fuel	0%	2%	5%	10%	Actual Avg. Annual Growth Rate 1995-2005
World Oil	75	45	31	22	1.7%
Canada Oil	200	80	48	31	2.6%
USA Oil	45	33	24	18	−2.1%
World Gas	75	46	32	22	2.6%
Canada Gas	55	38	27	20	1.6%
USA Gas	35	27	20	15	−0.2%
World Coal	1200	160	84	50	2.5%
Canada Coal	1800	180	92	54	−1.5%
USA Coal	1500	170	88	52	0.3%

Table 2-6 shows the results of applying Eq. [2-8] to determine resource lifetimes of oil, gas, and coal for annual growth rates of 2%, 5% and 10% in production. Also tabulated for reference are the no-growth (constant production) lifetimes from Table 2-5, and the actual average annual growth rates[6] in production from 1995 to 2005. It is apparent that even a modest growth rate of 2% has an enormous effect on resource lifetime, and a growth rate as large as 10% is devastating. You might even be wondering whether the coal lifetimes are correct — it does seem surprising that a no-growth lifetime of 1800 years can be reduced to only 180 years with only 2% annual growth. However, consider the following: at 2% growth per year, the doubling time is 35 years; this means that

[6] Source: BP Statistical Review of World Energy 2006

after 35 years of 2% growth, coal is being produced twice as fast, and if the growth were then to cease, the resource would last only about 800 years instead of 1600 years. Only 35 years of 2% growth has effectively wiped out 800 years of coal resources. And the growth does not stop — after another 35 years, production doubles yet again, and this process continues until all the coal is depleted, in only 180 years.

It is clear that if fossil fuels are to last a long time, their production cannot be allowed to grow exponentially. However, the last column in Table 2-6 indicates that *world* production of oil, gas and coal has been growing recently. Although US oil production has been declining by 2.1% annually, US oil *consumption* from 1995 to 2005 increased with an average annual growth rate of 1.6% as a result of imports. During the same time period, US natural gas production and Canadian coal production also declined at an average annual rate of 0.2% and 1.5%, respectively..

The Future

Is exponential growth a good model for the future? Probably not. It would be more reasonable to expect that as a resource begins to dwindle, its price will rise and production will fall, and this type of scenario is explored in the Hubbert model of resource production in Chapter 3. Nevertheless, a knowledge of exponential growth is important in understanding how production usually behaves when the resource is still very abundant, and how quickly a finite resource would disappear if production were to continue to increase exponentially.

EXERCISES

2-1. The growth rate of the world's population is about 1.2% annually (down from 2.0% in 1972). In the year 2006 the population was about 6.5 billion.
(a) Determine the growth constant and doubling time which correspond to the present growth rate.
(b) If the population continues to grow at 1.2% per year, what will the world's population be in 2030?
(c) If the average mass of a person is 70 kg, how long would it take until the mass of people equals the mass of Earth, assuming population growth of 1.2% per year? The mass of Earth is 5.98×10^{24} kg.

2-2. Between 2000 and 2005, production of natural gas in Mexico grew with an average growth rate of 2.0% per year. Determine (in your head) approximate values of the growth constant and doubling time.

2-3. Oil production in Italy increased from 2.6 million tonnes in the year 1986 to 5.2 million tonnes in 1995, that is, production doubled in 9 yr. Assuming exponential growth, determine (in your head) approximate values of the growth constant and annual percent growth rate for Italian oil production.

2-4. Imagine that a savings account was established at the time of Christ, 2000 years ago, with a deposit equivalent to 1¢. Assume interest, compounded annually, at 5%. Calculate the present value of the savings account.

2-5. The following data represent exponential growth of a quantity N as a function of time t.

t (yr)	0	1	2	3	4	5	6	7	8
N	2.85	4.50	6.90	9.00	14.9	22.3	35.5	51.3	90.0

(a) Plot $\ell n\, N$ vs. t, using semilog paper. (A sheet is provided in Appendix II.)
(b) Draw a straight line to fit the points as well as possible, and then use two well-separated points on your line to determine the growth constant.

PROBLEMS

2-6. Figure 2-8 shows natural gas production[7] in Norway from 2000 to 2005. The black squares indicate the actual data points, and the line is a computerized linear best-fit to the points.
(a) In 2004 what was the natural gas production in Norway?
(b) Convert your answer in (a) to exajoules.
(c) Using the line on the graph, determine the growth constant for Norwegian natural gas production.
(d) Assuming exponential growth at the rate determined in (c), when would natural gas production in Norway reach 400 million tonnes of oil equivalent?

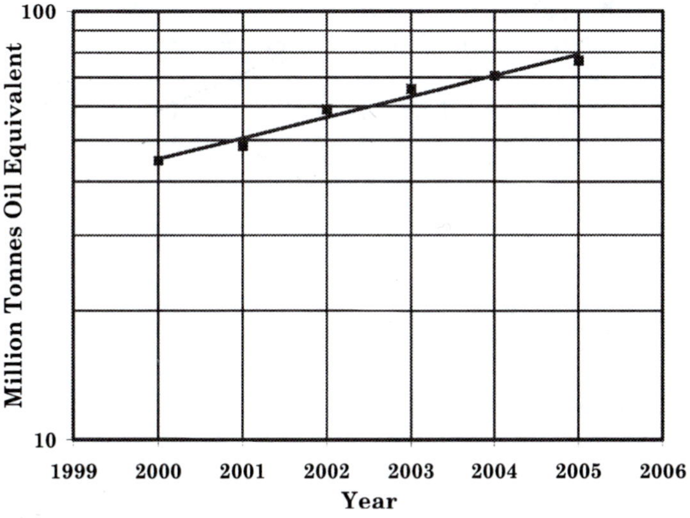

Figure 2-8 Natural Gas Production in Norway

2-7. How long would world oil resources last if production were to continue to grow at 1.7% per year (the average growth rate for 1995-2005)? Use appropriate data from Tables 2-3 and 2-4.

2-8. From 1990 to 1996, the growth in coal production in China was approximately exponential. The 1990 production was equivalent to 542 million tonnes of oil, and the 1996 production was equivalent to 681 million tonnes of oil.
(a) Determine the growth constant and the average annual % growth rate.
(b) Determine the doubling time of Chinese coal production, assuming constant exponential growth at the 1990-1996 rate.
(c) Chinese coal resources were approximately 18000 quads at the end of 1996. Determine the lifetime of coal resources in China if:
(i) production remains constant at the 1996 rate
(ii) production continues to increase exponentially at the 1990-1996 rate.

[7] The unit, million tonnes of oil equivalent, is often used to quantify production or consumption of energy sources other than oil.

2-9. The production of a resource in a particular country is increasing at 12% per year. If production continues to increase exponentially at this rate, the resource will be depleted in 47 yr. Suppose that the quantity of resource is suddenly found to be 3.0 times what was previously known. How many years will it take to deplete this larger resource, assuming exponential growth at the 12% annual rate?

ANSWERS

2-1. (a) 0.012 yr^{-1}, 58 yr (b) 8.7 billion
(c) 2.5×10^3 yr, a short time in the history of humanity

2-2. 0.020 yr^{-1}, 35 yr

2-3. 0.08 yr^{-1}, 8% per yr

2-4. $\$2.4 \times 10^{40}$

2-5. (b) 0.40 yr^{-1} to 0.42 yr^{-1}

2-6. (a) 70 million tonnes of oil equivalent (b) 3.0 EJ (c) 0.12 yr^{-1} (d) 2018 or 2019

2-7. 48 yr

2-8. (a) 0.038 yr^{-1}, 3.9% (b) 18 yr (c) (i) 6.6×10^2 yr (ii) 86 yr

2-9. 57 yr

CHAPTER 3

THE HUBBERT MODEL OF RESOURCE PRODUCTION

In Chapter 2, two possible models of future fossil fuel production were explored: constant production, and exponential growth. It was argued that exponential growth seemed to be a good model, based on historical and present data, for the production of a resource when the resource is still abundant. However, it is not reasonable to expect that production of, say, oil will continue to increase until the moment when all the oil is depleted. As supplies start to dwindle, it is more probable that prices will increase and production will drop. Hence, rather than having production that increases continuously, it is more likely to have a levelling-off and then a decline. This model was suggested in the 1950s by Dr. M. King Hubbert, a geologist working for Shell Oil in Texas. At that time, his work was given little credence by the petroleum industry, but US domestic oil production has followed his predictions very well, with a peak in the mid-1970s and now a decline. (Because of imports, US *consumption* has been able to increase.)

3.1 INTRODUCTION TO THE HUBBERT MODEL

To gain an overall appreciation of the Hubbert model, consider a typical pattern of resource production. Important factors are the rate of discovery of the resource, the rate of production, and the size of the reserves at any time. Figure 3-1 illustrates a plausible graph of the cumulative quantity of resource produced, Q_P, vs. time, t, for a finite resource such as oil, gas, or coal. The resource is first utilized slowly, and then more rapidly as technologies are developed to make better use of the resource's potential. However, the rate of production (that is, the slope of the curve) eventually declines as the resource begins to "run out,"

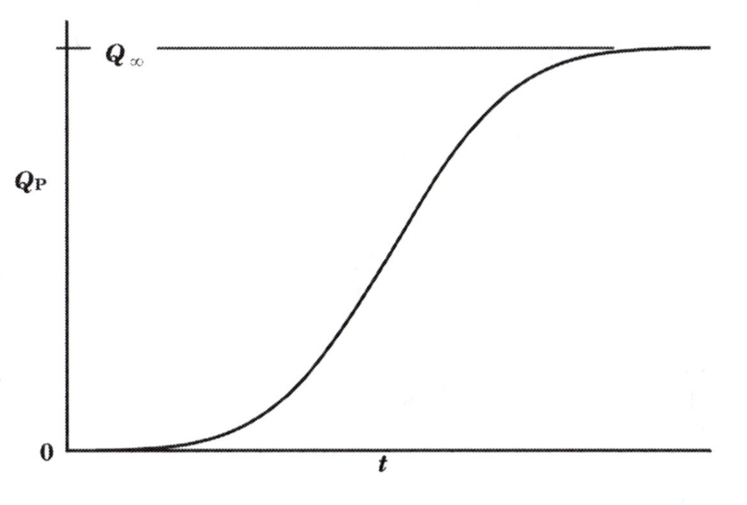

Figure 3-1 A graph of cumulative quantity of resource produced, Q_P, vs. time, t. Q_∞ represents the total quantity of the resource.

and ultimately the cumulative production asymptotically approaches a value that equals the total quantity of the resource that was ever available. In Fig. 3-1 this latter quantity is represented as Q_∞. A curve as smooth as the one shown would not occur in actual practice, of course; there would be bumps and wiggles along it, depending of the state of the economy, international conflicts, etc. Nonetheless, it is reasonable to assume that a graph of long-term cumulative production vs. time would show the general "S-shape" of Fig. 3-1.

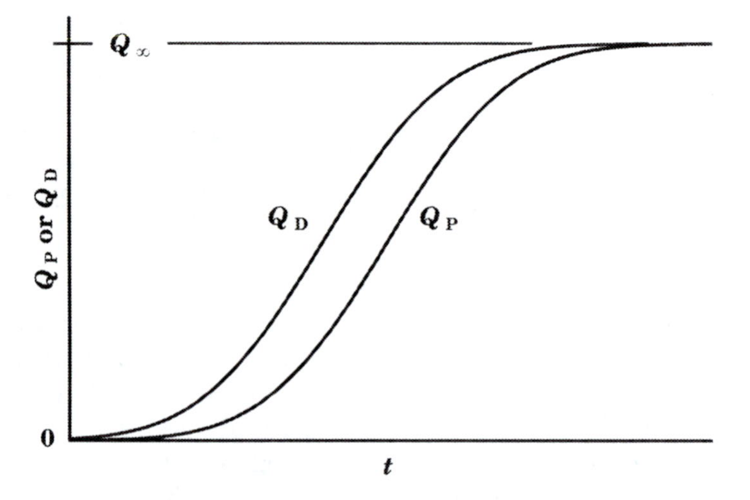

Figure 3-2 Quantity of resource discovered, Q_D, and quantity produced, Q_P, vs. time t. The Q_D curve precedes the Q_P curve.

A graph of the cumulative quantity of resource *discovered* will have a similar S-shape, but precedes the production curve. Figure 3-2 shows both the cumulative quantity of resource produced, Q_P, and the cumulative quantity of resource discovered, Q_D. The Q_D curve precedes the Q_P curve by a fixed amount of time.

At any given time the quantity of resource that is known to be available is simply the difference between the cumulative amount discovered and the cumulative amount produced. This available quantity represents the reserves, represented as Q_R:

$$Q_R = Q_D - Q_P \qquad \textbf{[3-1]}$$

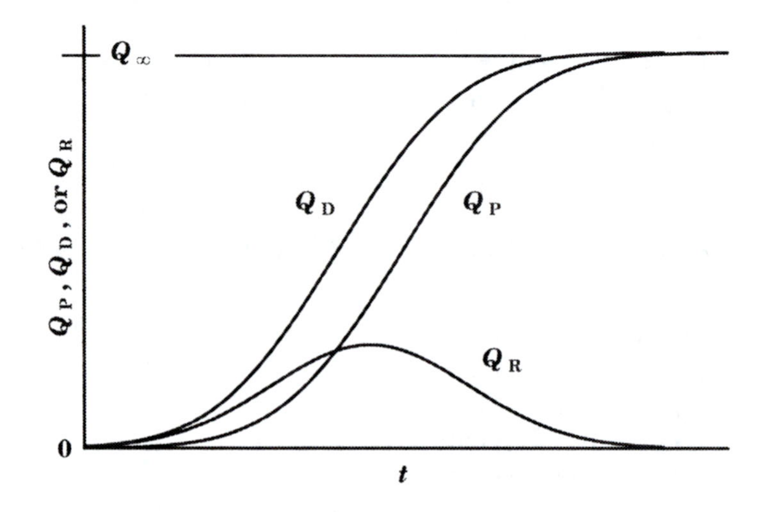

Figure 3-3 The quantity of reserves, Q_R, at any given time is the difference between the quantity discovered, Q_D, and the quantity produced, Q_P.

The three quantities, Q_P, Q_D, and Q_R, are illustrated as functions of time in Fig. 3-3. You might wish to confirm by direct measurement on the graph that, for any given time, $Q_R = Q_D - Q_P$.

Thus far the *cumulative* amounts of resource produced and discovered, and their difference (the reserves), have been considered. It is also useful to investigate the time rate of change of each of these quantities.

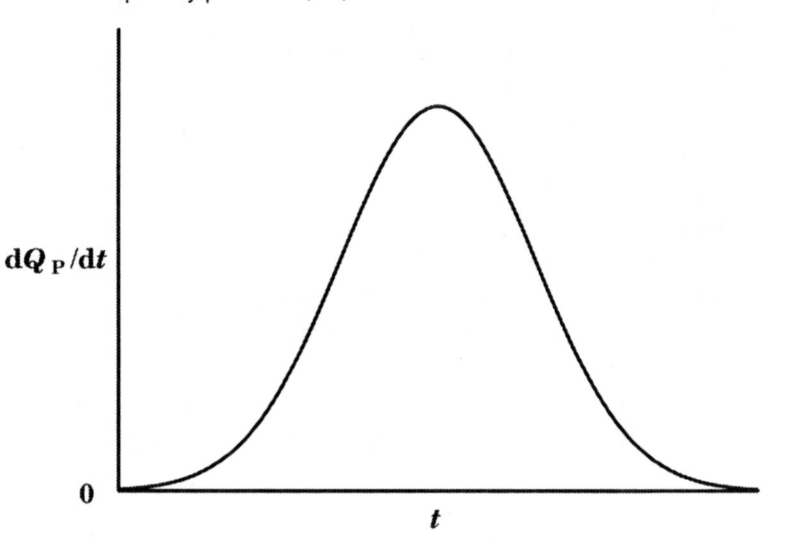

Figure 3-4 The rate of production of a resource, dQ_P/dt, vs. time, t.

First consider the rate of change of the cumulative quantity produced, that is, dQ_P/dt, which is the slope of the curve of Fig. 3-1. This slope is zero at the beginning (bottom) of the curve, reaches a positive maximum in the middle, and approaches zero again at the end (top). This slope, that is, the rate of production, is shown in Fig. 3-4. In Chapter 2, rate of production (often expressed in exajoules/yr) was given the symbol N, and that notation is continued here:

$$N = \frac{dQ_P}{dt} \qquad \textbf{[3-2]}$$

The rate at which resource is discovered (dQ_D/dt) is the slope of the curve of Q_D vs. t in Fig. 3-2, and has the same shape as the graph of rate of production vs. time (Fig. 3-4), but precedes it in time. Figure 3-5 illustrates both the production rate ($N = dQ_P/dt$) and the discovery rate (dQ_D/dt).

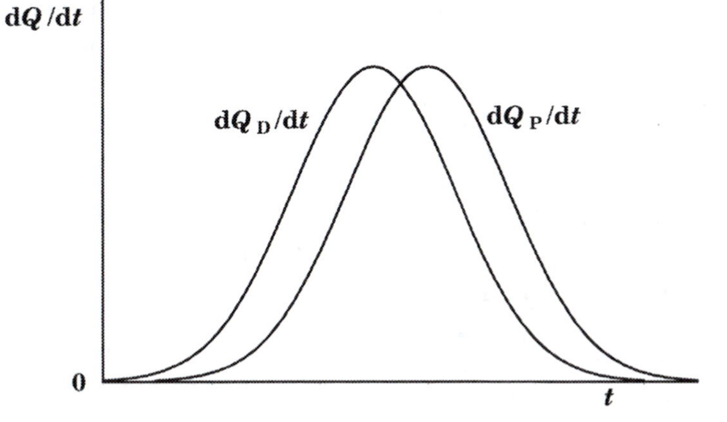

Figure 3-5 The production rate, dQ_P/dt, and the discovery rate, dQ_D/dt, of a resource.

Now consider the rate of change of reserves, dQ_R/dt. The graph of Q_R vs. t shown in Fig. 3-3 begins with a slope of zero, has a positive slope until the quantity Q_R reaches its peak, at which time the slope is again zero, and then the slope is negative (that is, the reserves are decreasing); the slope eventually becomes zero again at large time-values. Figure 3-6 shows a graph of dQ_R/dt vs. t, along with dQ_P/dt and dQ_D/dt.

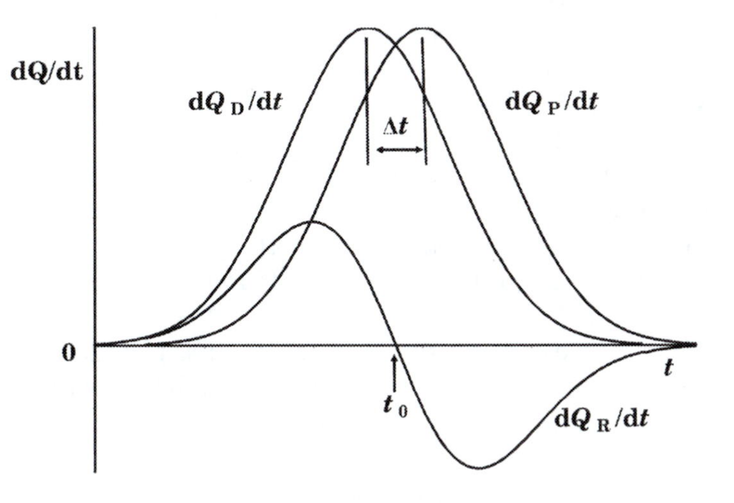

Figure 3-6 The rate of change of reserves, dQ_R/dt, has a value of zero at time t_0. Δt is the lag time between discovery and production (see text).

The three curves in Figure 3-6 together illustrate a very interesting property: as the curve of dQ_R/dt passes through its central value of zero (that is, as the reserves are at their maximum), the curve of production rate going up crosses the curve of discovery rate going down. This statement can be justified mathematically by taking the time derivative of Eq. [3-1]:

$$\frac{dQ_R}{dt} = \frac{dQ_D}{dt} - \frac{dQ_P}{dt}$$

Therefore, when $dQ_R/dt = 0$, it follows that $dQ_D/dt = dQ_P/dt$, that is, the discovery rate equals the production rate.

Furthermore, from the symmetry of the curves, this particular time is half-way (in time) between the peak of the discovery rate curve and the peak of the production rate curve. This observation provides a method of predicting in advance when the peak of the production rate will occur[1]. All that is required is a knowledge of (1) the lag time, Δt, between discovery rate and production rate, which is straightforward to obtain by plotting a graph of these quantities vs. time, and (2) the time, t_0, at which $dQ_R/dt = 0$, which can be obtained from a graph of dQ_R/dt vs. t. (See Fig. 3-6.) Then the peak in production can easily be predicted to occur at a time of $t = t_0 + \Delta t/2$, that is, one-half lag time after the time when $dQ_R/dt = 0$.

In the 1960s Hubbert plotted dQ_R/dt vs. t for US domestic oil production and found that the graph crossed the t-axis around 1961. A plot of this graph, including data up to 1979, is shown in Fig. 3-7; this plot corresponds only to the central, approximately linear, region of the graph of dQ_R/dt shown in Fig. 3-6. When Hubbert plotted his graph, the lag time Δt was known to be about 12 yr, and so he estimated that US oil production would peak in about the year 1967 (i.e., 1961 + 12/2). US production actually peaked in 1973, and so Hubbert's prediction was not far off the mark.

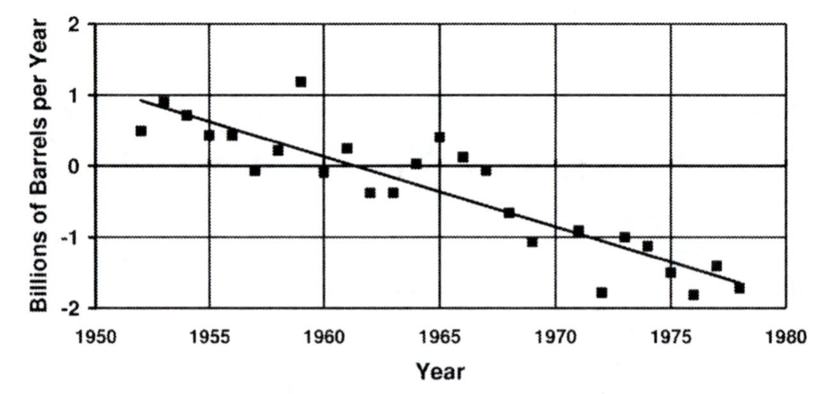

Figure 3-7 Graph of dQ_R/dt, in billions of barrels per year, for US oil production. The black squares are data points, and the line is a computerized fit to the data.

The next section focuses on the graph of production rate ($N = dQ_P/dt$) as a function of time, and uses Hubbert's model to predict not only when the production rate will peak, but also how much resource will be left at any time in the future.

3.2 THE MATHEMATICS OF THE HUBBERT MODEL

Figure 3-4 in the previous section shows a graph of production rate of a resource vs. time; this rate increases, then levels off and declines in a symmetrical way. A

[1] A different method of predicting this peak time will be discussed in Section 3.2.

convenient mathematical function that follows this pattern is the *Gaussian*, or *normal*, function (Fig. 3-8).[2]

To describe the production of a resource by such a function, three parameters are of particular importance:

- the time T_M when the function has its maximum value (peak)
- the height of the peak, that is, the maximum production rate N_M (the production rate at time T_M)
- the "width" of the function. There is a convention for specifying the width: the *standard deviation*, represented by σ. There are a number of ways to describe σ, but probably the easiest is to say that σ is the width (measured from the peak in either direction) at which the value of the function — the production — is 61% of the peak value (0.61 N_M).

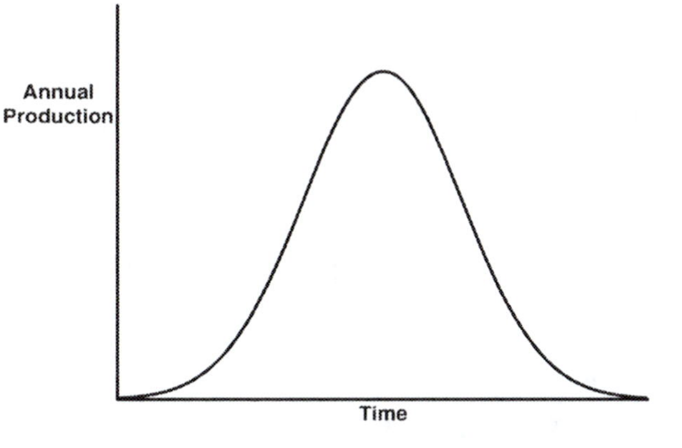

Figure 3-8 A Gaussian function.

The parameters T_M , N_M , and σ are illustrated in Fig. 3-9.

The full mathematical expression of the production rate N (or dQ_P/dt) as a Gaussian function of time t is:

$$N = N(t) = N_M e^{-\frac{(t-T_M)^2}{2\sigma^2}} \qquad \textbf{[3-3]}$$

Notice that when $t = T_M$, this equation becomes $N = N_M\, e^0 = N_M$, as it should. As well, when the time t differs from the mid-time T_M by σ, i.e., when $|t - T_M| = \sigma$, then $N = N_M\, e^{-1/2} = 0.61\, N_M$, as mentioned above.

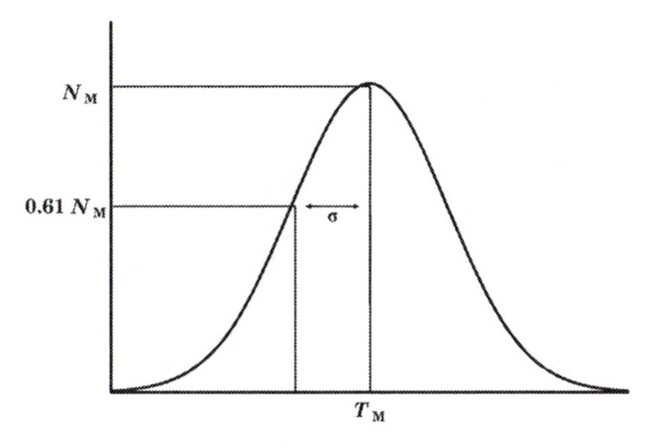

Figure 3-9 Parameters T_M, N_M, and σ of a Gaussian function.

The total resource, including that already produced and that remaining, is the area under the entire curve, which is the integral of the function; as in Section 3.1 this total quantity of resource is represented as Q_∞ . (Note that in Section 2.7, Q_T represented only the quantity of resource remaining.)

$$Q_\infty = \int_{-\infty}^{+\infty} N(t) = \int_{-\infty}^{+\infty} N_M e^{-\frac{(t-T_M)^2}{2\sigma^2}}\, dt \qquad \textbf{[3-4]}$$

There is a convenient relation between the area, Q_∞ , under the Gaussian function, and the quantities N_M and σ:

[2] A function such as the Lorentzian function also could be used to model the rise and fall of production rate.

$$N_{\mathrm{M}} = \frac{Q_{\infty}}{\sigma\sqrt{2\pi}} \qquad \textbf{[3-5]}$$

Equation 3-5 will not be derived here, but its usefulness will be demonstrated in Example 3-2. When working with Gaussian functions, it is convenient to define another variable, z:

$$z = \frac{|t - T_{\mathrm{M}}|}{\sigma} \qquad \textbf{[3-6]}$$

With this definition of z, Eq. [3-3] simplifies to:

$$N = N_{\mathrm{M}} e^{-z^2/2} \qquad \textbf{[3-7]}$$

As illustrated in Example 3-1 below, the variable z indicates how many standard deviations a particular time t is away from the peak-time T_{M}.

EXAMPLE 3-1

If $z = 2$, what are the corresponding values of the time t, in terms of T_{M} and σ?

SOLUTION

Using Eq. [3-6]: $z = 2 = \frac{|t - T_{\mathrm{M}}|}{\sigma}$ or $2\sigma = |t - T_{\mathrm{M}}|$

There are two possible values of t, depending on whether the quantity $t - T_{\mathrm{M}}$ is positive or negative. If $t - T_{\mathrm{M}} > 0$, that is, if $t > T_{\mathrm{M}}$ (time t is past the peak-time T_{M}), then $|t - T_{\mathrm{M}}| = t - T_{\mathrm{M}}$, and hence $2\sigma = t - T_{\mathrm{M}}$, which gives: $t = T_{\mathrm{M}} + 2\sigma$.

If $t - T_{\mathrm{M}} < 0$, that is, $t < T_{\mathrm{M}}$ (time t has not yet reached the peak-time), then $|t - T_{\mathrm{M}}| = T_{\mathrm{M}} - t$ and therefore, $2\sigma = T_{\mathrm{M}} - t$, giving $t = T_{\mathrm{M}} - 2\sigma$.

Thus, if $z = 2$, then $t = T_{\mathrm{M}} \pm 2\sigma$.

Example 3-1 shows that if $z = 2$, then the time t is 2 standard deviations away from T_{M} (Fig. 3-10). If z were 3.67, then t would be 3.67 standard deviations away from T_{M}. In effect, z is just a different way of specifying time t; for any arbitrary value of z, time $t = T_{\mathrm{M}} \pm z\sigma$.

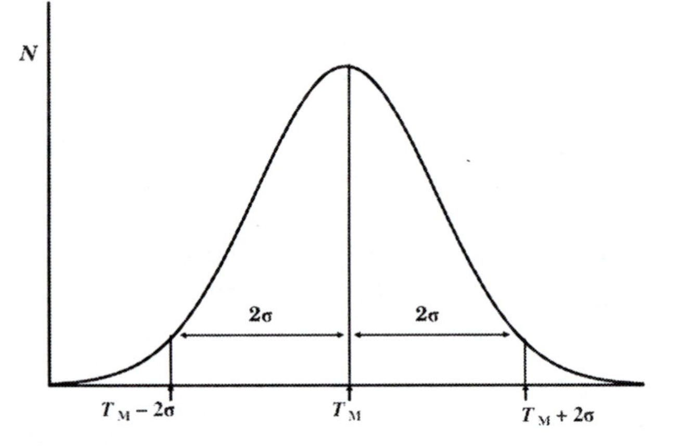

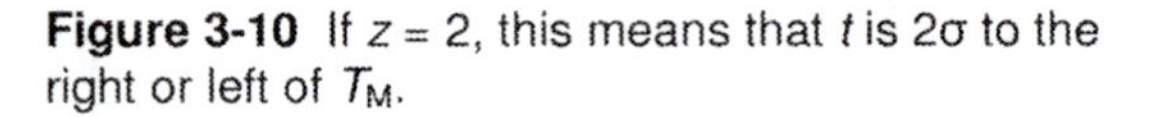
Figure 3-10 If $z = 2$, this means that t is 2σ to the right or left of T_{M}.

Before proceeding to another example using a Gaussian function, one of its most useful features needs to be considered. Regardless of the specific values of T_{M} , N_{M} , and σ, the area under

Table 3-1
Integral of the Gaussian Function vs. z

This table gives the integral of the Gaussian function from $T_M - z\sigma$ to $T_M + z\sigma$, shown as the shaded area to the right, as a decimal fraction of the total area under the function.

z	.00	.01	.02	.03	.04	.05	.06	.07	.08	.09
0.0	.0	.0080	.0160	.0239	.0319	.0399	.0478	.0558	.0638	.0717
0.1	.0797	.0876	.0955	.1034	.1113	.1192	.1271	.1350	.1428	.1507
0.2	.1585	.1663	.1741	.1819	.1897	.1974	.2051	.2128	.2205	.2282
0.3	.2358	.2434	.2510	.2586	.2661	.2737	.2812	.2886	.2961	.3035
0.4	.3108	.3182	.3255	.3328	.3401	.3473	.3545	.3616	.3688	.3759
0.5	.3829	.3899	.3969	.4039	.4108	.4177	.4245	.4313	.4381	.4448
0.6	.4515	.4581	.4647	.4713	.4778	.4843	.4907	.4971	.5035	.5098
0.7	.5161	.5223	.5285	.5346	.5407	.5467	.5527	.5587	.5646	.5705
0.8	.5763	.5821	.5878	.5935	.5991	.6047	.6102	.6157	.6211	.6265
0.9	.6319	.6372	.6424	.6476	.6528	.6579	.6629	.6680	.6729	.6778
1.0	.6827	.6875	.6923	.6970	.7017	.7063	.7109	.7154	.7199	.7243
1.1	.7287	.7330	.7373	.7415	.7457	.7499	.7540	.7580	.7620	.7660
1.2	.7699	.7737	.7775	.7813	.7850	.7887	.7923	.7959	.7995	.8029
1.3	.8064	.8098	.8132	.8165	.8198	.8230	.8262	.8293	.8324	.8355
1.4	.8385	.8415	.8444	.8473	.8501	.8529	.8557	.8584	.8611	.8638
1.5	.8664	.8690	.8715	.8740	.8764	.8789	.8812	.8836	.8859	.8882
1.6	.8904	.8926	.8948	.8969	.8990	.9011	.9031	.9051	.9070	.9090
1.7	.9109	.9127	.9146	.9164	.9181	.9199	.9216	.9233	.9249	.9265
1.8	.9281	.9297	.9312	.9327	.9342	.9357	.9371	.9385	.9399	.9412
1.9	.9426	.9439	.9451	.9464	.9476	.9488	.9500	.9512	.9523	.9534
2.0	.9545	.9556	.9566	.9576	.9586	.9596	.9606	.9615	.9625	.9634
2.1	.9643	.9651	.9660	.9668	.9676	.9684	.9692	.9700	.9707	.9715
2.2	.9722	.9729	.9736	.9743	.9749	.9756	.9762	.9768	.9774	.9780
2.3	.9786	.9791	.9797	.9802	.9807	.9812	.9817	.9822	.9827	.9832
2.4	.9836	.9840	.9845	.9849	.9853	.9857	.9861	.9865	.9869	.9872
2.5	.9876	.9879	.9883	.9886	.9889	.9892	.9895	.9898	.9901	.9904
2.6	.9907	.9909	.9912	.9915	.9917	.9920	.9922	.9924	.9926	.9929
2.7	.9931	.9933	.9935	.9937	.9939	.9940	.9942	.9944	.9946	.9947
2.8	.9949	.9950	.9952	.9953	.9955	.9956	.9958	.9959	.9960	.9961
2.9	.9963	.9964	.9965	.9966	.9967	.9968	.9969	.9970	.9971	.9972
3.0	.9973									
3.5	.9995									
4.0	.9999									

the curve from time $T_M - \sigma$ to time $T_M + \sigma$ is always 0.6827 Q_∞, that is, this area is always 68.27% of the total area under the curve. Similarly, the area from $T_M - 2\sigma$ to $T_M + 2\sigma$ is always 95.45% of the total, i.e., 0.9545 Q_∞. In general, the area from $T_M - z\sigma$ to $T_M + z\sigma$ as a fraction of the total area depends only on z.

As will soon be apparent, this property is very convenient when dealing with resource production. Table 3-1 shows the area under the curve from T_M - $z\sigma$ to $T_M + z\sigma$ (as a decimal fraction of the total area) as a function of z. The value of z is given in the left-hand column (to one decimal place) and the top row (to the second decimal place). As practice, use the table to confirm that:

- for $z = 1$, the fractional area is 0.6827
- for $z = 2$, the value is 0.9545
- for $z = 1.46$, the value is 0.8557.

Using the Hubbert Model

Any Gaussian function can be characterized by the three parameters T_M, N_M, and σ. In order to determine these parameters, three independent pieces of information are needed as input. In Example 3-2 that follows, the three given quantities are:

- current production rate (N)
- cumulative production to date
- total resource (Q_∞)

EXAMPLE 3-2

In 2005, the world oil production (rate) was 168 EJ/yr, and the cumulative production to the end of 2005 was 6200 EJ. The resource remaining (from Table 2-3) is approximately 12000 EJ. Use the Hubbert model to predict

(a) the year of peak production (T_M)
(b) the peak production (N_M)
(c) the year when 20% of the resource will remain
(d) the production in the year 2040.

SOLUTION

(a) First determine the total resource, Q_∞, which is the sum of the resource produced and the resource remaining:

Q_∞ = (6200 + 12000) EJ = 18200 EJ

Now find the value of z for the year 2005, and then determine σ and T_M.

At the end of 2005, the world had produced 6200 EJ of oil (the left-hand unshaded area in Fig. 3-11), out of a total of about 18200 EJ. As a fraction of the total, this is 6200/18200 = 0.341, or 34.1%. Because of the symmetry of the Gaussian function, in the future as oil production is dwindling there will be another 34.1% produced, represented by the right-hand unshaded area. The two unshaded areas make up 0.682 of the total, leaving 0.318 in the centre (shaded). Now that this central shaded area is known, z can be found. Look in the body of Table 3-1 for the entry closest to 0.318, and read off the z-value, 0.41.

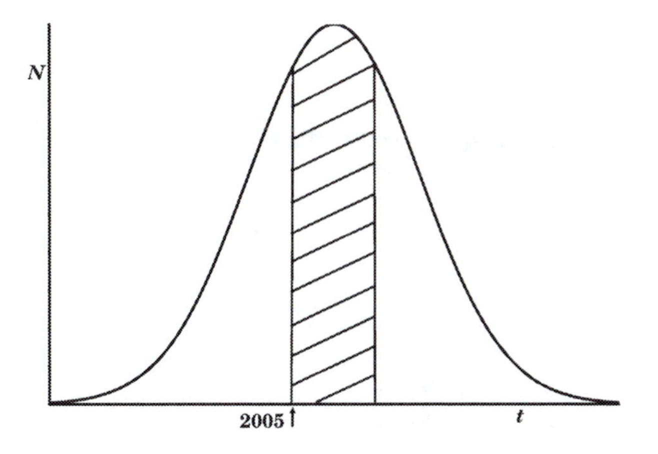

Figure 3-11 By the end of 2005, the world had consumed 6200 EJ of oil (the left-hand unshaded area), out of a total of about 18200 EJ.

Now use Eq. [3-7]: $N = N_M e^{-z^2/2}$

Into this equation, substitute the expression for N_M from Eq. [3-5]:

This gives: $$N = \frac{Q_\infty}{\sigma\sqrt{2\pi}} e^{-z^2/2}$$

Re-arrange to solve for σ: $$\sigma = \frac{Q_\infty}{N\sqrt{2\pi}} e^{-z^2/2}$$

Use this equation to calculate σ; for the year 2005, N = 168 EJ/yr and z = 0.41; and since Q_∞= 18200 EJ:

$$\sigma = \frac{18200 \text{ EJ}}{(168 \text{ EJ/yr})\sqrt{2\pi}} e^{-(0.41)^2/2}$$
$$= 39.7 \text{ yr}$$

Now, knowing σ, z, and t, use Eq. [3-6] to find T_M: $$z = \frac{|t - T_M|}{\sigma}$$

For the year 2005, $t < T_M$, since less than half of the total oil has been produced at this time. Therefore, $|t - T_M| = T_M - t$.

Hence, $$z = \frac{T_M - t}{\sigma} \quad \text{or} \quad z\sigma = T_M - t$$

Solving for T_M : $T_M = t + z\sigma = 2005 + (0.41)(39.7) = 2021$

Hence, the Hubbert model predicts that world oil production will peak in about the year 2021.

(b) To determine the peak production N_M , simply use Eq. [3-5]:

$$N_M = \frac{Q_\infty}{\sigma\sqrt{2\pi}} = \frac{18200 \text{ EJ}}{(39.7 \text{ yr})\sqrt{2\pi}} = 183 \text{ EJ/yr, or } 1.8\times10^2 \text{ EJ/yr (to 2 sig. digits)}$$

Thus, the peak production predicted by the Hubbert model is about 1.8×10^2 EJ/yr.

(c) The situation when 20% of the oil remains is illustrated in Fig. 3-12. The answer required is the time t at the beginning of the right-hand unshaded region, which contains 0.20 of the total area. By symmetry, the left-hand unshaded region also contains 0.20 of the total area, leaving 0.60 as the shaded area in the centre. From Table 3-1, this corresponds to z = 0.84. Now use Eq. [3-6] to find t:

$$z = \frac{|t - T_M|}{\sigma}$$

For the time t when only 20% remains, $t > T_M$ since more than half of the oil has been produced. Therefore, $|t - T_M| = t - T_M$, and

$$z = \frac{t - T_M}{\sigma} \quad \text{or} \quad z\sigma = t - T_M$$

Thus, $t = T_M + z\sigma$
$= 2021 + (0.84)(39.7)$
$= 2054$

The Hubbert model predicts that 20% of world oil will remain in about the year 2054.

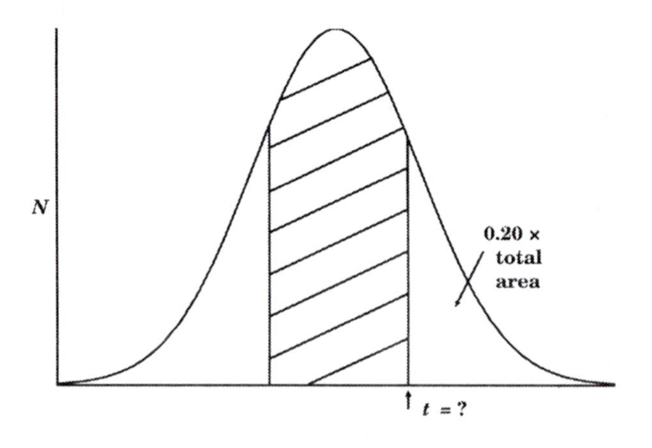

Figure 3-12 Determining when only 20% of world oil will remain.

(d) To find the production in the year 2040, first determine the z-value for that year:

$$z = \frac{|t - T_M|}{\sigma} = \frac{|2040 - 2021|}{39.7} = 0.48$$

Now use Eq. [3-7]: $N = N_M e^{-z^2/2} = 183\, e^{-(0.48)^2/2} = 1.6 \times 10^2$ EJ/yr

Therefore, according to the Hubbert model, the production will be approximately 1.6×10^2 EJ/yr in the year 2040.

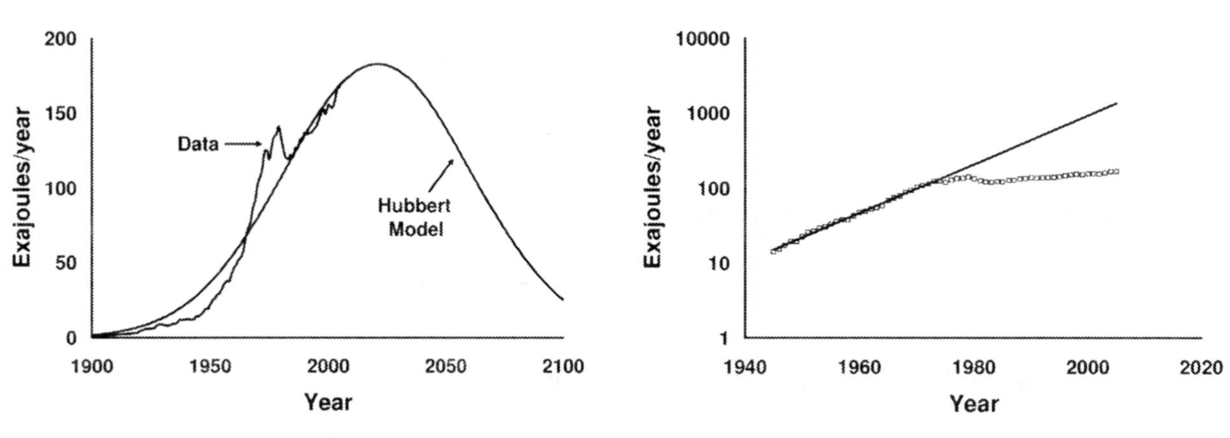

Figure 3-13 Hubbert model for world oil production.

Figure 3-14 World oil production from 1945 to 2005. Data points are shown by open squares (□). The straight line has a slope $k = 0.078$ yr^{-1}.

It is interesting to use the results of the above example to plot a graph of world oil production vs. time (Fig. 3-13). The smooth bell-shaped Gaussian curve has been determined from calculations of production N for various times t, just as was done in part (d) of Example 3-2 for the year 2040. The irregular curve shows actual data points for world oil production. What this graph shows quite clearly is that world oil production increased rapidly from 1900 to 1975, and has been following a somewhat erratic pattern since then. For the past few years the data curve follows the Hubbert model quite well.

Figure 3-14 shows the same world oil production data (but only from 1945 to 2005) replotted on a semilog graph. The straight line on this graph has a slope of 0.078 yr^{-1}, and fits the data extremely well between 1945 and 1973. This indicates that world oil production was increasing exponentially with a growth of about 8% per year during this period. Since 1973, the line and the data points have very poor agreement, demonstrating that the exponential growth phase of world oil production has ended.

The calculations in Example 3-2 were done with the assumption that the remaining world oil resources are 12000 EJ. However, as pointed out in Chapter 2, there is much uncertainty in quantities of resources. The value of 12000 EJ is in the middle of the range of estimated remaining resources, which could be as low as 7000 EJ or as high as 17000 EJ according to the US Energy Information Administration.[3] Using 7000 EJ as the value for remaining resources, the year of peak production becomes 2007, and using 17000 EJ, the peak year is predicted to be 2033. Regardless of which value is used for oil resources, the Hubbert model predicts that world oil production will peak in the next few decades. At that time one half of all the world's conventional oil resources will be depleted, and attention will have to be turned toward harnessing other sources of energy with minimal environmental effects.

3.3 FUTURE ENERGY PRODUCTION

Chapters 2 and 3 have briefly explored various scenarios for future production of fossil fuel resources: constant production, exponential growth, and the Hubbert model. For any one fossil fuel (or any finite resource), it is likely that its early production will grow exponentially, and then will level off and decline, following a curve such as a Gaussian. World oil has been considered in Example 3-2, and natural gas and coal could be treated in a similar manner.

But what about total energy production, which historically has shown exponential growth with very few downturns? What will it do in the future?

This question is extremely difficult to answer, because production of individual energy sources and total production of energy depend on a large number of factors:

- population
- economic activity (GDPs of major countries)
- the price of energy
- availability of energy (e.g., a strike by coal-miners can stop the supply of coal)

[3] J.H. Wood et al, "Long-Term World Oil Supply Scenarios," US Energy Information Administration, August 18, 2004

- technological advances[4]
- development of alternative sources of energy[3]
- government policies
- world political stability
- public attitude

Many of these factors are inter-related — for example, public attitude can affect government policies, which in turn can affect the price of energy and economic activity. At present, many people have strong concerns about the environmental effects of energy production, and might be willing to pay more for energy if environmental problems could be reduced. These concerns might well affect the future choice of sources, the price of energy, and the willingness of government, public and private utilities, and private industries to reduce environmental pollution and encourage greater energy efficiency.

In the long term, neither population nor energy production can continue to increase indefinitely — Earth's environment simply would be unable to withstand the pressures.

EXERCISES

3-1. Use Fig. 3-15 for this exercise.
(a) If area $2B$ represents 75% of the total area under the curve, what is the corresponding z-value?
(b) If area $2A$ represents 30% of the total area under the curve, what is the corresponding z-value?
(c) If area B represents 30% of the total area under the curve, what is the corresponding z-value?
(d) If area A represents 19% of the total area under the curve, what is the corresponding z-value?
(e) If area B is twice area A, what is the corresponding z-value?
(f) If $z = 1.19$, then what percentage of the total area under the curve is represented by $2B$?
(g) If $z = 1.96$, then what percentage of the total under the curve is represented by A?

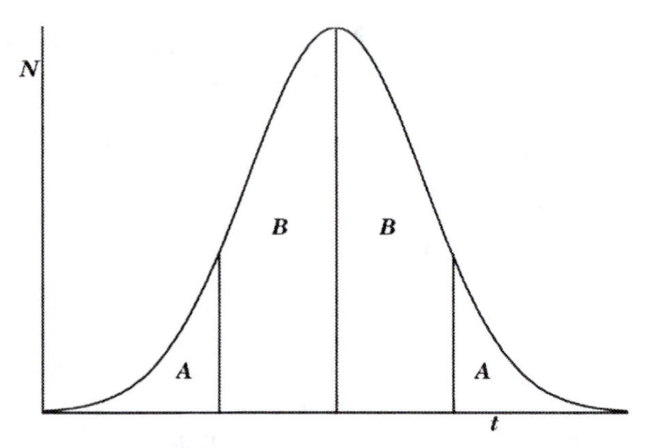

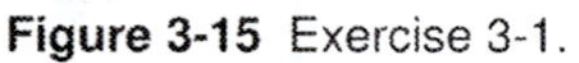

Figure 3-15 Exercise 3-1.

3-2. A Gaussian function is defined by $N = N_M e^{-\frac{(t-T_M)^2}{2\sigma^2}}$ with the following parameters:
$T_M = 1992$ (yr); $\sigma = 45$ yr; and $N_M = 78$ quads/yr.

(a) If $t = 2015$ (yr), determine N. (b) If $N = 37$ quads/yr, determine t.

---

[4] Recent technological advances and alternative energy sources, as well as energy conservation, are discussed in later chapters.

PROBLEMS

3-3. The following are data for world natural gas production:
cumulative production to end of 2005 = 3850 EJ,
production (rate) in 2005 = 105 EJ/yr,
approximate resource remaining at end of 2005 = 6000 EJ.

Apply the Hubbert model to world natural gas production to predict the year of peak production.

3-4. The following are data for Canadian oil production:
cumulative production to end of 2005 = 30.9 billion barrels,
production (rate) in 2005 = 145 million tonnes/yr,
approximate resources remaining at end of 2005 = 1200 EJ.

Apply the Hubbert model to Canadian oil production to predict:
(a) the year in which the production will peak,
(b) the production rate in exajoules/year in that peak year,
(c) the year in which only 10% of the resource will remain,
(d) the production rate in exajoules/year in the year 2025.

3-5. Figure 3-16 shows the production and consumption of oil in the USA from 1900 to 2005. Imports account for the difference between production and consumption. It is apparent that production has passed its peak and is now declining.
(a) Use the Hubbert model to predict when US oil production will decline to 5.0 EJ/yr. Use the following data:
Production (rate) in 2005
= 13.1 EJ/yr.
Cumulative production to end of 2005 = 1195 EJ.
Approximate resource remaining at end of 2005 = 600 EJ.
(b) Use a spreadsheet or graphing program to plot a graph of the Gaussian function with the parameters N_M, T_M, and σ determined in part (a). Does your graph compare well with the graph of actual data in Fig. 3-16? (Estimates of remaining resources are highly uncertain and unreliable. Try repeating this problem with a value of 250 EJ for the resources and see if your graph gives a better fit.)

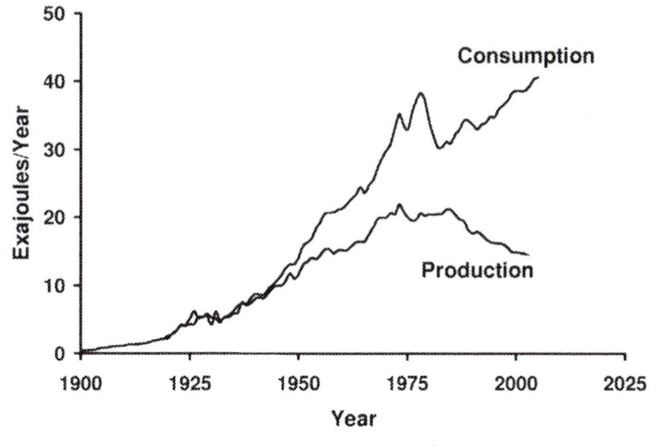

Figure 3-16 US Production and Consumption of Oil. (Problem 3-5)

ANSWERS

3-1. (a) 1.15 (b) 1.04 (c) 0.84 (d) 0.88 (e) 0.97 (f) 76.60% (g) 2.50%
3-2. (a) 68 quads/yr (b) 2047 or 1937
3-3. 2015
3-4. (a) 2059 (b) 12 EJ/yr (c) 2120 (d) 9.0 EJ/yr
3-5. 2056

CHAPTER 4

EFFICIENCY OF ENERGY CONVERSIONS

Imagine how wasteful it would be to discard more than half of the available energy when converting one form of energy to another. Unfortunately, that's exactly what we do in many energy-converters, such as coal-burning electric power plants and automobiles. Why don't engineers design these machines to operate more efficiently? As will be seen in this chapter, there is a fundamental law of nature that dictates the high level of waste energy in such devices.

4.1 TEMPERATURE, THERMAL ENERGY, AND HEAT

The study of the efficiency of energy conversions is an intriguing one, but it is one with several specific terms and distinctions. In particular, there are differences between the specific meanings of temperature, thermal energy, and heat — synonyms in common speech — which are important in the study of efficiency.

Although the term *temperature* appears frequently in day-to-day life, its precise meaning is somewhat difficult to define. Mathematically, the temperature T of an ideal gas is defined through the *ideal-gas law*:

$$PV = n\mathrm{R}T \qquad \textbf{[4-1]}$$

where P and V are the pressure and volume, respectively, of n moles of an ideal gas, that is, a gas at sufficiently low density that the interactions between the gas molecules can be ignored. The temperature in Eq. [4-1] is measured on the absolute temperature scale, for which the unit is the kelvin (K). The quantity R is the universal gas constant:

$$\mathrm{R} = 8.351\ \mathrm{J{\cdot}K^{-1}{\cdot}mol^{-1}} \qquad \textbf{[4-2]}$$

Although Eq. [4-1] defines temperature, it does not provide much of an intuitive feeling for the meaning of temperature. It is useful to think of temperature as being related to the average translational[1] kinetic energy ($\frac{1}{2}mv^2$) of molecules: as temperature increases, the average molecular kinetic energy increases, that is, the molecules move more rapidly; as temperature decreases, average kinetic energy decreases and the molecules move more slowly. In fact, there is a quantitative relation between average molecular kinetic energy and temperature:

---

[1]The term "translation" refers to a change in position of the centre of mass of an object. Thus, translational kinetic energy (KE) is the KE associated with the centre-of-mass motion, i.e., the motion of the object as a whole. This contrasts with, say, rotational KE, which is associated with rotational motion of the object around the centre of mass. An object can have zero translational KE, but still have rotational KE; an example is a wheel spinning on a fixed axle.

$$\left(\frac{1}{2}mv^2\right)_{AVG} = \frac{3}{2}k_B T \quad [4\text{-}3]$$

where m is the mass of one molecule, v is its speed, and k_B is Boltzmann's constant:

$$k_B = 1.381\times10^{-23}\ \text{J/K} \quad [4\text{-}4]$$

Equation [4-3] shows that average molecular kinetic energy is linearly related to temperature T; this relationship is valid for typical gases, liquids, and solids at room temperature. (For solids, the atoms vibrate in place, but still have a kinetic energy associated with the vibration.) At low enough temperatures, quantum energy effects make Eq. [4-3] invalid.

Whereas temperature is related to the *average* kinetic energy of a collection of molecules, *thermal energy* (sometimes called *internal energy*) is the *total* energy of all the molecules. Thus, thermal energy depends on the number of molecules as well as on the average molecular energy. A large number of molecules at low temperature can have more thermal energy than a small number of molecules at high temperature, simply because there are more of them. Thermal energy includes not only translational kinetic energy associated with the centre-of-mass motion of molecules, but also energy within the molecules themselves, such as rotational and vibrational kinetic energy, and electric potential energy.

The term *heat* is often used loosely to mean thermal energy, but there is in fact a difference between these two concepts. Heat refers to a *transfer of energy* from one object to another, as a result of a difference in temperature between the objects. As an example, consider a hot cup of coffee on a table in a room — as the coffee cools, its thermal energy decreases, heat is transferred from the coffee to the table and air in the room, increasing the thermal energy of the table and air. It is not correct to refer to the heat of the coffee, nor the heat of the table.

When heat is received by an object, all of it does not necessarily go into increasing the object's thermal energy. For example, a gas might be heated and hence expand to move a piston, thus doing work [2] on the piston. In this case, some of the heat received by the gas goes into increasing the thermal energy of the gas, and the rest provides the work done.

4.2 SPECIFIC HEAT AND LATENT HEAT

Suppose that the same amount of heat is applied to different objects. How much does the temperature of each object increase? This depends in part on the mass of the object; if heat is applied to a very massive object, it will experience a smaller temperature

[2] Recall the definition of work: if a force of magnitude F is applied to an object, which undergoes a displacement of magnitude Δr, then the work W done by the force on the object is given by $W = F\Delta r\cos\theta$, where θ is the angle between the force and displacement. If the force and displacement are in the same direction, $\theta = 0$, and hence $W = F\Delta r$. The SI unit of work is the joule (Exercise 4-1).

increase than a less massive object of the same material. The temperature change also depends on what the object is made of. If the molecules have many modes of motion (such as translation, rotation, vibration), then the heat can go into increasing the energies of each mode by a small amount, and the temperature is not raised very much. On the other hand, if we consider applying heat to something such as a monatomic gas, which has only translational kinetic energy of its atoms to make up its thermal energy, then all the heat goes into increasing the energy of only one mode of motion, and the temperature increase is larger. This difference between the internal modes of molecular motion manifests itself through different *specific heats* (or *specific heat capacities*) of objects, discussed below.

When an amount of heat Q is applied to an object having mass m, then the object undergoes an increase in temperature ΔT. The quantities Q, m, and ΔT are related by:

$$Q = mc\Delta T \qquad \textbf{[4-5]}$$

where c is the specific heat of the object, defined as the amount of heat, per unit mass and per unit change in temperature, required to change the temperature of a substance. The SI unit of specific heat is joule·kilogram^{-1}·kelvin^{-1} (J·kg^{-1}·K^{-1}). Equation [4-5] can be re-arranged as:

$$\Delta T = \frac{Q}{mc}$$

From Eq. [4-5b] it is easy to see that if the same amount of heat Q is applied to various objects having the same mass, then the object having the largest specific heat will experience the smallest increase in temperature. Notice also from Eq. [4-5b] that if objects made of the same material receive the same heat Q, then the object with the largest mass will undergo the smallest temperature change. Although we have been discussing application of heat to produce temperature increases, Eq. [4-5] works just as well for removal of heat and corresponding decreases in temperature.

Table 4-1: Some Specific Heats

Substance	Specific Heat (J·kg^{-1}·K^{-1})
Lead	128
Mercury	139
Copper	387
Aluminum	895
Ice (–10°C)	2.22×10^{3}
Ethyl alcohol	2.43×10^{3}
Water	4186

Table 4-1 lists the specific heat for a few substances. The values tabulated were determined at room temperature (20°C) and one atmosphere of pressure, unless otherwise noted. (The specific heat of a material depends somewhat on both temperature and pressure.) Notice the large value of specific heat for water; this makes water an excellent coolant, since a relatively small mass of water can absorb a large quantity of heat without undergoing a large increase in temperature.

Thus far, it has been assumed that heating or cooling an object will result only in an increase or decrease of temperature. However, the object might undergo a change of

Table 4-2: Some Latent Heats of Fusion and Vaporization

Substance	Melting Temperature (°C)	Latent Heat of Fusion (J/kg)	Boiling Temperature (°C)	Latent Heat of Vaporization (J/kg)
Mercury	–39	1.1×10^4	357	2.96×10^5
Lead	327	2.3×10^4	1744	8.58×10^5
Nitrogen	–210	2.6×10^4	–196	2.00×10^5
Hydrogen	–259	5.8×10^4	–253	4.55×10^5
Water	0	3.33×10^5	100	2.26×10^6
Aluminum	660	4.0×10^5	2467	1.05×10^7

state, that is, it might melt, solidify, vaporize, or condense. For a solid of mass m to liquefy (or a liquid of mass m to solidify), heat Q must be added (or removed):

$$Q = mL_F \qquad \textbf{[4-6]}$$

where L_F is the *latent heat of fusion* of the material. (Sometimes the word "latent" is omitted.) Notice that there is no ΔT in Eq. [4-6]; changes of state occur with no temperature change.

For a liquid to vaporize through the addition of heat, or a vapour to condense by the removal of heat, there is a similar relationship:

$$Q = mL_V \qquad \textbf{[4-7]}$$

where L_V is the *latent heat of vaporization.* Table 4-2 gives some typical heats of fusion and vaporization.

4.3 FIRST LAW OF THERMODYNAMICS (CONSERVATION OF ENERGY)

In leading up to a discussion of energy conversions and efficiency, it is useful to discuss the *first law of thermodynamics.* This law states that energy is conserved (although it can be converted from one form to another). We have already used this law implicitly; for example, in Section 4.1 it was stated that when a gas is heated and expands against a piston, the heat energy provided equals the increase in thermal energy of the gas plus the work done by the gas on the piston.

As another example, consider the situation shown in Fig. 4-1. Room-temperature water of mass m rests in a tank at the top of an incline (Fig. 4-1 (a)); a valve is then opened at the bottom of the tank, allowing the water to run down the incline into a tank at the bottom (Fig. 4-1 (b)). The gravitational potential energy of the water has decreased by mgh, where h is the vertical distance that the centre of mass of the water has moved, and has been converted into increased thermal energy of the water. The water is warmer. Conservation of energy gives:

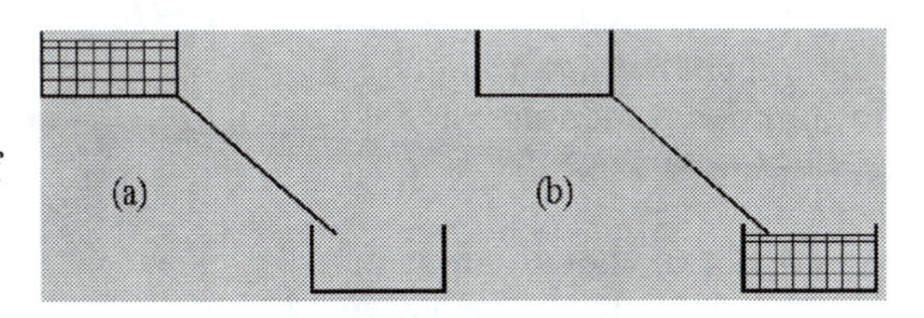

Figure 4-1 (a) Room-temperature water is at rest in a tank. (b) The water has flowed down the incline to the lower tank, with a resulting increase in temperature.

$$mgh = mc\Delta T$$

Consider now Figure 4-2, which shows hot water flowing up an incline, converting thermal energy into gravitational potential energy, and becoming cooler as it goes uphill. This uphill flow is, of course, impossible, but it is not conservation of energy that makes it so. In order to explain the non-reversibility of the water flow, another law of nature must be invoked: the second law of thermodynamics.

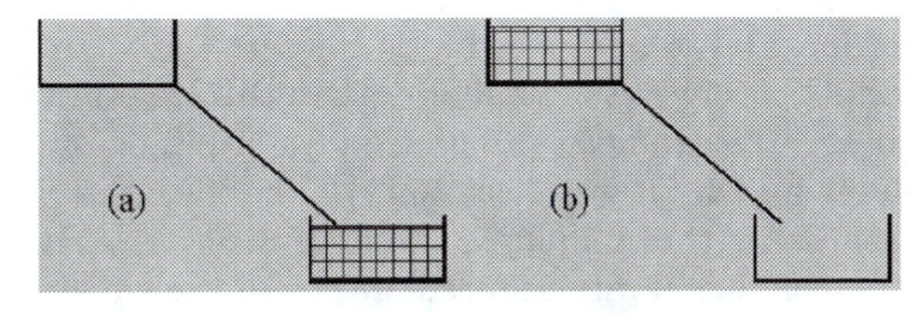

Figure 4-2 (a) Hot water is at rest in the lower tank. (b) Conservation of energy would allow the water to flow to the higher tank, converting thermal energy to gravitational PE.

4.4 THE SECOND LAW OF THERMODYNAMICS

The second law of thermodynamics can be stated in many ways; probably the simplest is that disorder is more probable than order. It is easy to generate disorder from order (just think of your room!), but much more difficult to create order from disorder. The thermal energy of the hot water in the lower tank in Fig. 4-2 (a) is disordered; the water molecules are moving randomly in all directions. It is highly unlikely, essentially impossible, that the molecules would all move spontaneously in the same direction up the incline in an ordered way. On the other hand, gravitational potential energy is an example of ordered energy (the vertical direction is a specific direction and gives order to this energy), and it is simple to convert this ordered energy to disordered thermal energy by allowing water to run down the incline.

These ideas can now be applied to an important example of energy production: a fossil-fuel electric power plant (Fig. 4-3). Fuel is burned to vaporize water in the boiler (label 1, Fig. 4-3), producing steam at high pressure and temperature. The steam passes over blades of a turbine (2), imparting energy to the turbine and causing it to spin. The turbine turns an electric generator (3), which produces an electric current in the external wires. Because the steam loses energy in spinning the turbine, the pressure and temperature of the steam decrease (4). The low-temperature steam is cooled and condenses back to water (5), to begin the process over again. The condensation is usually done by passing the steam through pipes that are cooled by water pumped from a nearby lake or river.

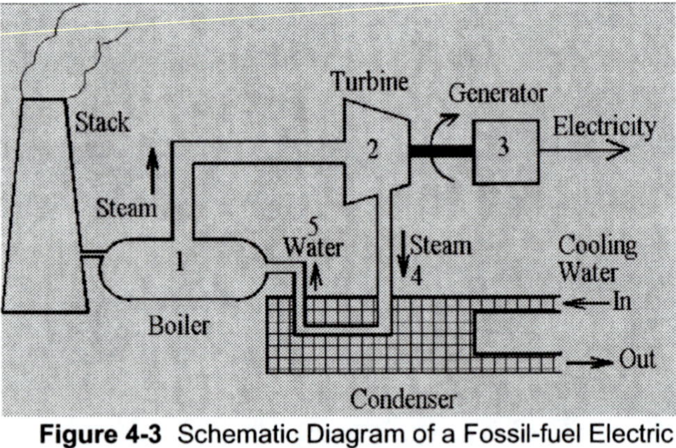

Figure 4-3 Schematic Diagram of a Fossil-fuel Electric Power Plant.

The energy of the hot steam is disordered thermal energy. However, the spinning turbine and the electric current have energy that is ordered. The electric plant is performing a conversion of disordered energy into ordered energy, which goes against the natural tendency from order to disorder, and it is impossible to convert all the disordered thermal energy in the steam into useful ordered kinetic energy of the turbine. We can now understand the function of the condenser (perhaps you were wondering why it was necessary). The heat that is discarded into the cooling water increases its thermal energy and its disorder. The disordered energy of the hot steam is partly converted into ordered energy of the turbine, and partly into disordered energy of the cooling water. If heat were not discarded to the cooling water (or elsewhere), the electric plant could not function — it cannot convert all the disordered energy of the steam into ordered energy. Not surprisingly, almost all the ordered energy of the spinning turbine can be converted into ordered electric energy.

It will be demonstrated in Section 4.6 that the cooling in the condenser improves the power-plant efficiency, which depends on the temperature of the steam produced by the boiler, and on the temperature of the cooling water in the condenser. In addition, from a practical point of view, if the water that is being returned to the boiler were not condensed, it would require all the energy output from the turbine to pump the waste steam from the turbine back to the boiler. This waste steam is at low pressure and the boiler is at high pressure, and the waste steam would not flow to the boiler without pumping. In order to move liquid water from the condenser to the boiler, a pump (not shown in Fig. 4-3) is also required. However, since the water has been condensed, its volume is roughly 1000 times smaller than the corresponding volume of steam, and hence the energy required to pump it to the boiler is about 1000 times smaller than to pump the same mass of steam. (The energy required here can be expressed as: energy = (difference in pressure) × volume.)

4.5 EFFICIENCY

Our discussion of the operation of a thermal electric plant in the previous section indicated that not all the energy available from the burning fuel is converted to useful electric energy. How much *is* converted? To answer this question, we turn to the concept of *efficiency*:

$$\text{efficiency} = \frac{\text{useful energy or work out}}{\text{total energy or work in}} \qquad [4\text{-}8]$$

Writing this with symbols:

$$\eta = \frac{\text{Useful } E_{\text{OUT}}}{\text{Total } E_{\text{IN}}} \qquad \textbf{[4-8b]}$$

where efficiency is represented by η (the lower-case Greek letter "eta"). Efficiencies are very often quoted as percentages:

$$\eta = \frac{\text{Useful } E_{\text{OUT}}}{\text{Total } E_{\text{IN}}} \times 100\% \qquad \textbf{[4-8c]}$$

A modern fossil-fuel electric plant has an efficiency of about 40%, that is, only about 40% of the energy in the coal is converted to useful electric energy. Much of the rest is discarded as waste energy to the cooling water, some is lost up the smokestack, and some to inevitable heat losses in steam pipes and friction in the turbine and generator. It is important to keep in mind that the main reason this efficiency is so low is that the electric plant is converting disordered energy into ordered energy. Older thermal electric plants have efficiencies in the range of 30% to 35%.

Table 4-3 lists typical efficiencies of various energy-converting devices. High efficiencies are achieved by devices that convert ordered energy to ordered energy (hydroelectric plants, electric generators, etc.), or ordered energy to disordered energy (electric heaters). Notice that automobiles, which — like thermal electric plants — convert disordered thermal energy to ordered mechanical energy, have an efficiency of only about 14%. Most of the energy that you put into your gasoline tank is wasted as heat expelled by the radiator and exhaust pipe, and in engine friction. The automobile is discussed in detail in Chapter 17.

Table 4-3: Typical Efficiencies

Device	Efficiency
Fossil-fuel electric power plant	40%
Hydroelectric plant	95%
Automobile	14%
Home gas furnace	85%
Electric generator	98%
Large electric motor	95%
Wind generator	55%
Electric heater	100%
Fluorescent light	20%
Incandescent light	5%

The common incandescent light has an efficiency of only 5%; although both electric energy and light energy have high order, the conversion from electricity to light occurs through the intermediary of disordered thermal energy. About 95% of the energy used by an incandescent lightbulb is dissipated as waste heat. The recent interest in compact fluorescent bulbs is a result of their higher efficiency, about 20%; while this is not high, it is nonetheless four times better than the incandescent efficiency. A 60-W incandescent bulb (i.e., one that uses 60 W of electric power) can be replaced by a 15-W fluorescent bulb to give the same amount of light.

A Variation on a Unit — MWe

The total thermal power provided by a fossil-fuel electric plant is typically of the order of 2500 MW. At an efficiency of 40%, this produces about 1000 MW of electrical power (0.40 × 2500 MW). In order to distinguish this useful electrical power from the total power, the electrical power is often written with a unit of MWe, where the "e" stands for electric.

EXAMPLE 4-1

A coal-burning electric power plant produces 950 MWe at an efficiency of 38%. What mass of coal (having energy content 28 MJ/kg) is required annually to operate this plant?

SOLUTION

First, use the definition of efficiency (Eq. [4-8c]) to calculate the total power input to the plant. Since power P is just the rate of energy use, power can be inserted into the definition of efficiency in place of energy.

$$\eta = \frac{\text{Useful } P_{\text{OUT}}}{\text{Total } P_{\text{IN}}} \times 100\%$$

Hence,

$$\begin{aligned}\text{Total } P_{\text{IN}} &= \frac{\text{Useful } P_{\text{OUT}}}{\eta} \times 100\% \\ &= \frac{950 \text{ MW}}{38\%} \times 100\% \\ &= 2.50 \times 10^3 \text{ MW} \\ &= 2.50 \times 10^9 \text{ W}\end{aligned}$$

Now determine the total energy E produced in one year, by multiplying the power P (in watts, or joules per second) and the time t (the number of seconds in a year):

$$\begin{aligned}E &= Pt \\ &= (2.50 \times 10^9 \text{ W})(365 \times 24 \times 60 \times 60 \text{ s}) \\ &= 7.88 \times 10^{16} \text{ J}\end{aligned}$$

To calculate the mass of coal required, divide this total energy by the energy content per kilogram of coal:

$$\frac{7.88 \times 10^{16} \text{ J}}{28 \times 10^6 \text{ J/kg}} = 2.8 \times 10^9 \text{ kg}$$

Thus, the amount of coal required per year is 2.8×10^9 kg, or 2.8 million tonnes. (That's a lot of coal, equivalent to almost 100 railroad-car loads per day!)

Capacity Factor

The solution to Example 4-1 assumed that the power plant would be running all year without stopping. However, electrical plants are often shut down for maintenance —

either regular or unexpected — or because the demand for electricity is low. In addition, if demand is low, a plant might be operating at less than full capacity, that is, a 1000-MWe plant might produce only 600 MW for some period of time. These factors are incorporated into an annual *capacity factor*, which is defined as:

$$\text{capacity factor} = \frac{\text{actual electrical energy (E.E.) produced in one year}}{\text{maximum E.E. producible in one year running at full capacity}}$$

Suppose that a plant's capacity factor is 65% for a given year. This means that its output is only 65% of the output that it could have produced if it had been running at 100% capacity all year. A capacity factor of 65% could mean that the plant was operating at its rated (maximum) power output for 65% of the time, and was completely shut down for 35% of the time. Or it could mean that it operated every day, but at only 65% of its maximum power. A capacity factor of 65% could also arise from any one of a number of combinations of shutdowns and operations at less than 100% power. Whatever the combination, a capacity factor of 65% means that the plant is using only 65% of the fuel that would be used by a plant with a capacity factor of 100%.

Nuclear plants are intended to run with a capacity factor of nearly 100%, because of the high construction cost and low fuel cost. As well, startup procedures at these plants are time-consuming. Fossil-fuel plants have lower capacity factors; these plants are cheaper to build than nuclear plants, but the fuel is more expensive, and startup is straightforward.

4.6 HEAT ENGINES

The automobile and the thermal electric plant are examples of a general type of machine called a *heat engine*, which is any device that converts heat energy into mechanical work. The second law of thermodynamics can be written specifically for heat engines: *no device can be constructed that, operating in a cycle (like an engine), accomplishes only the extraction of heat energy from some source and its* ***complete*** *conversion to work.*

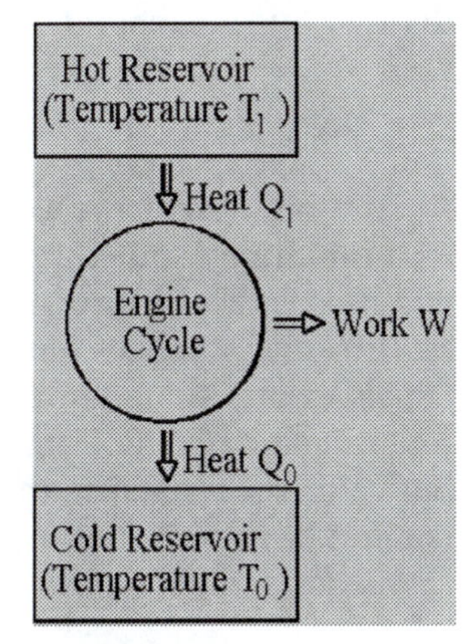

Figure 4-4 Schematic Diagram of a Heat Engine.

Figure 4-4 shows the general structure of a heat engine. Heat energy Q_1 is extracted from a thermal reservoir at high absolute temperature T_1, some useful work W is done, and heat Q_0 is discarded to a reservoir at low absolute temperature T_0. The efficiency of a heat engine is given by:

$$\eta = \frac{\text{Useful } E_{\text{OUT}}}{\text{Total } E_{\text{IN}}} = \frac{W}{Q_1}$$

By the first law of thermodynamics, the input heat Q_1 must equal the sum of the useful work W and the discarded heat Q_0, that is, $Q_1 = W + Q_0$, and hence $W = Q_1 - Q_0$. Therefore, the expression for η can be rewritten as:

$$\eta = \frac{Q_1 - Q_0}{Q_1}$$

$$\text{or} \quad \eta = 1 - \frac{Q_0}{Q_1} \qquad \textbf{[4-9]}$$

Although Eq. [4-9] is a correct expression for the efficiency of a heat engine, it is not particularly useful. In order to develop a more useful equation, the concept of entropy must be introduced.

Entropy

We have already discussed disordered and ordered energy. *Entropy* is a measure of disorder. One statement of the second law of thermodynamics is that in any energy transfer or conversion within a closed system, the entropy of the system must increase (or in rare cases, remain constant). In a thermal electric plant, if all the disordered thermal energy of the hot steam were to be converted solely to electric energy, then there would be a decrease in the total entropy, in violation of the second law.

Is it possible to be more quantitative about entropy? It is certainly beyond the scope of this book to go into the fundamental, philosophical roots of the concept and its mathematical formulation in terms of the randomness, or microscopic level of disorganization, of a system. However, the macroscopic approach to entropy of the original thermodynamicists can be adopted. They stated that if heat is added to a system, then the change in entropy (ΔS) of the system is the heat energy received divided by the absolute temperature of the system[3]:

$$\Delta S = \frac{Q}{T} \qquad \textbf{[4-10]}$$

At first sight, there appears to be no relation between this definition and the basic one of randomness, but there are some semblances. The heat input Q is obviously an input of a disorganized form of energy, since it consists of random molecular motions. Thus, heat input increases entropy and conversely loss of heat decreases the entropy of an object. But why is temperature in the denominator? In very cold materials, there is not much molecular motion, so a given quantity of heat energy will stir things up quite noticeably, increasing the disorder a great deal. But if the same quantity of heat energy is added to a very hot object (already having a huge amount of disorder), the entropy increase is much smaller.

When an amount of heat Q is extracted from an object at temperature T, the change in entropy is negative:

---

[3] This statement is valid as long as the system is so large that the addition of heat Q increases the temperature T by only an infinitesimal amount.

$$\Delta S = \frac{-Q}{T}$$

Equation [4-10] for ΔS will now be applied to the specific case of a heat engine in which an amount of heat Q_1 is converted to work W and the remainder Q_0 is discarded. The system is considered to be the engine plus the two thermal reservoirs. The engine runs in cycles, and at the end of a cycle, it is in the same state in which it began; that is, there is no change, and $\Delta S = 0$ for the engine alone. The entropy change due to the heat flowing out of the hot reservoir at temperature T_1 is $\Delta S_1 = -Q_1/T_1$, and the change in entropy due to heat Q_0 flowing into the cold reservoir at temperature T_0 is $\Delta S_0 = Q_0/T_0$. Since there is no change in entropy associated with the organized mechanical work W, the total entropy change is

$$\Delta S = \frac{-Q_1}{T_1} + \frac{Q_0}{T_0}$$

According to the second law of thermodynamics, the overall entropy of a system can never decrease, that is, $\Delta S \geq 0$.

Hence,
$$\frac{-Q_1}{T_1} + \frac{Q_0}{T_0} \geq 0 \quad \text{or} \quad \frac{Q_0}{T_0} \geq \frac{Q_1}{T_1}$$

Re-arranging,
$$\frac{Q_0}{Q_1} \geq \frac{T_0}{T_1} \qquad \textbf{[4-11]}$$

Return to Eq. [4-9], which gives the efficiency η of a heat engine: $\eta = 1 - \frac{Q_0}{Q_1}$

Substituting relation [4-11] into this:

$$\eta \leq 1 - \frac{T_0}{T_1} \quad \text{(heat engine)} \qquad \textbf{[4-12]}$$

Relation 4-12 is an extremely useful expression for the efficiency of a heat engine, in terms of only the temperatures of the thermal reservoirs. The equality sign in [4-12] holds only in the case of what is known as a reversible heat engine (for which $\Delta S = 0$), and gives the maximum possible efficiency of a heat engine:

$$\text{Max. } \eta = 1 - \frac{T_0}{T_1} \quad \text{(heat engine)} \qquad \textbf{[4-13]}$$

Unfortunately, a reversible heat engine (also known as a "Carnot[4] engine") is impractical — it can operate only if the temperature difference (T_1 - T_0) is infinitesimal, or, in the case of a finite temperature difference, the engine requires an infinite amount

[4] Sadi Carnot (1796-1832) was a French physicist and engineer who in 1824 described a hypothetical, idealized heat engine that has the maximum possible efficiency consistent with the second law of thermodynamics.

of time in which to work. Nonetheless, Eq. [4-13] serves as a convenient expression for the maximum conceivable efficiency of a heat engine. Real heat engines will always have efficiencies less than this, because of the irreversible transfer of heat from the hot reservoir to the cold reservoir, which have a finite difference in temperature, and because of energy dissipated in the engine by friction, imperfect insulation, etc.

Equation [4-13] can be used to determine the maximum possible efficiency of a fossil-fuel electric plant, as shown in Example 4-2 below.

EXAMPLE 4-2

In a fossil-fuel electric plant, such as that shown in Fig. 4-5, the temperature of the hot steam is typically about 500°C (773 K), and the low-temperature reservoir is the cooling water from a lake or river with a temperature of roughly 20°C (293 K). What is the maximum efficiency that is theoretically possible in such a plant?

SOLUTION

The maximum possible efficiency is given by Eq. [4-13]:

$$\text{Max. } \eta = 1 - \frac{T_0}{T_1} = 1 - \frac{293\text{ K}}{773\text{ K}} = 0.62\text{, or } 62\%$$

Hence, the maximum efficiency theoretically possible for a typical fossil-fuel electric plant is 62%.

The answer in Example 4-2 above shows that even if one could build a reversible heat engine operating between the temperatures encountered in a typical thermal electric plant, its efficiency at converting disordered energy to ordered energy would only be 62%. Such a heat engine would still discard 38% of the available heat energy to the low-temperature reservoir. As can be seen from Eq. [4-13], the efficiency could be increased by increasing T_1 or decreasing T_0. However, temperature T_1 is limited by the strength of the boiler materials, and T_0 is just the natural temperature of the cooling waters. Actual efficiencies of fossil-fuel electric plants are approximately 40%.[5]

Figure 4-5 The maximum efficiency theoretically possible from a typical fossil-fuel electric plant is only about 62%. *Photo courtesy of Ontario Power Generation.*

[5] A better estimate than the Carnot efficiency (Eq. 4-13) for the maximum efficiency of a heat engine can be obtained by assuming that the engine is operating at its maximum power output. The result is $\eta = 1 - (T_0 / T_1)^{1/2}$ = 38% for the temperatures given in Example 4-2. (Ref. F.L. Curzon and B. Ahlborn, "Efficiency of a Carnot Engine at Maximum Power Output," Am. J. Phys. **43**(1), 22-24 (1975).)

Nuclear electric plants have lower efficiencies than fossil-fuel plants, only about 30%. In a nuclear power plant a reactor produces heat to generate steam at high temperature and pressure. This steam is then used in the same way as in a fossil-fuel plant to spin turbines. In order to have high efficiency, one wants to have the core of the reactor at the highest possible temperature, but this results in unique demands on the structural materials and the cooling fluid, which must maintain their integrity under the twin stresses of temperature and radiation damage. Consequently, T_1 has to be kept lower than in fossil-fuel plants, giving a lower efficiency.

4.7 REFRIGERATORS

Refrigerators are one example of a heat-moving device, that is, something that transfers heat from one place to another. A refrigerator moves heat from the inside of the appliance to the outside; the heat is expelled to the air in the room via coils at the rear of the refrigerator. Heat naturally is transferred from hot objects to cold ones; refrigerators perform the reverse (unnatural) function, and in order to do so, need an external input of energy, normally provided by electricity. Another version of the second law of thermodynamics can be stated specifically for refrigerators as: no device can be constructed that (operating in a cycle) accomplishes *only* the transfer of heat from a cooler source to a hotter one.

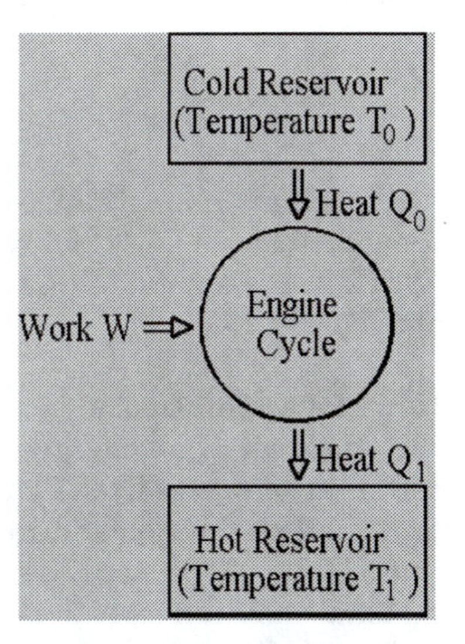

Figure 4-6 A Refrigerator. Heat Q_0 is extracted from a low-temperature reservoir by using external work W, and heat Q_1 is expelled to a high-temperature reservoir.

Figure 4-6 illustrates the basic operation of a refrigerator. Heat Q_0 is extracted from a reservoir at a low absolute temperature T_0 by input of external work W. Heat Q_1 is expelled to a reservoir at high absolute temperature T_1. By the first law of thermodynamics, the sum of the heat extracted and the external work must equal the heat expelled:

$$Q_0 + W = Q_1 \quad \textbf{[4-14]}$$

The "efficiency" of a refrigerator is the ratio of the useful heat transferred to the external work provided. (The reason for the quotation marks will be given shortly.) Since the function of a refrigerator is to cool the interior of the appliance, the *useful* heat transferred is the heat Q_0 removed from the interior. Thus, the "efficiency" is

$$"\eta" = \frac{Q_0}{W} \quad \textbf{[4-15]}$$

From Eq. [4-14], the work is $W = Q_1 - Q_0$, and substitution of this expression for W into Eq. [4-15] gives

$$"\eta" = \frac{Q_0}{Q_1 - Q_0} = \frac{1}{\frac{Q_1}{Q_0} - 1}$$

By using an analysis involving entropy (similar to that used in Section 4.6 for heat engines), it can be shown (Problem 4-12) that

$$\frac{Q_1}{Q_0} \geq \frac{T_1}{T_0}$$

Substituting this relation into the expression for "efficiency,"

$$\text{"}\eta\text{"} \leq \frac{1}{\frac{T_1}{T_0} - 1} \qquad \textbf{[4-16]}$$

Hence, the maximum possible "efficiency" of a refrigerator is

$$\text{Max. "}\eta\text{"} = \frac{1}{\frac{T_1}{T_0} - 1} \qquad \textbf{[4-17]}$$

A typical maximum "efficiency" of a refrigerator is now calculated. Assume temperatures of $T_0 = 263$ K (i.e., –10°C) in the freezer compartment, and $T_1 = 293$ K (20°C) in the surrounding room. Simple substitution into Eq. [4-17] gives an "efficiency" of 8.77, or 877%.

How can an "efficiency" be greater than 100%? What has been calculated is not actually the efficiency of an energy *conversion*, but rather a parameter that indicates how well a refrigerator *moves* heat. The value of 8.77 indicates that a refrigerator could move 8.77 units of heat energy for an input of one unit of external work. It should now be clear that although the word "efficiency" has thus far appeared, it is not efficiency at all; it is instead something which is commonly called the *coefficient of performance (C.O.P.)*, that is, "η" = C.O.P. Replacing "η" with C.O.P. in [4-16]:

$$\text{C.O.P.} \leq \frac{1}{\frac{T_1}{T_0} - 1} \quad \text{(refrigerator)} \qquad \textbf{[4-18]}$$

For completeness, and as a reminder of the fundamental meaning of the C.O.P. of a refrigerator, Eq. [4-15] is rewritten here, with "η" replaced by C.O.P.:

$$\text{C.O.P.} = \frac{Q_0}{W} \quad \text{(refrigerator)} \qquad \textbf{[4-19]}$$

Actual mass-produced refrigerators have a C.O.P. slightly less than 1, and the most efficient refrigerators have a C.O.P. of about 2. Notice that C.O.P.s are normally not quoted as percentages. The C.O.P. could be improved easily by using thicker insulation, improving door seals, improving the efficiency of the motor and compressor, and placing the motor and compressor (both of which get hot during operation) above the cool cabinet so that the heat from these devices will rise into the room instead of into the cabinet. A number of sensible design features of refrigerators in the 1950s, such as thicker insulation, were discarded in the era of cheap energy in the 1960s.

How a Refrigerator Works

The operation of a real refrigerator is rather simple. A fluid is alternately compressed and allowed to expand, and in between these processes the fluid either absorbs or gives off heat. Until recent years, a fluid commonly used in household refrigerators was freon (CCl_2F_2), one of the chlorofluorocarbons (CFCs) responsible for the decline of ozone in the upper atmosphere. CFCs also contribute to global warming and climate change (Chapter 6). Now hydrogenated freons are being used that break down much more rapidly in the environment if they are accidentally released from a refrigerator. In industry, ammonia is frequently employed as the working fluid.

The workings of a refrigerator cycle begin at the expansion valve shown in Fig. 4-7. To the right of the valve, the fluid is a warm liquid at high pressure. When a sensor in the refrigeration box indicates that cooling is required, the expansion valve (also known as the throttling valve) is opened. The valve has only a very small opening, and liquid sprays slowly through it into the lower-pressure region to the left. Because of the pressure decrease, some of the liquid vaporizes, thus removing energy from the liquid, and the temperature falls. The liquid is now cold, with a small quantity of it having been vaporized. This cold liquid flows through coils in the refrigeration box, removing heat from the contents of the box. The liquid vaporizes during this process, and the coils that it is travelling through are called the evaporator. The fluid is now a warm, low-pressure vapour, which passes into the compressor. This device is run by an electric motor and compresses the vapour, heating it somewhat, and the fluid becomes a hot, high-pressure vapour. It passes through coils (usually at the rear of the refrigerator), giving up heat to the surroundings. In the process, it condenses (hence the name "condenser" in Fig. 4-7) to form a warm liquid, which is ready to re-enter the cycle.

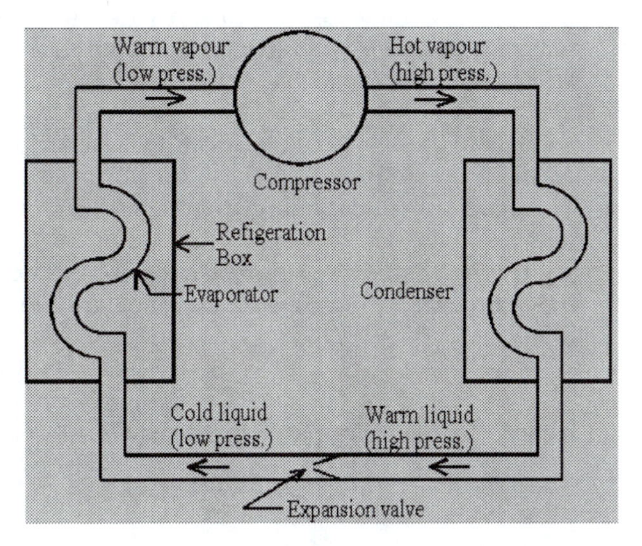

Figure 4-7 Operation of a Refrigerator.

When a refrigerator is operating, the heat given off at the condenser is appreciable — put your hand on the coils and feel it. A little attention to these coils can make a refrigerator work more efficiently. If these coils are dirty, the efficiency of heat transfer from the condenser decreases, and so it is useful to clean the coils periodically. As well, if the heat from the coils cannot easily escape from the region surrounding the refrigerator, some of it will be conducted back into the refrigeration box, requiring further refrigeration cycles. Home refrigerators are often placed in a recess in a wall, hampering heat escape; it makes good sense to allow for ample air circulation around the refrigerator, especially at the top, since hot air rises. Ideally, it would be best to allow the heat to vent through the rear wall, heating another room.

EXAMPLE 4-3

A refrigerator that has a C.O.P. of 1.2 consumes 0.50 kW·h (1.8 MJ) of electrical energy while operating over a particular period of time. During this time, (a) how much heat energy (in megajoules) is removed from the refrigerator compartment, and (b) how much heat energy (in megajoules) is dissipated by the condenser?

SOLUTION

(a) The C.O.P. of a refrigerator is given by Eq. [4-19]: C.O.P. = Q_0 / W. Given quantities are: C.O.P. = 1.2, and external work W (the electrical energy provided) = 1.8 MJ. The energy removed from the cold refrigerator compartment is Q_0, and is given by a simple re-arrangement of the above equation:

$$Q_0 = \text{C.O.P.} \times W = 1.2\,(1.8\text{ MJ}) = 2.2\text{ MJ}$$

Thus, 2.2 MJ of heat energy is removed from the refrigerator compartment.

(b) By the first law of thermodynamics, the heat Q_1 dissipated by the condenser must equal the sum of the external work provided and the heat extracted, as stated in Eq. [4-14]:

$$Q_1 = W + Q_0 = (1.8 + 2.2)\text{ MJ} = 4.0\text{ MJ}$$

Hence, the heat dissipated by the condenser is 4.0 MJ.

4.8 AIR CONDITIONERS AND HEAT PUMPS

This section examines two devices — air conditioners and heat pumps — that are essentially identical in form and operation to refrigerators. An air conditioner is just a refrigerator that has the evaporator (where the cooling occurs) in the interior of a building (Fig. 4-8), and the condenser (where the heat is expelled) outside the building, or at least in a window where the heat can easily be sent outside. The C.O.P. of an air conditioner is defined in exactly the same way as that of a refrigerator:

$$\text{C.O.P.} = \frac{Q_0}{W} \leq \frac{1}{\dfrac{T_1}{T_0} - 1} \quad \text{(air conditioner)} \qquad \textbf{[4-20]}$$

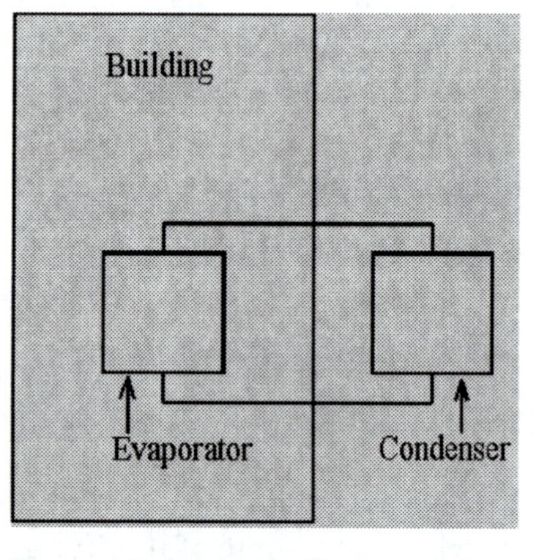

Figure 4-8 Schematic Diagram of an Air Conditioner.

Sometimes air conditioners or refrigerators carry a label specifying their energy-efficiency ratio (EER). This is the rate at which heat energy is removed in Btu per hour divided by the electrical power consumption in watts (an interesting, but unfortunate, combination of units).

A heat pump is virtually the same as an air conditioner, except that the placement of the evaporator and condenser are reversed (Fig. 4-9). The evaporator is placed outside a building, where it extracts heat from the air, ground, or groundwater; the condenser is positioned inside, where it is used to heat the building. The function of the heat pump is to provide heat to the interior, and hence the *useful* heat is the heat Q_1 transferred into the warm building. Therefore, the C.O.P. is the ratio of the heat Q_1 provided to the work W required. Using an analysis similar to that performed in Section 4.6 for a heat engine, the C.O.P. of a heat pump can be written (Problem 4-14) in terms of the absolute temperatures of the two heat reservoirs:

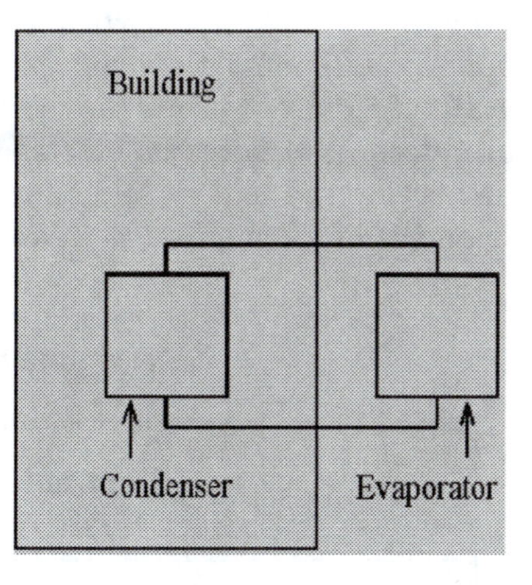

Figure 4-9 Heat Pump.

$$\text{C.O.P.} = \frac{Q_1}{W} \leq \frac{1}{1-\frac{T_0}{T_1}} \quad \text{(heat pump)} \qquad \textbf{[4-21]}$$

Many heat pumps serve double-duty as air conditioners. In the winter, the working fluid flows in a direction such that the condenser is inside the building, as shown in Fig. 4-9. In the summer, a simple flick of a switch causes the compressor to drive the fluid in the opposite direction. Thus, the evaporator becomes the condenser, and vice-versa, and the interior of the building is now cooled.

Suppose that a heat pump is operating with an exterior temperature of 0°C (i.e., T_0 = 273 K), and an interior temperature of 20°C (i.e., T_1 = 293 K). Simple substitution into expression 4-21 gives a maximum possible C.O.P. of 15. The actual C.O.P. of a typical heat pump working between these temperatures is about 3. This means that for 1 unit of electric energy as input, the pump can provide 3 units of heat energy to the building's interior. However, as the exterior temperature drops below about –10°C, the typical C.O.P. drops below 1, and it becomes more efficient simply to use an electric heater.

Notice in expression 4-21 that in order for a heat pump to have a high C.O.P., the temperatures T_1 and T_0 should be as close as possible. In contrast, consider the efficiency of a heat engine given earlier in expression 4-12:

$$\eta \leq 1-\frac{T_0}{T_1} \quad \text{(heat engine)}$$

For this efficiency to be large, T_0 should be small and T_1 large; that is, T_1 and T_0 should be as far apart as possible.

4.9 SYSTEM EFFICIENCY

Is it more efficient to heat a home using electric heaters or an oil furnace? At first glance, the answer might appear to be obvious: electric heaters convert electrical energy to thermal energy with an efficiency of 100%, whereas an oil furnace is only about 75% efficient. However, to give a more complete answer, we should consider the overall efficiency of producing and delivering the electricity and the oil. If the electricity is produced in a coal-fired electrical plant, then the energy consumed in mining and transporting the coal should be taken into account. The *system efficiency* includes the efficiencies of *all* such steps in an energy-conversion process.

Table 4-4
Approximate System Efficiency for Electric Heat and Oil Heat

Electric Heat			**Oil Heat**		
Step	Step efficiency	Cumulative efficiency	Step	Step efficiency	Cumulative efficiency
Coal-mining	96%	96%	Oil extraction	97%	97%
Transporting coal	97%	93%	Transporting oil	97%	94%
Electricity generation	40%	37%	Heating	75%	70%
Electricity transmission	90%	34%	System efficiency		70%
Heating	100%	34%			
System efficiency		34%			

Table 4-4 shows the efficiency of each step for electric heat (assuming a coal-burning power plant) and oil heat. The cumulative efficiency is the *product* of the various step efficiencies, and the system efficiency is the final cumulative efficiency. The system efficiency of 70% for oil heat turns out to be larger than that (34%) of electric heat, indicating that it more efficient to heat with oil than with electricity from a coal-fired plant. Such system efficiencies are useful when comparing alternative ways to perform the same task.

EXERCISES

4-1. Use the definition of work ($W = F\Delta r\cos\theta$) to show that the unit of work is $kg{\cdot}m^2/s^2$, which is a joule. (You might find it handy to recall Newton's second law of motion: $F = ma$.)

4-2. How much heat is required to vaporize 7.8 kg of water, initially at 25°C? Tables 4-1 and 4-2 provide useful data.

4-3. If a 60-W incandescent bulb is replaced by a 15-W fluorescent bulb, how much less electrical energy (in kilowatt·hours) is consumed in a week by the fluorescent bulb, operating for 12 h each day?

4-4. Suppose that an ocean-thermal-energy-conversion device were developed to produce work from the temperature difference between the warm surface waters of the ocean and the colder deeper waters. If the surface and deep-water temperatures are 22°C and 5°C respectively, what is the maximum theoretical efficiency of such an engine?

4-5. What is the maximum C.O.P. that is theoretically possible for an air conditioner operating between a room at 21.0°C and exterior air at 31.0°C?

4-6. A heat pump that has a C.O.P. of 2.4 consumes 0.75 kW·h (2.7 MJ) of electrical energy while operating over a certain period of time. During this time, (a) how much heat energy (in megajoules) is transferred into the high-temperature reservoir, and (b) how much heat energy (in megajoules) is transferred out of the low-temperature reservoir?

4-7. The C.O.P. of a refrigerator or an air conditioner is given by C.O.P. = Q_0/W, whereas the C.O.P. of a heat pump is given by C.O.P. = Q_1/W. Why are these expressions different?

PROBLEMS

4-8. A satellite in orbit could derive its energy from a very strong radioactive source of alpha particles. Suppose that 30 W (two significant digits) of electrical power are to be produced with 5.0% efficiency by conversion of the (kinetic) energy of the alpha particles. If the energy of each alpha particle is 5.5 MeV, how many particles must be emitted by the source per second?

4-9. It takes 2.0 million tons of coal per year to feed a 1000-MWe steam-electric plant. Assuming that the plant has a 60% capacity factor, what is its efficiency? Coal has an energy content of 22 MBtu per ton. Assume two significant digits.

4-10. The latent heat of fusion of ice is 333 kJ/kg. In the freezer compartment of a refrigerator, a tray holds 12 cubes of water each of mass 8.0 g. If the C.O.P. of the refrigerator is 3.3, how many kilowatt·hours of electrical energy would be required to run the refrigerator to change the water cubes (at 0°C) into ice cubes (at 0°C)?

4-11. A refrigerator is operating at a C.O.P. of 0.98, which is 13% of the maximum C.O.P. theoretically possible. The temperature of the freezer compartment of the refrigerator is –12°C. What is the temperature (to the nearest degree Celsius) of the surrounding room?

4-12. Use the second law of thermodynamics and the concept of entropy to show that for a refrigerator, $Q_1/Q_0 \geq T_1/T_0$. (The symbols are defined in Section 4.7.)

4-13. A particular heat pump uses 1.0 kW of electrical power. Every second, it is able to remove 2.5×10^3 J of energy from a low-temperature reservoir. What is its C.O.P.?

4-14. Derive expression 4-21 for the C.O.P. of a heat pump.

4-15. Calculate the system efficiency of a fluorescent light powered by electricity from a coal-fired power plant. Use Tables 4-3 and 4-4 as necessary.

4-16. A 15-W compact fluorescent bulb produces the same amount of light as a 60-W incandescent bulb, because the fluorescent bulb is more efficient at converting electrical energy to light energy. The purchase price of a compact fluorescent bulb is much more than that of an incandescent bulb, but the fluorescent bulb costs less over its lifetime because of its greater efficiency, and because it lasts ten times longer than an incandescent bulb. Using the information given below, determine the payback time (to the nearest month) for a 15-W compact fluorescent bulb, that is, determine how much time is needed before the extra purchase cost of the fluorescent bulb has been recovered by savings. Assume that the compact fluorescent bulb is "on" for five hours per day, and that the cost of electrical energy is 10¢ per kW·h.

Price of 60-W incandescent bulb: \$0.50
Price of 15-W compact fluorescent bulb: \$10.00
Lifetime of 60-W incandescent bulb: 1000 h
Lifetime of 15-W compact fluorescent bulb: 10,000 h

4-17. The world's nuclear electrical capacity in the year 2000 was 350 GWe. If all the waste heat from operating these nuclear plants for one year were used to melt ice (initially at 0°C) from the Antarctic ice cap, what would be the increase in the depth of the oceans? Assume that 70% of the surface area of Earth is covered with water, and that the nuclear power reactors operate at an efficiency of 30%. The radius of Earth is 6400 km.

ANSWERS

4-2. 2.0×10^{7} J
4-3. 3.8 kW·h
4-4. 5.8%
4-5. 29
4-6. (a) 6.5 MJ (b) 3.8 MJ
4-8. 6.8×10^{14}
4-9. 41%
4-10. 2.7×10^{-3} kW·h
4-11. 23°C
4-13. 3.5
4-15. 7%
4-16. 13 months
4-17. 0.21 mm

CHAPTER 5

HEALTH AND ENVIRONMENTAL IMPACTS OF FOSSIL FUELS

Almost every day newspapers carry stories about air pollution, or the greenhouse effect, marine oil spills, fires or cave-ins in coal mines, acid rain, or burning oil wells or gas pipelines. All of these serious environmental problems are associated with the use of fossil fuels. The greenhouse effect has become so important that it will be considered on its own in the next chapter, within the context of the energy balance of the Earth's atmosphere. In this chapter we will deal with issues related to extracting and transporting fossil fuels, thermal pollution due to the waste heat from burning fossil fuels (especially for production of electricity), and air pollution, including acid rain, smog and particulate.

Did You Know?
The world's first producing oil field was near Petrolia, Ontario, which brought in the first gusher in 1858. The photograph shows a pump (of Canadian design) still in operation in the Petrolia field.

5.1 EXTRACTION, TRANSPORT and REFINING

Extraction

Fossil fuels must first be removed from the ground. In the case of oil and gas wells, the drilling and extraction of the fuels produces local disruptions in ecosystems, although in most areas the impact is rather minor, comparable to that of any small industry. Of course, there are examples of situations where the effects are significant, such as the long-range air and water pollution that resulted from the burning of the oil wells in Kuwait in 1991. As well, in regions such as the Arctic and offshore, spills and blowouts can have major and lasting effects on plant and animal life, and extra care must be taken to minimize the possibility of serious accidents. Nowadays there is concern about the potential impact of pipelines upon large-scale migration patterns, such as those of the vast caribou herd in the Canadian Arctic.

In the case of coal, both underground and surface (strip) mining have disruptive consequences. Underground mining can lead to land subsidence, and the risk of fires and explosions is always present. Strip mining produces complete disfigurement of the landscape, unless costly measures are taken to reclaim the land afterwards. In both methods, water seepage reacts with sulphur compounds in the coal to produce sulphuric acid, which leaks into rivers and lakes. In addition, mining is an extremely hazardous occupation: in the USA between 1991 and 1999, an average of 93 people was killed annually in coal mining accidents. In 2005, 6000 Chinese coal miners were killed by accidents,[1] and government figures put the number of miners suffering from pneumonoconiosis at 600,000, a figure that increases by 70,000 each year. This "black lung disease" is known to mining communities around the world; it is a result of breathing coal dust for many years. The lungs become damaged to such an extent that breathing becomes extremely difficult, and the miners become susceptible to a host of respiratory ailments and problems.

Transport

Once out of the ground, the oil, gas, and coal need to be transported, either to a site where further processing occurs, or to an end-use site. Coal is usually transported by ship or train, (although it may be as local as shown in Fig. 5-1) and environmental effects are usually minimal. However, the transport of oil by ship occasionally has disastrous consequences – we have all heard of major spills due to oil tankers running aground and producing local havoc in fragile coastal environments.

Figure 5- 1 Local Delivery of Coal "Pucks" in China

Although these accidents make major news stories, tankers actually put more oil into the oceans in another way — by flushing ballast water. After delivering oil to a destination, tankers are filled with water as ballast for the return trip. Upon completion of this trip, the water (containing oily residues from the tanks) is discharged into the sea. It was estimated[2] in 1981 that ballast water accounts for about 20% of the oil in the oceans, whereas tanker accidents introduce only about 3%. Other major sources of oil[2] reaching the oceans include river runoff (25%), unburned hydrocarbons from the

[1] The Independent, 29 July 2006

[2] The Global 2000 Report to the President, *Entering the Twenty-First Century*, A Report Prepared by the Council on Environmental Quality and the Department of State, US Government Printing Office, Washington, D.C., (1981).

atmosphere (10%), natural seepage (10%), and flushing of ballast water from fuel tanks (not to be confused with oil-cargo tanks) in ships (10%).

Oil and gas are also transported by pipeline, which occasionally have leaks and/or explosions. Problems such as these are particularly troublesome in areas that are environmentally sensitive, such as the Arctic. Corrosion of the pipelines is the main issue. There was considerable concern in 2006 when a pipeline from Alaska, supplying 8% of overall US oil production, had to be shut down for testing due to corrosion which had eaten more than two-thirds of the way through the wall in particular spots. It was revealed that sludge had not been scrubbed out of the line for 14 years. A similar line ruptured earlier in that year, spilling over 200,000 gallons of oil.

Table 5–1:
Energy Used in Freight Transport

Mode	Energy cost kJ/(tonne· km)
Ship	70 – 400
Pipeline	70 – 900
Railroad	150 – 700
Truck	800 – 1500
Aircraft	6000 – 18000

As some countries run low in domestic natural gas, attention is turning to a massive expansion in the shipping of liquid natural gas (LNG) from places such as Russia. This requires the building of massive terminals to receive the tankers and de-pressurise the fuel so that it can then be distributed on land through pipelines. The need for the highest safety standards is obvious.

Transporting a fossil fuel requires some energy, of course, and the further the fuel must be moved, the more energy is required. Ships provide the most energy-efficient mode of transport [3], followed closely by pipelines and railroads (Table 5-1). Trucks use more energy, and aeroplanes considerably more.

Refining

Oil, of course, is refined to create the various end products that have so many uses. High safety standards are required here. Nevertheless accidents happen occasionally, one example being the explosion in Texas in 2005 which killed 15 workers and injured 180.

There is no question of refining natural gas, but attention is turning to the conversion of coal to liquid and gaseous fuels, and here again safety must be considered as a primary concern.

---

[3] R.F. Hemphill, *Energy Conservation in the Transportation Sector*, in Energy Conservation and Public Policy, J.C. Sawhill, ed., Prentice-Hall, (1979), p. 92

5.2 THERMAL POLLUTION

We saw in Chapter 4 that most of the thermal energy produced in a fossil-fuel or nuclear electrical plant is not converted to electricity, but instead is discharged as waste heat to water (or the atmosphere). As more and more electrical generating stations come into service, this thermal pollution becomes an increasing problem.

In nuclear plants, all of the waste heat is removed by cooling water; at a typical efficiency of 30%, the cooling water is removing heat energy 2 or 3 times the station's electrical rating. In fossil-fuel plants, a significant portion (about 10%) of the available energy goes up the stack as hot gases and is lost to the atmosphere; the demand for cooling water is considerably less than at a nuclear plant both for this reason and because of the higher efficiency. In the developed countries, coal plants have efficiency 37 - 42% and natural gas plants 43 - 49%. [4] In China, the world's largest consumer of coal, the efficiency is only 33%.

In many plants, the cooling water is drawn from a lake, river, or ocean, used once and then returned with its temperature increased by about 10°C. (In most areas of North America and Europe, there are regulations requiring that the rise in temperature in "once-through cooling" be no larger than this.) The temperature increase produces changes in the overall aquatic ecology in the surroundings, encouraging some organisms and discouraging others. Most types of fish are particularly sensitive to temperature changes and can remain healthy only within rather narrow temperature limits, which vary from species to species. Temperature affects fishes' appetites, digestion, growth, behaviour, spawning, and longevity. As well, the warmer water contains less dissolved oxygen, which can affect the viability of aquatic organisms. When an electrical plant is closed down for maintenance or other reasons, fish and other creatures that thrived in the warmth of the cooling waters are suddenly thrust into a colder environment and might not be able to survive.

Heat energy is not the only pollutant. Upon intake the water is usually treated with chemicals to prevent build-up of slime and to kill various kinds of larvae, etc. After the water is used for cooling, it is discharged along with the chemicals into some larger body of water. (Fish are prevented from being taken into the cooling system by the use of screens.)

EXAMPLE 5–1

A coal-burning electrical plant produces 1000 MWe at 40% efficiency. What minimum volume of water is needed per second to remove the waste heat, if the water undergoes a temperature increase of 10°C? (Assume two significant digits in all given data and neglect the heat energy dissipated up the chimney.)

---

[4] *Energy Technology Perspectives*, International Energy Agency, (2006), p.179

SOLUTION

Consider a time period of 1 s. Since a watt is a joule per second, and the electrical power produced is 1000 MW, then the electrical energy produced in 1 s is 1000 MJ. To determine the waste heat generated in 1 s, we first use the efficiency to determine the total energy input to the plant.

$$\text{efficiency} = \frac{\text{useful (electrical) energy}}{\text{total energy in}}$$

$$\text{Thus, total energy in} = \frac{\text{useful (electrical) energy out}}{\text{efficiency}} = \frac{1000\ \text{MJ}}{0.40} = 2500\ \text{MJ}$$

The waste heat energy is just the difference between the total energy and the electrical energy:

$$\text{Heat energy} = (2500 - 1000)\ \text{MJ} = 1500\ \text{MJ}$$

Now determine the mass of water required by using Eq. [4-5]:

$$Q = mc\Delta T \quad \text{giving} \quad m = \frac{Q}{c\Delta T} = \frac{1500 \times 10^6\ \text{J}}{(4186\ \text{J} \cdot \text{kg}^{-1} \cdot \text{K}^{-1})(10\text{K})} = 3.6 \times 10^4\ \text{kg}$$

Since the density of water is 1.0×10^3 kg/m^3, the volume of water required is:

$$\text{volume} = \frac{\text{mass}}{\text{density}} = \frac{3.6 \times 10^4}{1.0 \times 10^3\ \text{kg/m}^3} = 36\ \text{m}^3$$

Hence, 36 m^3 of water are required every second.

The volume of water determined in the above example is huge! On a daily basis, it works out to 3.1×10^6 m^3 or 3.1 billion litres, and this is only for one electrical plant. In Problem 5-9 at the end of this chapter, we see that if once-through cooling were used in all fossil-fuel and nuclear plants in the USA, then the annual cooling water requirement would be around 1/5 of the total water available as runoff from precipitation. Of course, some water can be used more than once at different locations along a river, but nonetheless the water requirement for cooling is enormous.

It is fairly obvious that in regions without access to large water bodies, once-through cooling is not practical. Instead, closed-cycle cooling is used in which the same water is used over and over again. There are three basic approaches: cooling ponds, wet cooling towers, and dry cooling towers.

Cooling Ponds

In regions where land costs are low, artificial lakes or ponds can serve to provide cooling water for an electrical plant. These ponds are of such size that the entire volume flows through the plant in a week or two, and the heat received by the water is dissipated mainly by surface evaporation to the atmosphere. In order to provide ample surface for evaporation, the ponds need to be quite large — several square kilometres. Of course a continual water supply is needed to replenish the evaporative loss, but this requirement is much less than for once-through cooling. In addition, the artificial lake has recreational and even fishery potential.

Wet Cooling Towers

These hyperboloid-shaped towers (Fig. 5-2), having both a height and diameter of 100 to 200 m, are widely used in Europe. The warm cooling water from the condenser in the electrical plant is introduced at the top of the tower (Fig. 5-3); it is either sprayed in as a mist of droplets or enters in bulk to fall on a series of baffles that break it up. In either case the result is that a large surface area of water is exposed to air; the consequent evaporation cools the remaining water prior to re-circulation to the condenser. The air in the tower naturally rises as a result of being heated, and is replaced by cooler air from below in this *natural-draft* arrangement. As with cooling ponds, wet cooling towers require makeup water, but only 2% to 4% of that needed for once-through cooling.

Figure 5-2 Cooling Towers at a Nuclear Power Station in France

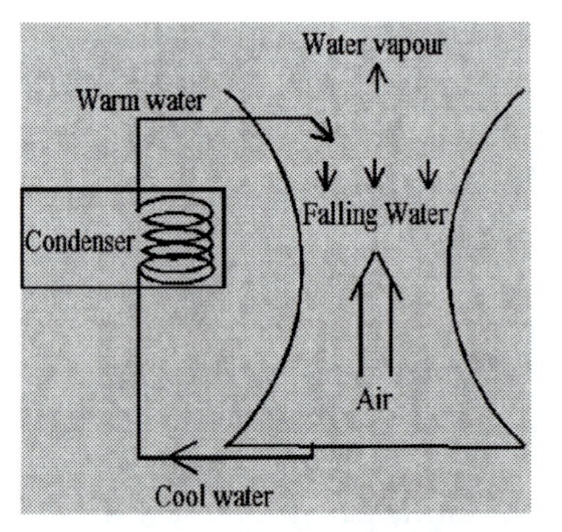

Figure 5-3 Schematic Diagram of a Wet Cooling Tower

The large size of wet cooling towers can be reduced by using fans to force much larger volumes of air into contact with the falling water. In these *forced-draft* cooling towers, the fans use about 4% of the station's energy output.

Both types of wet cooling tower share a major environmental drawback in that by inducing evaporation, they introduce large amounts of water vapour into the atmosphere. At the very least, the towers are a constant source of clouds and introduce humidity into the local environment. As well, chemicals are added to the water to prevent organic growths and corrosion, and these chemicals are released into the atmosphere.

Dry Cooling Towers

Dry towers do not rely on evaporation, but instead allow the warm cooling water to pass through a large heat exchanger (like a huge car radiator) inside the cooling tower. Air passes up the tower past this exchanger by either natural or forced draft. The environmental problems of introducing moisture and chemicals into the atmosphere are removed, but the cost is two to three times than for a wet tower and up to 8% of the station's energy is needed to operate a forced-draft dry tower. This is the obvious option for a location where there is not sufficient water for a wet tower or cooling pond.

Thermal Enhancement

It should now be evident that fossil-fuel (and nuclear) electrical plants produce large quantities of waste heat. In some situations, it would make sense to heat buildings with this energy, or to provide low-temperature thermal energy for industries. In some parts of the world, co-generation (production of both steam and electricity for industry) and district heating (using waste heat from electrical production to heat a community) are much more common than in North America. District heating is discussed in more detail in Chapter 16.

5.3 THE URBAN HEAT-ISLAND

Thermal pollution is not just the waste heat dumped into the environment at a generating station. Most of the energy of fossil fuels consumed by society finds its way eventually as heat energy to the environment; the exception is energy “stored” in manufactured materials. When averaged over the globe, the heat generated by society is a minute fraction of the energy reaching us from the Sun; however, in urban areas our energy consumption is highly localized, and as a result the temperature of urban regions is often greater than in the surrounding districts. This increase in temperature is due not only to heat generated in buildings and by motor vehicles, but also to increased absorption of solar energy because of the large quantities of concrete and asphalt. As well, an important cooling mechanism — the evaporation of water — is reduced in cities because rainwater quickly drains away instead of being held by soil and vegetation. In the summer, there is an increasing use of air conditioning in buildings and vehicles; heat energy is extracted from interiors, and expelled to the local atmosphere, producing higher air temperatures. This results in increased use of the air conditioners, which warms the outdoor air even more. As well, air conditioners use electricity, which has thermal pollution associated with its generation.

Many examples of this urban heat-island effect have been documented. At the Municipal Airport within the city limits of Edmonton, Alberta, the average annual temperature in the mid-1990s was 1.5°C higher than at the International Airport, 20 km to the south. [5]

5.4 AIR POLLUTION

> "Forget six counties overhung with smoke,
> Forget the snorting steam and piston stroke,
> Forget the spreading of the hideous town;
> Think rather of the pack-horse on the down,
> And dream of London, small and white and clean,
> The clear Thames bordered by the gardens green."
> — William Morris (1834-1896), *Prologue, The Wanderers, 1.1*

The use of fossil fuels — first coal, then oil and gas — has increased the general standard of living dramatically. However, it has also fouled the air so much that in many major cities the air is unhealthy for much of the year. When there were few people on this planet, the rate at which pollutants were sent into the air was low enough that the atmosphere remained relatively clean. Now that there are over six billion of us, air pollution is a serious problem. Here we consider the various kinds of air pollutants, technologies available to reduce them, and progress so far.

Air pollution from fossil fuels can be categorized into five main types, five of which will be addressed in the remainder of this chapter:

- particulate matter (PM)
- sulphur dioxide (SO_2)
- nitrogen oxides (NO_X)
- volatile organic compounds (VOCs)
- carbon monoxide (CO).

These substances, released directly from the fossil fuels when burned or processed, are *primary* pollutants. Table 5-2 shows the relative contributions from different human activities (anthropogenic contributions) to the overall annual releases of these primary pollutants in Canada. Of course there are also "open sources' including agriculture, wind-blown soil and road dust, and forest fires; for PM these dominate over human-made emissions, but the opposite is true for the other pollutants. The damage that the

[5] *The State of Canada's Environment*, Government of Canada, Ottawa, (1996) p.12-12

primary pollutants cause is often due to other compounds — *secondary* pollutants — formed as a result of chemical reactions involving the primary ones. For example, both

Table 5-2: Percentage Breakdown of Air Pollutants by Source in Canada for the Year 2000 [6]

Source	PM	PM2.5	SO_2	NO_x	VOC	CO
Industry	61.9	39.7	67.6	24.9	40.9	12.3
Electricity production	13.5	6.3	27.2	11.6	0.1	0.3
Residential + commercial combustion (incl. wood)	12.8	30.9	1.6	3.1	6.5	6.6
Transport	9.2	20.6	3.5	60.2	29.9	80.7
Miscellaneous and incineration	2.5	2.6	0.1	0.2	22.7	0.2

SO_2 and NO_X undergo reactions to form acids, which fall to the earth elsewhere as acid rain. Sub-micron particles in industrial PM agglomerate to form particles of a few microns diameter which can be carried great distances by wind. Because air can move quickly and easily over hundreds and thousands of kilometres, pollutants generated in one area can affect locations far away. Many types of air pollution present global problems that require international cooperation for their solution.

Particulate Matter (PM)

Particulate matter consists of a wide variety of particles — dust, smoke, droplets, that are small enough to remain airborne for a considerable time. The so-called "fine component" (described as PM2.5, i.e., having diameter < 2.5 μm) can be inhaled deeply into the respiratory tract and can result in impaired breathing and exacerbation both of pre-existing pulmonary problems such as asthma and of heart disease. Particles of this size reduce visibility and scatter sunlight before it reaches the ground, and hence also play a role in atmospheric physics that is opposite to the greenhouse effect (next chapter). In addition, they can carry damaging materials such as sulphuric acid to surfaces of buildings, accelerating their deterioration. We will return to human health effects in more detail in section 5-5. Of course, wind-blown dust from soil and roads (PM – particles of all sizes) far exceeds anthropogenic PM2.5, but it is not a health concern.

There are control devices that the majority producers – industry and electric utilities – can instal: *electrostatic precipitators* and *bag houses.* As smoke and dust particles move, they acquire electric charges by rubbing with each other; in electrostatic precipitators, the particles pass through a metal tube containing a central rod at an electric potential of up to 100000 volts relative to the tube. This means that the tube and rod have opposite charges, and each charged particle is attracted to either the tube or rod, where

[6] Criteria Air Contaminants Emission Summaries for 2000, Environment Canada

the particle's charge is neutralized. As the particles collect, they agglomerate and become large enough to settle out. A bag house is like a large vacuum cleaner, where a long (about 15 m) narrow cloth bag is used to filter out the particles from air blown through the bag by large fans.

Sulphur Dioxide (SO_2)

All fossil fuels contain some sulphur, with decreasing amounts from coal through oil to natural gas; in each case the sulphur content is highly variable. On burning, much of it is released as SO_2, which is a colourless gas with a strong pungent odour. Canada is in a unique situation as regards SO_2; its high level of hydro-electric power plants means that utility emissions are low compared to countries that burn a lot of coal; but its well-developed mining and smelting industry contributes strongly to the industrial emissions. In many other countries, e.g., the USA, where there is a heavier reliance on using coal to produce electricity; as a result 67% of US emissions in 1999 were from electric power plants. Around the globe, coal is the source of more than half the SO_2 produced. [7]

Sulphur dioxide itself is very toxic to plants, even in very low concentrations (< 0.1 parts per million by volume – ppmv), and can cause respiratory irritation in humans at concentrations of 1 to 2 ppmv. Additional concerns are presented by SO_2 because of the acids that it forms. When released into the atmosphere, some of it oxidizes to SO_3, which combines with water to form sulphuric acid (H_2SO_4). In addition, some combines directly with water to form sulphurous acid (H_2SO_3). These acids fall to the earth either as acid rain or by adhering to particulate matter.

The easiest way to reduce SO_2 emissions is to switch to a fuel with less sulphur. High-sulphur coal can be replaced with low-sulphur coal, or with oil or gas. However, such a change cannot often be accomplished cheaply. It might be necessary to import the low-sulphur fuel over long distances at some expense, and furnaces might need to be retrofitted for a new fuel. There might also be political problems — imagine importing low-sulphur coal into a region rich with high-sulphur coal, and thus putting coal miners out of work. It is also possible to clean coal before burning. Some of the sulphur is in the form of iron pyrite (FeS_2), which has a higher density than the rest of the coal. If the coal is crushed, and then agitated with water, air, oil, and a small amount of surfactant (a wetting agent), the iron pyrite settles out and can be removed; this process is called *oil flotation*. Cleaning coal in this way is cheaper than removing SO_2 from the flue gas, but it does not remove sulphur that is organically bound, and typically removes only up to 30% of the sulphur.

A common, but more expensive, approach is to use *scrubbers* to remove SO_2 by spraying a mixture of water and lime ($Ca(OH)_2$) or limestone ($CaCO_3$) down through the flue gas passing upward in the smokestack. The SO_2 combines with the lime or limestone to

---

[7] W. Fulkerson et al., Sci. Am. 263, no. 3, p. 130 (Sept. 1990)

produce $CaSO_3$, which can be removed as a solid from the water. Scrubbers can remove up to 99 of the sulphur. Other alternatives for controlling SO_2 that might make significant contributions in the future are fluidized-bed combustion and coal gasification, discussed in Chapter 18.

Nitrogen Oxides (NO_X)

Whenever air is heated above 500°C, nitrogen oxide (NO) is formed from nitrogen and oxygen, regardless of what has provided the heating — automobile engines, home furnaces, coal-fired electric plants, etc. In the atmosphere, NO is oxidized to produce nitrogen dioxide (NO_2); whereas NO is colourless, NO_2 is a brownish gas. Once formed, nitrogen dioxide can produce nitric acid (HNO_3) via more than one chemical route, and this acid settles to the earth as acid rain or on particulate matter. Nitrogen dioxide is also one of the primary pollutants in photochemical smog, discussed later in this chapter. In Canada, over half of the NO_x comes from the transportation sector (Table 5-2), primarily motor vehicles. Globally, most of the NO_X emissions are produced by burning fossil fuels, with about 30% from coal-burning. [8]

Because motor vehicles produce huge amounts of NO_X, much attention has been paid to decreasing their NO_X emissions. Various generations of catalytic converters have been developed to chemically reduce NO_X in exhaust gases to N_2; they remove about 75% of the NO_X and they also have the function of oxidizing carbon monoxide and any un-burned hydrocarbons (more about these pollutants later in this chapter).

Emissions of NO_X from coal-burning power plants can be decreased by lowering combustion temperatures through use of newer burners that produce a cooler flame, or through injection of steam into the combustion chamber. Of course, these alternatives have expenses associated with them. For example, it cost Ontario Hydro $13 million in 1989 to replace all 320 burners at its largest coal-fired plant with a cooler-burning model. [9]

Acid Rain

One of the pollution problems of which the public is most aware is acid rain. We have already discussed how SO_2 and NO_X produce damaging acids that can fall to the earth in rainwater. Natural rainwater is slightly acidic with a pH of about 5.6, as a result of carbon dioxide dissolving in it to produce carbonic acid (H_2CO_3). However, as a result of SO_2 and NO_X emissions, rain with a pH between 3.5 and 4.5 regularly falls in many areas of the world.

---

[8] Ref. 7, pp. 129-130.
[9] *"Curbing Acid Rain"*, Ontario Hydro Publication G0096, March (1989) Ref. 3, p. 12-23

The effects of acid rain are many: damage to forests, crops, soils, lakes and rivers, fish, stonework, etc. In the case of forests and crops, it is often difficult to point the finger of guilt solely at acid rain, because there are often other possible causes of damage, such as ground-level ozone. Nonetheless, there is much evidence to indicate that acid rain is the cause of extensive forest declines in Europe and on the higher slopes of mountains in the eastern USA. In Canada, there has been deterioration in sugar maple groves exposed to acidic deposition and other pollutants. Damage to vegetation and soil depends strongly on the natural ability of soil and bedrock to buffer or neutralize the acid. For instance, the granite bedrock and thin soils of the Canadian Shield have very low buffering capability, and acid rain falling in this region can be especially damaging.

The situation is more clear-cut with respect to the effects on lakes, rivers, and aquatic life. In regions with soils and rock of poor buffering ability, acid rain has decreased the pH of lakes and rivers. Healthy aquatic ecosystems have pH values above 6. As the pH approaches 6, crustaceans, insects, and some algal and zooplankton decline or vanish. As pH decreases from 6 to 5, fish populations disappear, with the most highly valued species (trout, salmon, etc.) dying at the higher pH values in this range. At a pH of 5, virtually all fish are dead. In Ontario alone, in the year 1990, there were more than 7000 lakes acidic enough that essentially all the fish had died, and another 12000 in which other forms of plants and animals had been affected. [10] Many lakes in the northeastern USA and Scandinavia cannot support fish. [11]

Acid rain also attacks buildings, corroding iron and steel, and especially destroying limestone, marble, and concrete. When artistic monuments made of stone are damaged, items of cultural heritage can be irretrievably lost.

Volatile Organic Compounds (VOCs)

Volatile organic compounds are released from any incomplete combustion of fuel and from evaporation of fuels and other organic liquids, and they play an important role in the production of photochemical smog. In Canada about 10% of the total VOC emissions come from open sources, mainly forest fires, while the rest are of anthropogenic origin; the main industrial sources are upstream oil and gas activities, and the miscellaneous portion has major contributions from solvents, dry-cleaning, and fuel-marketing.

For obvious reasons, most of the effort in controlling VOC pollutants has focused on vehicles. California has led the way over the years in setting vehicle emission standards — the serious air pollution problems in the Los Angeles area are well known. Thirty years ago, about half of the VOCs from a typical car came from the exhaust, and the other half from "blowby" (gas forced past the piston rings) and evaporation from the carburetor and gas tank. The blowby gas is now fed back into the engine, and the

---

[10] *Toronto Globe and Mail*, January (1990) reporting on a study by the Ontario Ministry of the Environment.
[11] Ref. 1, p. 165.

evaporation problem is controlled by better seals around the carburetor and tighter gas caps. The VOCs in the exhaust are now reduced by the catalytic converter, which oxidizes the VOCs to H_2O and CO_2.

Smog

The modern photochemical smog seen above many large cities is not at all the same as the classical smog associated since the time of Dickens with London, England. Classical smog — literally smoke and fog — is formed by PM from burning coal, and from moisture and SO_2. Classical smog was common in many industrial regions of the world in the 19th and early 20th centuries, and is still a problem in China, where economic development is fuelled by coal. Classical smog is usually worst in winter, when a lot of fuel is being burned. In December 1952, there was the infamous "killer smog" in London, with extremely high concentrations of PM and SO_2, resulting in a death rate two to three times normal for several days. About 4000 deaths were attributed to the smog, most of the victims being elderly people with respiratory problems. The smog problem in London was finally solved by placing restrictions on coal-burning.

Photochemical smog requires light, and thus is a problem in summer months in northern or southern regions of the world, and year-round in tropical regions. Also required are NO_X, VOCs, and warm temperatures (above about 18°C). We will not go into the detailed chemistry here; in brief summary, sunlight drives various temperature-dependent chemical reactions involving NO_X and VOCs to produce secondary pollutants, the most important of which is *ozone* (O_3). A typical smog is a mixture of ozone and particulates, with the latter imparting the colour.

Ozone is a powerful oxidizing agent that damages lungs and causes breathing difficulties, aggravation of asthma, increased severity of bronchitis and emphysema, and decreased lung function, all resulting in illness which can range to high severity and even death; active children and older people are particularly at risk. It is also very injurious to plants, so much so that the loss of agriculture and forest production in Ontario due to ozone is estimated at $280 million per year. [12] (Note that what is being discussed here is *ground-level* ozone. The decline in ozone concentration in the upper atmosphere is a completely different problem, initiated by chlorofluorocarbons (CFCs) and resulting in decreased absorption of harmful ultraviolet radiation from the Sun). Ozone is a serious problem in most urban centres in the world, but the city with the worst air quality is probably Mexico City, home to 20 million people. In 1996 the ozone air pollution index exceeded the official acceptable level (which is close to that recommended by the World health Organization) on 89% of days in the year, and for 20% of them the level was over twice the acceptable level. [13] Along some sidewalks in Mexico City, booths are set up where people can breathe oxygen for a few minutes by inserting money into a machine.

---

[12] Air Quality in Ontario, 2004 Report, Ontario Ministry of the Environment
[13] *Air pollution in Mexico City*, M. Yip and P. Madl, University of Salzburg (2002)

In Canada, ozone concentrations are highest in southern Ontario and Quebec, and are increasing (see below). The Ontario Medical Association [14] points out that during some hot summer periods of highly elevated ozone, over 50% can be attributed to ozone and its precursors arriving from the USA; at the same time, Ontario's ozone is being exported to Quebec, the Maritimes, New York and New England.

Ozone concentrations could be reduced if people would use mass transit instead of private automobiles. Many North Americans drive to work alone in a car, releasing much more NO_x and VOCs than if buses, trains, subways, or car-pools were used. Unfortunately, many people are accustomed to the convenience of the single-person vehicle, and are unlikely to switch to mass transit unless it is clearly a faster way to travel. Introduction of special traffic lanes for buses and vehicles with three or more persons, or road tariffs for single-person commuters, can help to alleviate the problem.

Carbon Monoxide (CO)

Carbon monoxide is formed whenever a fuel undergoes incomplete combustion. This gas is colourless, odourless, and highly toxic — it binds strongly to haemoglobin in the blood, preventing the haemoglobin from carrying oxygen. High concentrations can be a serious pollution problem in urban centres. In Canada, transportation is by far the most important source producing 80% of the emissions, with almost all of this contribution coming from road vehicles. Globally, the burning of tropical rain forests and savannas produces at least as much CO as does the burning of fossil fuels.[15] This is a strong indication of how rapidly the rain forests are disappearing.

As is the case with VOC and NO_x pollution, control of CO has concentrated on road vehicles. Catalytic converters oxidize most of the CO to produce CO_2 (which is a serious global pollutant on its own — see Chapter 6.)

5.5 LINKS BETWEEN AIR POLLUTION AND HUMAN HEALTH

A link between various health problems and extremely high particulate and SO_2 levels during severe air pollution episodes was well established by the 1970s. However, the suggestion of a possible cause-and-effect relationship between lung and heart illness and death and the more typical every day levels of these pollutants in the air was a much more controversial matter. Epidemiological research on this topic accelerated dramatically after 1990, when researchers became able to combine large regional

---

[14] *Ontario's Air: Years of Stagnation*: Ontario Medical Association (2001)
[15] R.E. Newell et al, Sci. Am. **261**, No.4, p. 82, Oct. (1989).

pollution databases with hospital databases on medical conditions and mortality. This was done for many different cities and regions around the world. Advanced statistical methods that control for other factors (e.g., diet, occupation, gender and weight) were weight) were developed to test whether the fraction of people becoming ill or dying exhibited a dependence on the PM2.5 or PM10 levels in the air.

Table 5-3: Relative Risk Associated with Exposure to PM2.5. *Data Extracted from Ref. 16.*

Cause of death	RR with 95% confidence range
All causes	1.06 (1.02 – 1.11)
Cardiopulmonary	1.09 (1.03 – 1.16)
Lung cancer	1.14 (1.04 – 1.23)
All other causes	1.01 (0.95 – 1.06)

The study we draw on here, sponsored by the American Cancer Society (Cancer Prevention Study II or CPS-II) involves lifetime tracking of 1.2 million adults living in over 150 metropolitan areas of the USA. Pope and colleagues [16] give an excellent summary of progress. Graphs of the fractional death rate due to heart and lung disease against PM2.5 concentration are linear with positive slopes, demonstrating a clear correlation. There is no evidence of any threshold below which the effects disappear. Researchers express the slopes as the Relative Risk (RR), i.e., the increase in death rate per 10 $\mu g/m^3$ increase in PM2.5. Table 5-3 shows the results.

The Ontario Medical Association [17] expressed its concern in 2002 over air pollution, particularly in the Windsor-Quebec corridor where high levels of PM2.5 and ground-level ozone occur most frequently in Canada. It pointed to a Canadian study where 2-4 % excess deaths from respiratory problems could be attributed to pollutant levels, and it commented on the increases in hospitalization and emergency room visits resulting from worsened air quality. The OMA recommended more stringent limits on SO_2 and NO_X emissions, and the imposition of more rigorous standards for vehicle emissions. It also suggested that action should be taken under the U.S. Clean Air Act as regards US emissions which damage the health of Canadians.

With the evidence of adverse health effects now firmly in hand, research is now turning towards the actual biological mechanisms. New work is directed at ascertaining what component of the PM2.5 particles is responsible? And what are the resulting cellular processes in humans that cause or exacerbate disease?

5.6 REDUCTION OF AIR POLLUTION

There are two main ways to estimate air pollution on a regional or country-

[16] C.A. Pope, R.T. Burnett, M.J. Thun, E.E. Calle, D. Krewski, K. Ito, G.D. Thurston, JAMA, 287, no. 9, p. 1132.
[17] Ontario Medical Association Ground Level Ozone Position Paper (2000)

wide scale. The first is to collect and/or estimate all possible data on emissions into the atmosphere; sometimes, as in the case of automobiles, this will be a complex task, depending on the variation of emission control equipment across the fleet. The second is to measure concentrations of pollutants in the air using networks of monitoring stations. It is then interesting to look at published emissions trends and air quality trends. Exact agreement between the two should not be expected: first, there are uncertainties in emissions estimates ;second, to take but one example, measured PM2.5 includes emitted PM2.5 together with PM2.5 that has formed from chemical and physical processes in the atmosphere.

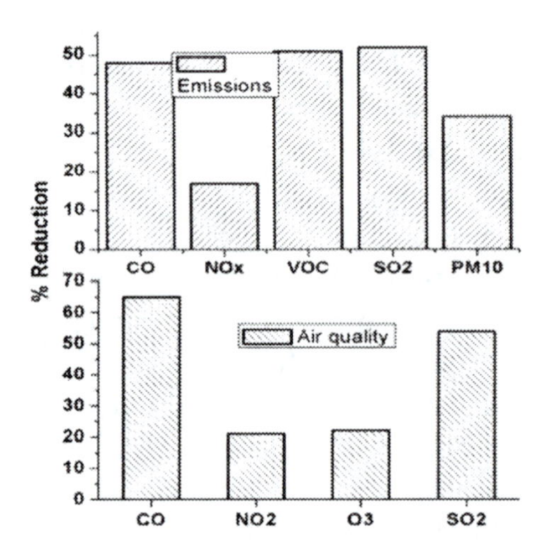

Figure 5-4 Air pollution reductions in USA from 1983 to 2002

The reductions in US emissions and in monitored air pollution [18] shown in Fig. 5-4 are remarkable, reflecting the various measures taken since the Clean Air Act of 1974. This was achieved in a period during which energy consumption increased by 42% and vehicle-km traveled by 155%!

Canada and the USA signed an Air Quality Agreement in 1991, and an important part of this is the Acid Rain Annex, which addresses SO_2 and NO_X emissions, particularly from electric power generation. In the period shown in Fig. 5-4, both countries reduced their SO_2 emissions by about 50%. In the USA, this reflects controls in electric utilities and desulphurization of diesel fuel; in Canada, it reflects reductions in both the mining and the utility sectors. The largest Canadian reductions were in the east, while emissions rose in the west as a result of "sour gas" processing, oil sands development, and burning coal for electricity.

In the 1983-2002 period shown in Fig. 5-4, US emissions of NO_X declined by 15% while measured NO_2 levels in the air fell by 21%. Graphs of NO_X emissions [19] from Environment Canada show very little reduction in the 1990-2002 period; however in Ontario[13] average measured concentrations of NO_2 fell by 35% between 1975 and 2004.

Considerable reductions have been achieved in the deposition of wet sulphate (acid rain) in eastern Canada and USA, reflecting the controls on SO_2 emissions put into place in the last two decades. Comparing the periods 1989-1991 and 1998-2000, reductions of up to 30% were seen over a large area of the eastern and north-eastern United States.

---

[18] National Air Quality and Emissions Trends Report 2003, USEPA
[19] U.S.-Canada Air Quality agreement: Progress Report 2004

However for wet nitrate deposition, very little change has been observed. At least some of the credit for the reduction in acid levels in Canada should go to the Canadian Coalition on Acid Rain, which quietly lobbied governments from 1981 until 1991. Having seen acid rain legislation approved in the USA in 1990, the group disbanded.

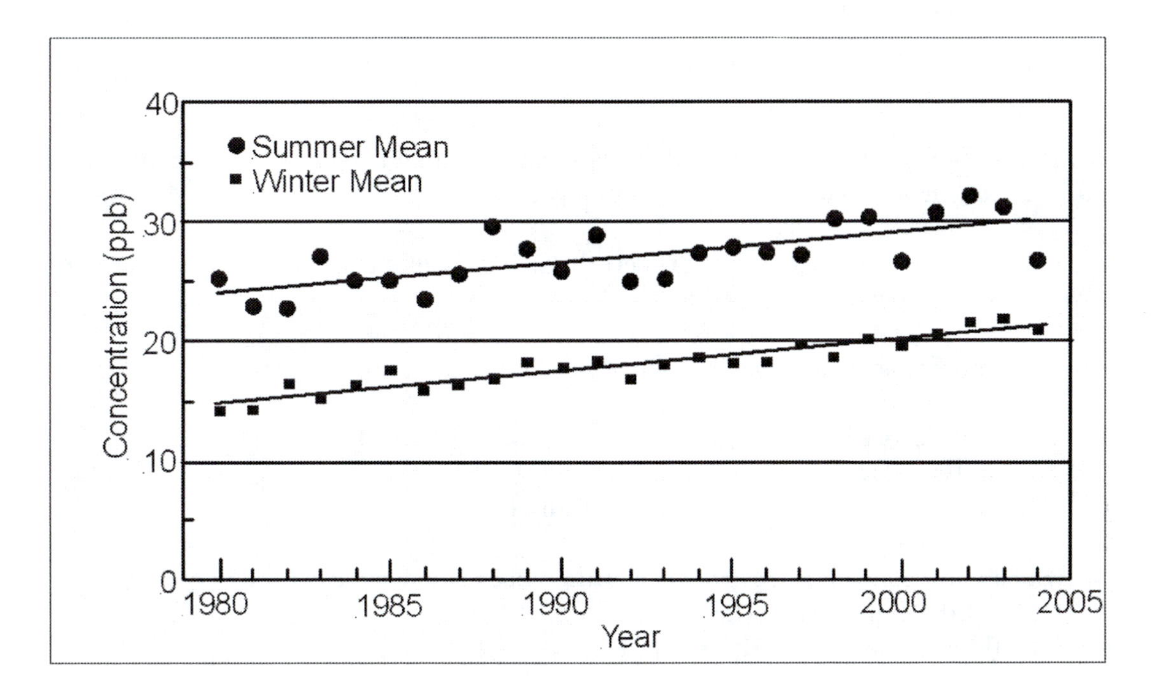

Figure 5-5 Twenty-five Year Trend of Seasonal Means of Ground-level Ozone Concentrations across Ontario [20]

Ground-level ozone remains an extremely significant issue in both the USA and Canada. The 22% decrease in the "1-hour" ozone concentration shown in Fig. 5-4 was achieved mostly in the 1983-1993 period, with only a 2% decrease in the 1993-2002 period. Indeed in the latter period the "8-hour" concentration shows a small increase. These last two observations are statistically compatible with a "no-change' situation. For Ontario, measurements of ozone concentrations [20] over the past 25 years, shown in Fig. 5-5, have a slowly increasing trend.

The Ozone Annex [21] to the Canada-USA Air Quality Agreement was signed in 2000, and focuses on NO_X and VOC emissions, these being the precursors of ozone and smog. It defines a Pollutant Emission Management area (PEMA) which covers Quebec and central and south Ontario, together with eighteen American States in the Great Lakes and north-east regions. The agreement requires a greatly expanded reporting of emissions. The commitments agreed upon by the partners include:

- stringent NO_X and VOC emission levels for vehicles;
- caps on NO_X emissions by utilities

[20] Air Quality in Ontario 2004 Report, Government of Ontario
[21] See Environment Canada website

- reduction of VOC emissions by the key sectors of iron and steel, base metal smelting and refining, cement, asphalt, pulp and paper, lumber and wood.

On the basis of these commitments, projections have been made for emission reductions over the period 1990-2010. For NO_x, these are 39% in both countries; for VOCs they are 35% in Canada and 46% in the USA.

EXERCISES

5-1. For a fossil-fuel electrical plant that is operating with an efficiency of 40%, what is the ratio of waste heat to electrical energy produced?

5-2. For a nuclear electrical plant that is operating with an efficiency of 30%, what is the ratio of waste heat to electrical energy produced?

5-3. Why is photochemical smog not a problem in Toronto in January?

5-4. Being as specific as possible, what is the major source (> 50%) of
(a) SO_2 in Canada? (b) SO_2 in the world?
(c) NO_x in Canada? (d) CO in Canada?

5-5. For which of the following types of air pollution does transportation provide at least 25% of the emissions in Canada?
(a) PM (b) SO_2 (c) NO_x (d) VOCs (e) CO

5-6. In a molecule of sulphur dioxide, what is the ratio of the mass of sulphur to the mass of oxygen? Refer to Appendix IV for molar masses.

5-7. What mass of sulphur is contained in 15000 tonnes of $CaSO_3$? Refer to Appendix IV for molar masses.

5-8. Canada is aligning its standards for pollutant emission by vehicles with those of the USA. The methodology is complex, allowing a car manufacturer to have 11 different emission levels among its products, but ensuring that the average across the "fleet" meets a defined standard that will drop in value over a period of time. The highest CO emission level in this arrangement for the 2004-2006 period is 2.6 g/km. What mass of carbon monoxide is emitted by a 2005 automobile if it travels 11500 km and its CO emission rate remains constant at the maximum permitted?

PROBLEMS

5-9. This problem compares the cooling water requirements of a nuclear plant producing electricity at 30% efficiency and a fossil-fuel electrical plant operating at 40% efficiency. Assume that the electricity production at the two plants is equal. Neglecting the heat that goes up the chimney in the fossil-fuel plant, and assuming two significant digits in given data, determine the following ratio: (cooling water required for nuclear plant)/(cooling water required for fossil-fuel plant).

5-10. (a) In the year 2004 the USA produced 1787 TW·h of electrical energy in conventional (i.e., fossil-fuel) thermal plants and 476 TW·h in nuclear plants. Assuming 30%

efficiency for nuclear plants and 40% for conventional thermal plants, determine the (annual) volume of cooling water required to cool these plants in once-through cooling if the cooling water undergoes a temperature increase of 10°C. (Neglect the heat lost up the chimney in conventional plants, and assume two significant digits in given data.)
(b) The total amount of precipitation in the USA is estimated to be about 5.6×10^{12} m^3 annually, of which approximately 30% is available as runoff. Confirm the statement made in the text that about 1/5 of the annual runoff would be required for once-through cooling of electrical plants.

5-11. For a typical 1000-MWe fossil-fuel electrical plant that has an efficiency of 40% and is cooled by a wet cooling tower, calculate the amount of water evaporated daily, assuming that the water is at 100°C and that all the waste heat removed from the plant is removed by evaporation.

5-12. In many devices involved in energy industries, a fluid (liquid or gas) flows through a pipe. If the cross-sectional area of a pipe is 2.0 m^2, and the fluid in it has a speed of 1.5 m/s:
(a)what is the volume flow rate in m^3/s? (i.e., what volume of fluid passes through any cross-section of the pipe every second?)
(b)what mass of fluid passes through any cross-section of the pipe every second, if the fluid has a density of 1.0×10^3 kg/m^3?

5-13. An old coal-fired steam-turbine electrical power plant that produces 800 MW of electrical power is located beside a lake. The overall efficiency of the plant is 34%.
(a) What is the total thermal power input to the plant?
(b) At what rate is waste heat discharged from the plant (in MW)?
(c)If the lake water used in the condenser is to be raised in temperature by no more than 6.0°C, what minimum volume of water must be made available per second?
(d) If the lake water enters the condenser through a pipe of diameter 8.0 m, what minimum speed must the water have in the pipe?

Additional Information for Problems 5-14 to 5-16:

Element	Molar Mass (g/mol)
C	12
O	16
S	32
Ca	40

5-14. In 2004, approximately 1 billion tonnes of coal were burned in China (one third of the world total). If each tonne of coal had an average sulphur impurity of 1.5% (by mass), how many tonnes of sulphur dioxide could have been put into the air in the absence of scrubbing devices?

5-15. A particular coal-fired electric power station uses 7.5×10^9 kg of coal per year to produce 950 MW of electrical power. The coal contains 2.4% sulphur by mass. If a scrubber is used to remove 85% of the sulphur from the exhaust gases in the form of $CaSO_3$, what mass of $CaSO_3$ is produced annually?

5-16. A coal-burning electric power plant produces 850 MWe at an efficiency of 37%. It burns coal of energy content 28 MJ per kg and sulphur content 1.7% by mass. What mass (in tonnes) of SO_2 is produced by this plant in a year? (1 tonne = 1000 kg)

ANSWERS

5-1. 1.5
5-2. 2.3
5-6. 1:1
5-7. 4000 tonnes
5-8. Approx. 30 kg
5-9. 1.6
5-10. **a.** 2.7×10^{11} m^3
5-11. 5.7×10^7 kg
5-12. **a.** 3.0 m^3/s, **b.** 3.0×10^3 kg/s
5-13. **a.** 2.4×10^3 MW, **b.** 1.6×10^3 MW, **c.** 62 m^3, **d.** 1.2 m/s
5-14. 60 million tonnes
5-15. 5.7×10^8 kg
5-16. 8.8×10^4 tonnes

CHAPTER 6

ELECTROMAGNETIC RADIATION AND THE GREENHOUSE EFFECT

The third method of heat transfer is radiation. The physics of the electromagnetic spectrum is reviewed here and applied to what has become known as the "Greenhouse Effect".

6.1 THE ELECTROMAGNETIC SPECTRUM

In Chapter 5 the two heat transfer processes of conduction and convection were discussed. In this section the third process *radiation* will be introduced. Whereas conduction and convection require the intermediary of matter to transport energy, radiation does not. The energy of the Sun comes to us by the process of radiation through essentially empty space.

Radiant energy consists of electromagnetic (E-M) waves of which visible light is only one part. Being waves they are characterized by a wavelength λ and a frequency f. The wavelength is the crest-to-crest distance of the waves (see Fig. 6-1.), and the frequency is the number of full wavelengths that pass any point (such as x_1 in Fig. 6-1) each second while the wave moves at a speed c. The relationship between these three quantities is:

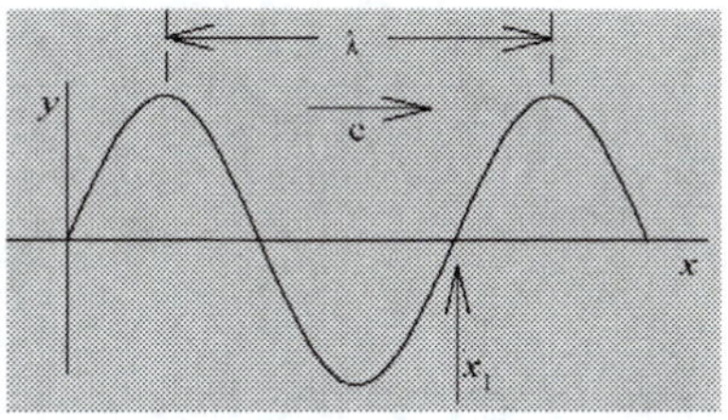

Figure 6-1 An E-M Wave Moving in the *X*-Direction With a Speed c.

$$c = f\lambda \qquad \textbf{[6-1]}$$

The speed of E-M waves is that of light, which in a vacuum is,

$$c = 2.998 \times 10^8 \text{ m/s}$$

If the wavelength is measured in metres then the frequency must have units s^{-1}. This unit is commonly called the *hertz*, (abbreviated Hz) and means *cycles per second*. Except for radio waves, it is very inconvenient to use metres to measure these wavelengths. Since the range of wavelengths is so very large, several different subsidiary units have been defined to keep the numbers simple. The total range of wavelengths define the *electromagnetic spectrum* and is described (along with the appropriate conversion factors) in Table 6-1.

This very wide range of wavelengths is shown diagrammatically in Fig. 6-2 where it is evident how narrow is the range of visible wavelengths relative to the whole E-M spectrum.

Table 6-1: The Approximate Wavelength Ranges of the Electromagnetic Spectrum

Type of wave	λ(m)	Conventional units
Radio waves	>0.1	>0.1m
Microwaves	10^{-1}–10^{-4}	100mm – 0.1mm
Infrared	10^{-4}–7×10^{-7}	100 μm – 0.7 μm
Visible light	7×10^{-7}–4×10^{-7}	700 nm – 400 nm
Ultraviolet	4×10^{-7}–10^{-8}	400 nm – 10 nm
X–Rays	10^{-8}–10^{-11}	10 nm – 0.01 nm
Gamma rays	$<10^{-11}$	<0.01 nm

Note 1mm = 10^{-3} m, 1μm = 10^{-6} m, 1 nm = 10^{-9} m

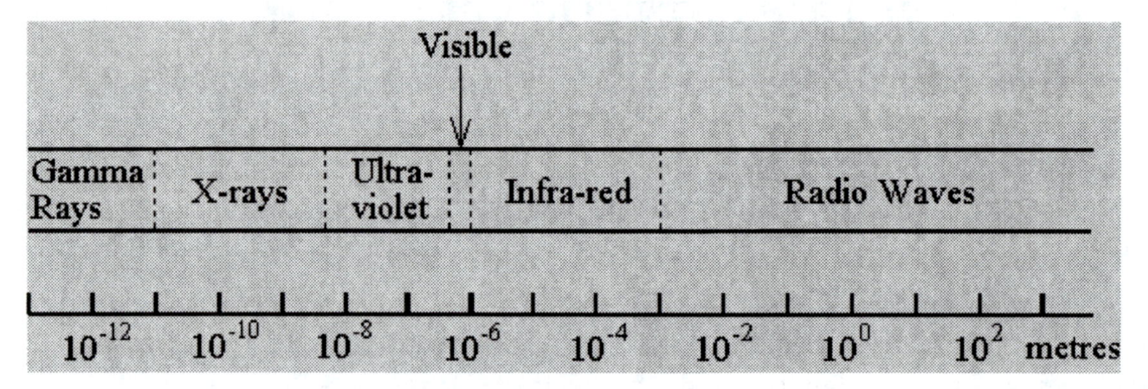

Figure 6– 2: The Electromagnetic Spectrum.

6.2 INVERSE-SQUARE LAW

It is important to know how the intensity of radiation varies with the distance away from the source of energy. Obviously it decreases as the distance increases; if you stand further from a fireplace you feel a smaller intensity of radiated heat. The quantity, *intensity,* is defined as the amount of radiant energy passing per unit time through a unit area or, the power per unit area (since power = energy/time).

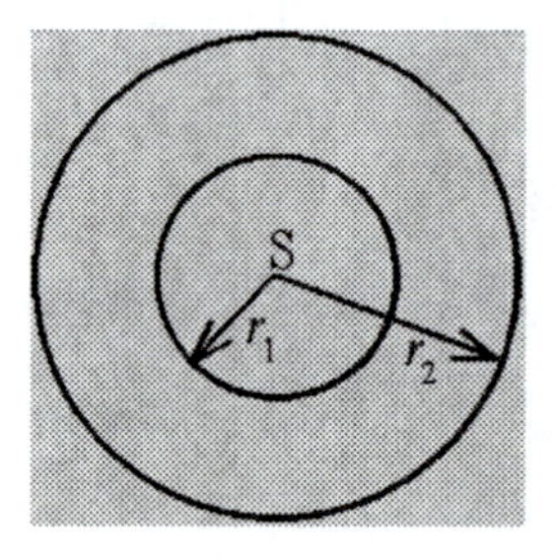

Figure 6-3 Inverse Square Law

In Fig. 6-3 the point S is a source of radiation, emitting P watts (joules/second). All of this power passes through the sphere of radius r_1 so that the intensity I_1 at a distance r_1 is

$$I_1 = P/4\pi r_1^2$$

The power which goes through this sphere must also pass through the sphere of radius r_2. The intensity I_2 at this new distance is

$$I_2 = P/4\pi r_2^2$$

Since the power is the same then

$$I_1/I_2 = (r_2/r_1)^2 \qquad \textbf{[6-2]}$$

This equation is the *inverse square law* which says that the intensity of radiation from a point source varies inversely as the square of the distance from the source. It holds for all types of radiation (e.g., sound, radioactivity, etc.) so long as the source can be considered point-like compared to all other dimensions.

EXAMPLE 6-1

The *solar constant* is the intensity of energy, supplied by the Sun, arriving at the top of the Earth's atmosphere; it is 1.40×10^3 W/m^2. The distance from the Sun to the Earth is 149.6×10^6 km and to Venus it is 108.2×10^6 km. What is the value of the solar constant on Venus?

SOLUTION

$I_1/I_2 = (r_2/r_1)^2$
$I_v/1.4\times10^3\ \text{W}\cdot\text{m}^{-2} = (149.6\ \text{M km}/108.2\ \text{M km})^2$
$I_v = 2.68\times10^3\ \text{W/m}^2$

6.3 BLACKBODY RADIATION

The next thing to consider is how the energy in the E-M radiation is distributed over this range of wavelengths. If the radiation from some hot object is passed through a glass prism, for example, it will be spread out according to wavelength, the shortest wavelengths being deviated the most by the prism and the longest wavelengths the least. The result is a distribution of energy vs. wavelength called the *blackbody spectrum*; this is shown in Fig. 6-4 for objects at temperatures of 1500 K, 2000 K, and 2500 K.

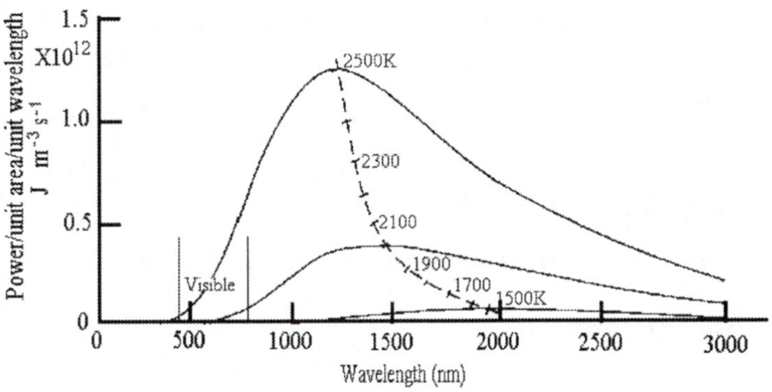

Figure 6-4 The Blackbody Spectrum

There are several important features of this figure:

1) The area under any curve is the total energy/unit area/unit time, that is the intensity, $I(T)$, radiated by the object at a temperature T.

2) Curves representing different temperatures never cross. This means that as the temperature of an object increases its energy output at every wavelength also increases. The way in which the total intensity increases with temperature is given by *Stefan's Law* which is,

$$I(T) \propto T^4 \qquad \textbf{[6-3]}$$

The fourth power of the temperature means that the total energy radiated by objects rises very rapidly as the temperature is increased. Eq. [6-3] can be expressed as an equality if a constant of proportionality is introduced. In the resulting equation,

$$I(T) = \sigma T^4 \qquad \textbf{[6-4]}$$

the constant σ is *Stefan's constant* and has the value 5.67×10^{-8} W·m^{-2}·K^{-4}.

An object which radiates according to Eq. [6-4] is called a "perfect blackbody". Most surfaces, however, radiate with less efficiency than the ideal black case. This is expressed in the equation by multiplying by another dimensionless parameter ε called the "average emissivity" (what it is averaged over is discussed in the next paragraph). Thus,

$$I(T) = \varepsilon\sigma T^4 \qquad \textbf{[6-5]}$$

The value of ε lies between 0 and 1. If it is 0 then the object doesn't radiate at all; if $\varepsilon = 1$ then the surface is said to be "perfectly black", hence the term "blackbody radiation". Real surfaces have values that vary with wavelength. For example, the metal tungsten, which is used as the filament in incandescent electric light bulbs, has an emissivity of 0.46 at a wavelength of 500 nm (yellow light), but only 0.38 at 1000 nm (infra red) and it is not a perfect blackbody ($\varepsilon = 1$) at any wavelength. The average emissivity in Eq. [6-5] is therefore the average over all wavelengths.

EXAMPLE 6-2

What is the total intensity $I(T)$ for a perfect blackbody at 2500 K?

SOLUTION

From Eq. 6-5 with $\varepsilon = 1$ (perfect blackbody),
$I(T) = 5.67 \times 10^{-8}$ W·m^{-2}·K^{-4} $\times (2500)^4$ K^4 $= 2.21\times10^6$ W/m^2

This is the area under the upper curve in Fig. 6-4. Can you think of a simple way to verify the order of magnitude of the value from the curve itself? [Hint: think of triangles.]

---

3) The wavelength of the peak of each curve moves from long wavelengths at lower temperatures to shorter wavelengths as the temperature is raised. If λ_m is the wavelength of the peak then it is related to the temperature of the radiating object by *Wien's Law*:

$$\lambda_m(\text{ nm }) = \frac{2.8972 \text{ x } 10^6 \text{ nm}\cdot\text{K}}{\text{T}(\text{K})} \quad \textbf{[6-6]}$$

The positions of the peak of the radiation curves are also shown on Fig. 6-4 as a dashed line. At a temperature of about 6000K, as we will see, the peak will appear near 500 nm; this is the temperature of the surface of the Sun and explains why sunlight appears white. It contains all the visible colours approximately equally.

EXAMPLE 6-3

What is the wavelength of the peak of the blackbody radiation curve for an object at a temperature of 2500 K?

SOLUTION

From Wien's law (Eq. [6-6])

$\lambda_m = 2.8972 \times 10^6$ nm·K/2500 K = 1159 nm

This is the wavelength of the peak of the upper curve in Fig. 6-4.

---

4) Figure 6-4 shows that most of the emission from objects at about 2000 K is in the infrared (heat) region. What visible light they emit is predominantly in the red; i.e. red-hot.

A proper theoretical understanding of the form of this radiation curve had to await the development of the Quantum Theory by Max Planck. In the quantum description, radiation is viewed as consisting of a rain of massless "particles" or *quanta* carrying the radiation energy and travelling at the speed of light. The energy of each quantum is proportional to the frequency of the radiation and is given by Planck's equation

$$E = \text{h}f \quad \textbf{[6-7]}$$

where h = 6.626×10^{-34} J s is *Planck's Constant.* (This equation will not be needed specifically in this chapter but will be important in the discussion of photo-voltaic devices in Section 15.4.)

The equation for the curve is *Planck's Radiation Equation*:

$$I(\lambda,\text{T}) = \frac{2\pi \text{h c}^2}{\lambda^5} \frac{1}{\text{e}^{\text{h c}/\lambda k_B \text{T}} - 1} \quad \textbf{[6-8]}$$

In this equation k_B is *Boltzmann's Constant.* When the well-known values for all constants are substituted, Eq. [6-8] can be written

$$I(\lambda,\text{T}) = \frac{3.746\times10^{-16}}{\lambda^5} \frac{1}{\text{e}^{1.46\text{x}10^{-2}/\lambda \text{T}} - 1} \quad \textbf{[6-9]}$$

For an object that is not perfectly black then Eq. [6-8] and [6-9] must be multiplied by the emissivity which is a function of the wavelength. Eq. [6-9] becomes,

$$I(\lambda,T)=\varepsilon(\lambda)\frac{3.746\times10^{-16}}{\lambda^5}\frac{1}{e^{1.46\times10^{-2}/\lambda T}-1} \qquad \textbf{[6-10]}$$

EXAMPLE 6-4

What is the intensity per unit wavelength I(λ,T) for a perfect blackbody at a temperature of 2500 K at the peak of its radiation curve? The wavelength (from Example 6-3) is 1159 nm.

SOLUTION

From Planck's Radiation Equation (Eq. 6-10) with ε(λ) = 1 (perfect blackbody),

$$I(\lambda,T)=\frac{3.746\times10^{-16}\,\mathrm{W\cdot m^2}}{(1159\times10^{-9})^5\,\mathrm{m^5}}\frac{1}{e^{1.46\times10^{-2}/(1159\times10^{-9}\times2500)}-1}$$

$$=1.17\times10^{12}\ \mathrm{W/m^3}$$

This is the peak value of the upper curve in Fig. 6-4.

6.4 SOLAR RADIATION

When a radiation detector is taken above the Earth's atmosphere by means of a rocket or satellite and the spectrum of the incoming radiation from the Sun is recorded, the upper curve of Fig. 6-5 is obtained. Except for a few features in the wavelength range 200 to 400 nm this curve is that of a blackbody at a temperature of 6000 K. This is one of the ways we know that the visible surface of the Sun (the photosphere) has a temperature of 6000 K.

If the radiation detector is operated at the Earth's surface, under the atmosphere, the lower curve of Fig. 6-5 is obtained. The curve retains the shape of a blackbody at 6000 K but is of lower overall intensity. About 31% of the incoming solar energy is scattered or reflected back into space by clouds, dust in the atmosphere, the Earth's surface, and by the atmospheric molecules themselves.

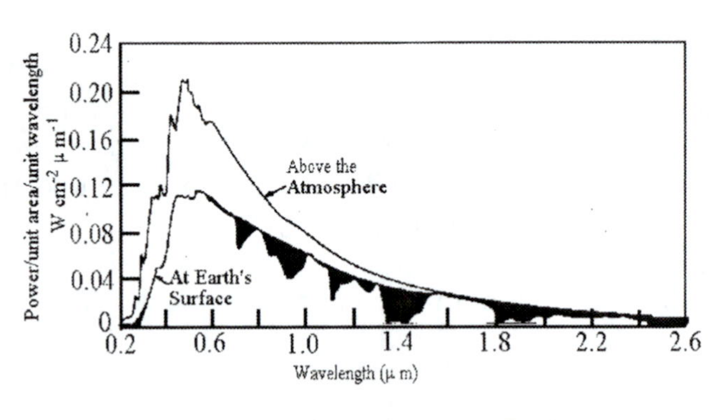

Figure 6-5 Solar Radiation at the Earth.

Astronomers call the fraction lost in this way the planetary *albedo* (α); for the Earth α = 0.31.

The Earth, as seen from space, is a very bright planet, that is, its albedo is large. This is caused by the large area of water on the surface and the highly reflective clouds in the atmosphere. For us, on the surface, this means that there is a smaller fraction (69%) of solar radiation which is actually absorbed. Some other planets are much darker when they are viewed from space, that is, their albedos are smaller. For example, the albedo of the Moon is 0.07, Mars is 0.16 and the rings of Uranus are as dark as coal dust with an albedo of 0.03.

The lower curve in Fig. 6-5 also has a number of regions in the infrared where the black body curve seems to be eaten away (shown darkened in Fig. 6-5). These are regions where the atmospheric molecules, particularly water (H_2O) and carbon dioxide (CO_2) strongly absorb radiation from the solar spectrum. About 20% of the incident solar energy is absorbed in this way. This absorption has a profound effect on the temperature and climate of the Earth, and is responsible for what is known as the *Greenhouse Effect*.

6.5 ENERGY BALANCE IN THE EARTH'S ATMOSPHERE

The Earth and its atmosphere receive energy in the form of solar radiation mostly at short wavelengths (because the Sun has a high temperature) and, as mentioned above, some is reflected or scattered out immediately ($\alpha = 0.31$) and the rest is absorbed. The Earth and the troposphere (lower atmosphere) also radiate energy but do so at long wavelengths (in the "far-infrared") since their temperature is much lower. The interchanges of energy in this system are complex and are shown schematically in Fig. 6-6 for an input of 342 W/m^2 of solar energy.[1]

Of these 342 units 107 units (77 + 30) are returned directly to space, accounting for the Earth's albedo ($\alpha = 107/342 = 0.31$). The remaining 235 units are absorbed: 67 by the atmosphere and 168 by the surface of the Earth. In addition to the direct solar input, the warm atmosphere radiates in all directions and contributes over twice as much energy to the Earth's surface (324 units). The surface, therefore receives 492 units (168 + 324). At equilibrium this energy must also be lost by the surface which occurs in three ways: 102 units are recycled back to the atmosphere by water evaporation and convection, and 390 units are re-radiated for a total of 492 units as expected. Energy balance must also prevail in the atmosphere. It receives 67units from the Sun, 350 of the 390 of the Earth's radiation, and 102 from evaporation and convection for a

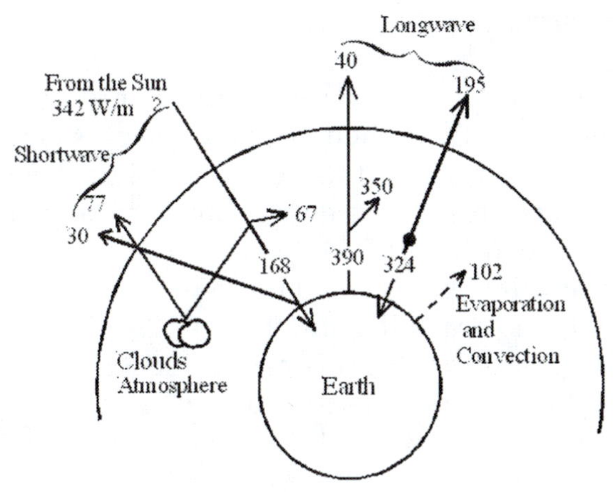

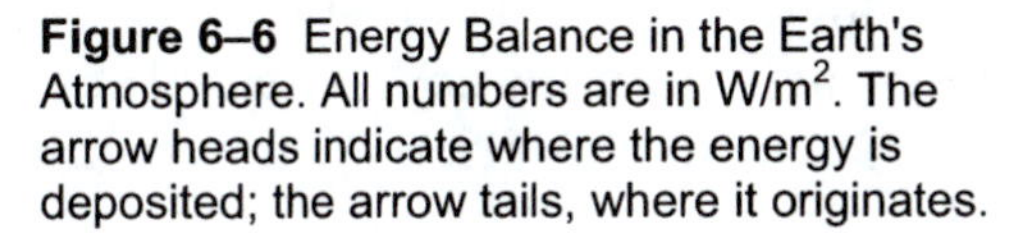
Figure 6–6 Energy Balance in the Earth's Atmosphere. All numbers are in W/m^2. The arrow heads indicate where the energy is deposited; the arrow tails, where it originates.

[1] The reason for this number will become clear in the next section.

total of 519 units. This is lost as 324 units radiated to the surface and 195 units radiated to space (total - 519).

This complicated system of exchanging energy back and forth among the surface, atmosphere, and space is in thermal equilibrium and each component has its characteristic temperature. Several important environmental questions follow from this: What would be the effect of disturbing any one of these cycles? Of particular concern these days is what would happen if the 350 units absorbed in the atmosphere from the 390 radiated from the surface were to increase because of an increased absorbing power of the atmosphere? How would the system adjust? Would any of the other cycles be affected?

6.6 THE GREENHOUSE EFFECT

Imagine that sunlight is falling on a black surface and that all of the energy is absorbed. The temperature of the surface will rise to some value, say T_1, where the rate at which energy is radiated away is equal to the rate at which the Sun provides it; equilibrium is established. The radiation curve of the surface at T_1 could be the lower curve in Fig. 6-7. If T_1 is around 300 K then the peak of the radiation curve is in the far infrared region and indeed all of the relevant wavelengths of Fig. 6-7 are in the infrared.

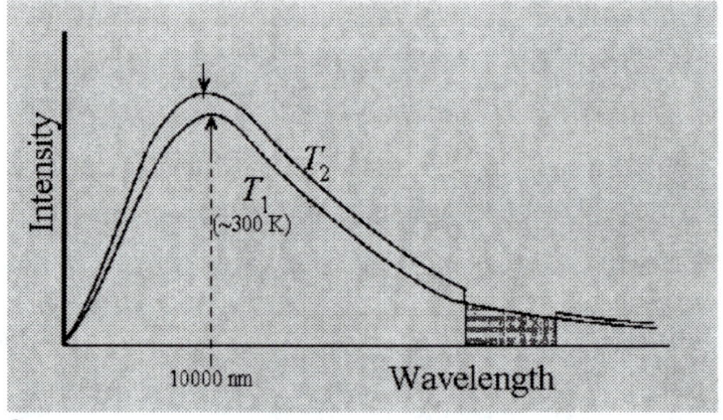

Figure 6-7 The Greenhouse Effect.

If a sheet of glass (or some other absorbing medium) is interposed between the sunlight and the surface then, because the glass is transparent in the visible (short wavelengths) but absorbs in certain regions of the infrared, the equilibrium is altered. In Fig. 6-7 this absorption is represented simply by the shaded region. The surface can no longer radiate away sufficient energy to maintain equilibrium and so its temperature will begin to rise. The temperature rises to a new value T_2 so that the excess energy represented by the region between the two curves is equal to the energy represented by the shaded region. At this stage equilibrium is re-established but necessarily at a higher temperature. This is the explanation of the greenhouse effect.[2]

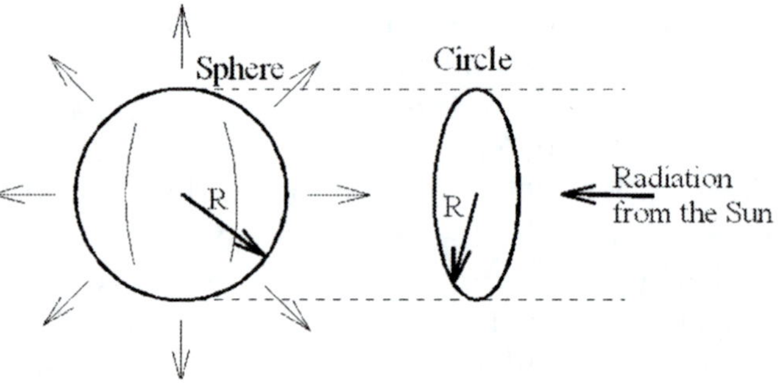

Figure 6-8 The Energy Balance of the Earth.

[2] Popular press explanations involving "trapped radiation" and other terms are obviously incorrect. The greenhouse effect is an equilibrium phenomenon depending on the relative transparency of the short and long–wavelength regions of the atmosphere.

6.7 THE EARTH'S GREENHOUSE EFFECT

In the Earth's atmosphere the gases, primarily H_2O and CO_2, provide both absorption and scattering of the incoming solar energy. If the intensity of the solar radiation at the top of the Earth's atmosphere is I_0 (W/m^2), then the total power incident on the atmosphere is that which flows through a circle of radius R, as shown in Fig. 6-8. This power can be written as the product of area and intensity $\pi R^2 I_0$, where R is the radius of the Earth (and therefore of the circle).

This radiation becomes rapidly distributed (albeit non-uniformly) over the whole spherical surface of the Earth (area = $4\pi R^2$) by the Earth's rotation, and atmospheric and oceanic circulation, so it is as if the top of the atmosphere were receiving a steady average power per square metre (the "insolation") of

$$I_0 \frac{\pi R^2}{4\pi R^2} = \frac{I_0}{4}$$

If we take $I_0 = 1.368\times10^3$ W/m^2, then the average insolation at the top of the atmosphere is $1.368\times10^3/4 = 342$ W/m^2, which is the value used in the previous section in the discussion of the Earth's energy balance.

A fraction, α, of the power arriving at the top of the atmosphere is reflected and scattered away and so the power delivered through the atmosphere is $\pi R^2 I_0(1-\alpha)$. The power radiated away by the planet at an effective temperature T_e is given by Stefan's Law (Eq. [6-4]) and is radiated from the entire surface of the Earth; it is therefore, $4\pi R^2\sigma T^4$. If P_{net} is the net power retained by the Earth's surface then

$$P_{net} = \pi R^2 I_0(1-\alpha) - 4\pi R^2\sigma T^4 \qquad \textbf{[6-11]}$$

and the net power per unit area is

$$I_{net} = \tfrac{1}{4} I_0(1-\alpha) - \sigma T^4 \qquad \textbf{[6-12]}$$

In a state of equilibrium $I_{net} = 0$. The value of I_0 at the mean distance of the Earth from the Sun is 1.368×10^3 W/m^2 and with $\alpha = 0.31$ this gives an effective temperature of the Earth's surface of 254 K. Note that we have made a major assumption in neglecting any absorbing effects of the atmosphere. The actual mean temperature of the Earth's surface is, however, 15 °C or 288 K. Thus the atmosphere accounts for a warming of the surface by 34K.

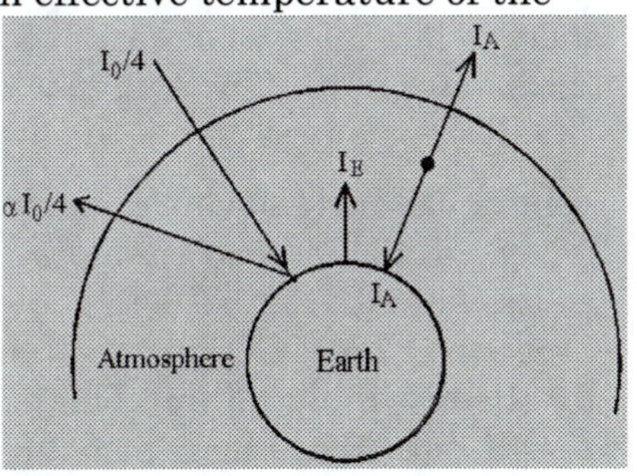

Figure 6-9 A simple Model of the Atmosphere.

We can introduce the absorptive effects of the atmosphere with a simple model shown in Fig 6-9. We assume that Earth receives per unit area $\tfrac{1}{4} I_0(1-\alpha)$ from the Sun as before and radiates an amount I_E which is completely absorbed by the atmosphere, since it is entirely far-infrared radiation. But the atmosphere is also a radiating blackbody and radiates an amount I_A equally to the Earth and to space. Since the total energy from the Sun must equal the total energy returned to space then

$$\tfrac{1}{4}I_0 = \tfrac{1}{4}\alpha I_0 + I_A, \text{ or } I_A = \tfrac{1}{4}I_0(1 - \alpha)$$

Applying the same reasoning to the atmosphere we have

$$I_E = 2I_A = 2(\tfrac{1}{4}I_0)(1 - \alpha) = \tfrac{1}{2}I_0(1 - \alpha) = \sigma T_E^4$$

Solving gives $T_E = 302$ K, which is close to the measured value of 288 K, and a great improvement on 254 K when the atmosphere is assumed to make no contribution.

The model can even be refined by having the atmosphere only absorb a fraction of the radiation from the Earth. This is left for Problems 6-17 and 6-18 where it can be seen that a value of T_E = 286 K is obtained, only 2 K below the measured value.

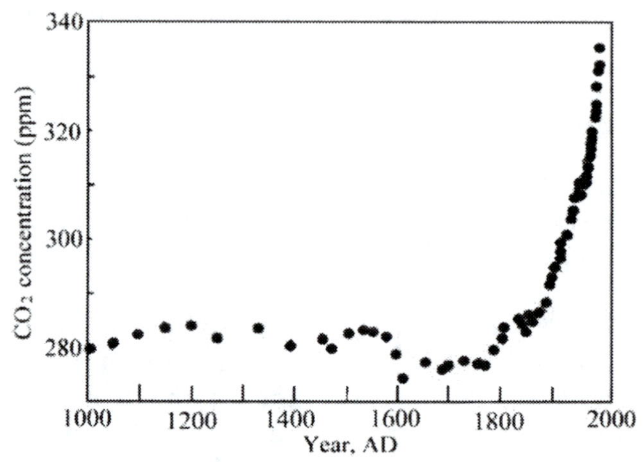

Figure 6-10 Atmospheric Carbon Dioxide since 1000 AD. Courtesy of V. Morgan et. al. Antarctic CRC and Australian Antarctic Division. See footnote 3.

Any planet with an atmosphere will experience a warming of its surface because of the greenhouse effect. Mars with its thin atmosphere, experiences a small effect. By contrast, on Venus, with a very dense atmosphere of CO_2, the greenhouse effect dominates producing temperatures of 700 K on the surface.

Until the 19th century the activities of people had very little effect on the equilibrium of the gases in the atmosphere. The Industrial Revolution and the rapid development of industry brought increasing emission of gases into the atmosphere. The evidence that the CO_2 concentration in the atom-sphere has increased is undeniable; the effects of its increase are more controversial. Figure 6-10 shows the concentration of CO_2 in the atom-sphere for the last millennium. The most recent figures are from measurements on gases in the contemporary atmosphere,

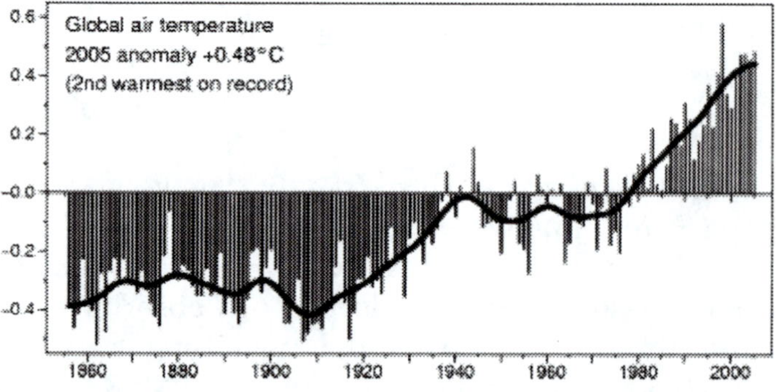

Figure 6-11 Global temperature since 1855. The reference line is the 1961-1990 mean. Courtesy: Climate Research Unit, School of Environmental Studies, Univ. of East Anglia see footnote. 4, 5 and http://www.cru.uea.ac.uk/

while data representing earlier times come from measurements on air-bubbles trapped in polar ice-caps or in glacial ice.[3]

The concentration of CO_2 remained unchanged at about 280 parts-per-million (ppm) until about 1800 and then started to increase. The increase, while gentle at first, has become very steep in the latter part of the twentieth century and has increased by over 20% from pre-industrial days.

Figure 6-11 shows the mean annual temperature of the Earth since 1850; the solid curve is the five year average.[4] That the temperature has risen by about one degree in this period is evident; to attribute that rise to the increase in greenhouse gases is more problematic. It is well known that there are temperature variations of the Earth on many time scales. Ice ages occur on the scale of tens of thousands of years and smaller fluctuations occur on shorter scales. For example, a 70 year period ending in 1715 became known as the "Little Ice Age" since it was a period of unaccustomed cold in northern Europe. It is also known that this same 70 years was a period in which there were very few sunspots; in other words, there was little solar *activity*.

There are scientists who are reasonably able to argue that the data of Fig. 6-11 are simply evidence of another variation in the Earth's temperature that has nothing to do with the greenhouse effect. Others, just as reasonably, argue that it is evidence of a major disturbance of the environment caused by our activities and refer in particular to the rapid rise since 1970. As more research is done and data accumulates the case put forward by the deniers becomes less and less tenable.

It is notoriously difficult to forecast seasonal or annual temperatures. The problem is that the Earth's atmosphere is so very complicated that detailed calculations and predictions tax the resources of the largest computers. Even if the calculations could be carried out, it is certain that all the interrelated processes in the atmosphere remain unknown or are imperfectly understood. There are many effects that might serve to decrease the greenhouse effect; these are *negative feedback* effects. For example, it is likely that the oceans of a warmer Earth would evaporate more water and thus there would be more cloud cover. An increased cloud cover would increase the Earth's albedo, α, in Eq. [6-12]; an increase in α would result in a decreased temperature T_e (see problem 6-8). In other words, the greenhouse effect may, to some extent, be self-correcting on a watery planet like the Earth. It is on grounds such as these that some scientists argue against predictions of catastrophic temperature changes in the short term. The rapidly increasing computing power now available makes the models more sophisticated and realistic. A recent study [5] clearly separates the natural and anthropogenic (man-made) effects, and ascribes the temperature rise of the last half of the 20th century to fossil fuel burning and other human activity. The result of such advances is that the majority of atmospheric scientists anticipate temperature rises of

---

[3] D.M. Etheridge, L.P. Steele, R.L. Langenfelds, R.J. Francey, J.–M. Barnola, V.I. Morgan. *Natural and anthropogenic changes in Atmospheric CO_2 over the last 1000 years from Antarctic ice and firn.* Journal of Geophysical Research, Vol. 101, (1996) pp 4115–4128.

[4] P.D. Jones, M. New, D.E. Parker, S. Martin, I.G. Rigor, *Surface air temperature and its changes over the past 150 years.* Reviews of Geophysics, Vol. 37, (1999) pp 173–199.

[5] P.A. Stott, S.F.B. Tett, G.S. Jones, M.R. Allen, J.F.B. Mitchell and G.J. Jenkins; *External Control of 20th Century Temperature by Natural and Anthropogenic Forcings.* Science Vol. 290 (2000) pp 2133–2137.

between 1 and 2 K in the next 50 years, and some predictions on more debatable assumptions do not rule out rises of up to 5K.[6]

6.8 OTHER GREENHOUSE GASES

Although the discussion so far has been in terms of CO_2 it is not the only contributor to the greenhouse effect. Other gases of human origin make about an equal contribution. The other important gases are methane (CH_4), chlorofluorocarbons (CFC), ozone (O_3), and nitrous oxide (N_2O). Table 6-2 summarizes their present concentration and rate of increase. (The table also shows the relative contribution of each gas to the greenhouse effect - this will be explained later).

Table 6-2: The Greenhouse Gases[7]

Species	Concentration (ppm)	Rate of Increase (% per year)	Greenhouse Contribution (%)
CO_2	350	0.5	60
CH_4	2	1	15
CFC	0.0008	NA**	12
O_3*	0.01- 0.05	0.5	8
N_2O	0.3	0.2	5

* in the troposphere (lower atmosphere) **This entry is essentially zero as a result of the Montreal Protocol severely curtailing the global use and release of CFCs

The increase in these gases is almost entirely a result of the activities of humans in the modern industrial society. The increase of CO_2 is not only due to the burning of fossil fuels but is enhanced by massive deforestation, as plants are a major factor in removing CO_2 from the atmosphere.

Methane has many sources: Major changes in agricultural practise, particularly increased growing of rice and increased herds of cattle have resulted in more atmospheric methane. Natural gas is largely methane and its production from drilled gas wells inevitably leads to leakage or incomplete combustion. Finally, methane is copiously produced in landfills (See Section 15.5).

CFCs found many industrial uses since their discovery, particularly in refrigeration, as a plastic foam expander, and as a driver gas in aerosol sprays. All of these uses result in atmospheric pollution. Since the adoption of the Montreal Protocol, however, the production and release of CFCs has been greatly reduced so essentially none is being added. It remains a greenhouse gas however since its lifetime in the atmosphere is very

[6] Intergovernmental Panel on Climate *Change Climate Change 2001: The Scientific Basis,* IPCC Secretariat, C/O World Meteorological Organization (2001)
[7] H. Rodhe, Science 248 p 1217 (1990)

long (see Table 6-3). In addition to the greenhouse effect caused by CFCs in the troposphere, the erosion of the protective layer of ozone by CFCs that diffuse to the stratosphere is a separate and very serious issue. Ozone is produced by sunlight in the stratosphere and is essential to our well being as it shields us from the harmful effects of the ultraviolet rays of the Sun. It is, however, a pollutant in the lower atmosphere. It is produced by a complex of chemical reactions involving the action of sunlight on the chemical smog that pollutes large urban areas.

N_2O is just one of many oxides of nitrogen that are produced when fossil fuels are burned with air. At the elevated temperature of the combustion process the nitrogen component of the air reacts with oxygen to produce nitrogen oxides. The hotter the combustion gases are, the more of these oxides that are produced. This is one of the unfortunate effects of increasing combustion temperature to increase the thermal efficiency of electric energy production; it increases the production of N_2O. The most important source of N_2O is the automobile engine so it is not surprising that it is a most serious problem in large traffic congested cities. In the presence of solar ultraviolet radiation and unburned hydrocarbons it takes part in a complex chain of reactions that is responsible, among other things, for urban smog. These gases are not equally effective in producing a greenhouse warming. In addition to the large effect of water, their relative contribution is given in the last column of Table 6-2.

If production of these gases were to stop, the effects would not disappear quickly. The methods by which the gases are removed are various and not at all well understood; this is particularly the case for CO_2. The oceans, plants, and chemical reactions and other unidentified agents remove them at various rates which have been estimated and are given in Table 6-3.

Table 6-3: Decay Time of Greenhouse Gases[8]
(Time to decay to 37% (1/e) of initial value)

Species	Time (yr.)
CO_2	120
CH_4	12
CFC	0.5 -260
O_3*	0.1
N_2O	114

* in the troposphere (lower atmosphere)

6.9 EFFECTS OF GREENHOUSE WARMING

As explained above, the calculations of future greenhouse warming remain estimates, owing to our imprecise understanding of the processes at work. For the same reason, predictions of climatic change are just as imprecise, perhaps more so. However it seems

[8] Ibid

certain that the global average temperature will rise by 1 to 5 K over the next half-century, and some consequences necessarily follow from this.

If it does nothing else, global warming will cause the ocean levels to rise. This need not be a result of an increased precipitation or melting of the polar ice-caps; the oceans will rise simply from the thermal expansion of water (see problem 6-16). Small though this effect might appear, it may account for a global average increase of 0.2 m in the sea level. (Local factors will also play a part in any given locality: changes in ocean circulation, distribution of salinity, etc.) One model calculation gives a rise in the north Atlantic of 0.4 m but no change (or even a small decrease) in the Ross Sea in Antarctica. Yet a rise of 0.4 m in the North Atlantic would be catastrophic in low-lying areas in, for example, the Eastern Seaboard of the USA, where many millions of people live in cities and much food is grown on the alluvial plains. A large alluvial-plain country like Bangladesh, which can already have half of its surface area flooded by rainfall runoff, could be devastated even by a small rise in sea level.

A further consequence of increased temperature will be increased ocean evaporation and the consequent increase of rainfall on the continental land masses. An increased snowfall might even make the polar ice-caps increase in size. An increased rainfall would make the flooding problem in Bangladesh even more severe, and models suggest it may be more important than the ocean rise itself. Paradoxically, increased precipitation and temperature does not necessarily mean increased soil moisture. Shorter winters will trap less moisture and longer summers will increase evaporation so the soil moisture could actually decrease. The large deserts around the Tropics of Cancer and Capricorn, such as the Sahara and the US southwest will increase, and semi-desert grasslands now used for grain production like the Canadian prairies will lose productivity.

Some agricultural scientists counter that the increase of CO_2 in the atmosphere will increase the rate of photosynthesis in plants, thus increasing productivity. In addition, they argue, increased plant coverage of the fertile regions of the Earth will increase soil-moisture retention. Thus they believe that, although the producing regions of the Earth may be redistributed, the effects may be globally self correcting or even positive! Arguments of this type have almost divided agriculturalists and climatologists on the "optimistic" and "pessimistic" sides of the greenhouse debate.

6.10 AMELIORATION OF THE GREENHOUSE EFFECT

From the lifetime data given in Table 6-3 it is obvious that nothing can be done to correct the greenhouse warming in a short time. Even if production of all greenhouse gases could be stopped immediately the effects would be felt for a century at least and longer if irreparable damage to the environment occurs. In addition, although the production takes place under the control of the various nations and their respective political systems, the atmospheric circulation knows no political boundaries. The problem is a global problem in a world that has not been used to solving global problems either peacefully or with dispatch.

Clearly a rapid phase-out of CFCs is necessary for more reasons than global warming. This is one area where some progress has been achieved. The so-called "Montreal Protocol" of 1990 called for a reduction of CFCs by 50% by the year 2000. This was speeded up, however, because of the discovery of an “ozone hole” (i.e. an area of depleted ozone) in the stratosphere over North America and Northern Europe in 1992.

The other greenhouse gases are almost all the result of our society's demand for energy. The utilization of energy cannot be stopped but methods must be found to minimize gas emission. First, and potentially easiest to address, is the efficiency with which energy is used. Switching from incandescent to fluorescent lighting, insulating houses, lowering winter home temperatures, raising summer air-conditioning temperatures, and building more fuel-efficient cars, are measures which, while variable in the size of their effect, will add together to reduce the rate of increase of green-house gases.

To make an absolute reduction of the emission of these gases is more difficult. To do this we must switch from our dependence on fossil fuels to a more diverse mixture of renewable and non-renewable resources. These are discussed in more detail in other chapters of this book.

To further complicate matters there is some theoretical modelling which predicts that if all of the world were to convert to renewable energy resources the result would be a period of global cooling before the pre-industrial temperature and atmospheric conditions were restored. This readjustment will take up to several centuries to complete and human influence will have very little effect on its amelioration.

Finally, the ultimate problem is population; people require energy, and energy produces greenhouse gases. Population control is a problem beyond the scope of this book, but its solution must involve all of humankind with all its knowledge including the scientific.

EXERCISES

6-1. While standing on a dock you count 20 wave crests (starting at zero) go by in 80 s. The distance between crests is 4 m. What is the wave speed?

6-2. What is the wavelength of the carrier wave of a radio station broadcasting at 740 kHz?

6-3. A small point-like source of radioactive material is placed 10 cm below a Geiger counter and 900 counts/s is recorded. The counter is raised to 30 cm; how many counts are now recorded in one second?

6-4. Two identical lead spheres are hung on thin threads in a vacuum. The temperature of one is 320 K and it radiates 6.0 W of thermal energy. The second sphere has a temperature of 325 K. How much heat does it radiate?

6-5. At what wavelength is the peak of the radiation curve of the first sphere in Exercise 6-4?

PROBLEMS

6-6. The solar constant at the mean position of the Earth is about 1.40×10^{3} W·m^{-2}; the mean distance from the Earth to the Sun is 149.6×10^{6} km. What is the total power developed by the Sun?

6-7. The distance from the Sun to Mars is 2.28×10^{8} km. Neglecting the planet's thin atmosphere, what is the solar constant at the surface of Mars?

6-8. The Earth orbits about the Sun in an elliptical orbit with a mean distance of 149.6×10^{6} km. The minimum distance (perihelion) is 147.0×10^{6} km and occurs about Jan. 4. The maximum distance is 152.1×10^{6} km and occurs about July 6. The average solar constant is about 1.40×10^{3} W/m^{2}; what is it on Jan. 4? What is the total percent change between Jan. 4 and July 6? Does this account for the seasons?

6-9. Using the result of Problem 6-6, calculate the surface temperature of the Sun. The radius of the Sun is 6.98×10^{5} km and Stefan's constant is 5.67×10^{-8} in SI units.

6-10. Using the solar constant given in Problem 6-6, calculate the mean total power intercepted by the Earth. The radius of the Earth is 6400 km.

6-11. Use the result of problem 6-10 to calculate the surface temperature of the Earth in the absence of the atmosphere and assuming that the surface is totally black, that is, the albedo is zero.

6-12. Repeat problem 6-11 assuming the correct albedo for the Earth of 0.31.

6-13. Show that an increase of the Earth's albedo by a small amount $d\alpha$ will produce a decrease in the surface temperature given by

$$dT = -71\, d\alpha/(1-\alpha)^{3/4}$$

If the albedo increases by 1%, what is the change in temperature assuming there is no atmosphere?

6-14. What is the ratio of the intensity per unit wavelength of a perfect blackbody at a wavelength of 500 nm to that at 1000 nm if the temperature is 3000 K?

6-15. Repeat question 6-14 for the case of tungsten (see pg 6-4).

6-16. The oceans of the world cover 3/4 of the earth's surface and have an average depth of 3.8 km. If the temperature of the oceans were to rise by 1 K how much would the sea level rise on average due to thermal expansion alone? Assume that the surface area of the oceans would not change. The thermal expansion of a volume V_0 of liquid to a volume V due to a temperature change T is given by $V = V_0(1 + \beta T)$ where β is the volume expansion coefficient. The value of β for water is 0.2×10^{-3} K^{-1} .

6-17. Energy Balance for the Earth with a Semi-transparent Atmosphere. Assume that the Earth receives, on average, a power of $\frac{1}{4}I_0$ from the Sun and has an albedo α. Assume that the Earth's atmosphere absorbs a fraction β of the far-infrared radiation from the Earth (and therefore transmits a fraction $(1-\beta)$ to space. Show that the intensity radiated by the Earth is

$$I_E = \frac{2(\frac{1}{4}I_0)(1-\alpha)}{2-\beta}$$

6-18. Measurements of the absorption spectra of the Earth's atmosphere in the far-infrared show that the atmosphere absorbs 75% of the blackbody radiation from the Earth. Calculate the average equilibrium temperature of the Earth's surface.

6-19. Figure 6-10 shows the concentration of CO_2 in the Earth's atmosphere over the past millennium. Figure 6-12 presented here is of the Earth's population for an even longer period. It could be instructive to plot a graph of the "population-specific" emission of CO_2: that is, how much CO_2 is each person responsible for? Has it changed over time?

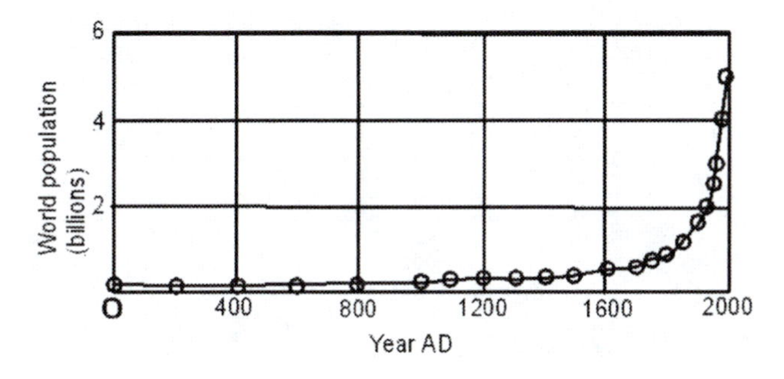

Figure 6–12 Earth's population for 2 millennia.

a) From the 2 graphs tabulate the total atmospheric CO_2 and the population in millions (the graphs scale better when the population is expressed in 10^6 units.) Smooth out the CO_2 data by drawing a curve through it.

b) Find a method from the data to estimate the natural concentration of CO_2 in the atmosphere and, assuming it to be constant in time subtract+ it from the measured concentrations. Presumably the remainder is the anthropogenic part.

c) Determine the "specific" anthropogenic CO_2 in ppm/10^6 persons. Is there evidence for an increasing anthropogenic part of the CO_2 concentration, or is the CO_2 concentration entirely a result of population increase?

ANSWERS

6-1. 1 m/s
6-2. 405 m
6-3. 100
6-4. 6.4 W
6-5. 9.05×10^3 nm
6-6. 3.94×10^{26} W
6-7. 0.60×10^3 $W{\cdot}m^{-2}$
6-8. 1450 W/m^2, +7%
6-9. 5800 K
6-10. 1.8×10^{17} W
6-11. 279 K
6-12. 255K
6-13. –0.29 K
6-14. 0.25
6-15. 0.30
6-16. 0.8 m
6-18. 286 K
6-19. Present anthro CO_2 per 10^6 population is about 10×10^{-3}.

CHAPTER 7

ELECTRICITY

Electricity is the form of energy that characterizes the modern era. It forms the basis of the industrial activity and the way of life of modern society. More energy resources are devoted to the generation of electricity than to any other single end use. In this chapter the fundamental laws of electricity are developed.

7.1 ELECTRIC CHARGE

Electric charge is a fundamental property of matter. One of the reasons that electric science developed later than mechanics was that most matter in bulk displays no obvious effect of electric charge; the electric effects are subtle and difficult to evoke. We now know, of course, that this is because electric charge exists in two forms called "positive" (+) and "negative" (–), and matter normally contains equal amounts of both, producing a null effect. In addition we know that "like" charges (i.e., + + or – –) repel each other, whereas "unlike" charges (+ –) attract.

Extensive experimentation in the 17th and 18th centuries elucidated the fundamental laws of the forces produced by stationary and moving charges on other stationary and moving charges, and the magnetic effects resulting from moving charges. Much of this fundamental electrical science will not be needed here as our primary concern is the production of electric current and potential, the flow of current through conducting materials, and its use as a carrier of useful energy.

In the 20th century, as the structure of the atom became better understood, it was realized that the electric neutrality of bulk matter was a result of several facts:

- The unit of electric charge, both positive and negative, is quantized. This means that charge does not exist in arbitrary amounts but occurs in integral multiples of some smallest finite amount called the *fundamental unit of charge.* To the highest precision of our measurement capabilities, the absolute values of the fundamental units of positive and negative charges are equal. In the SI system of units this charge has a magnitude of

$$e = 1.602\times10^{-19}\ \text{C} \qquad \textbf{[7-1]}$$

  where C stands for "coulombs.[1]"

- Ordinary matter is electrically neutral (equal numbers of positive and negative charges) right down to the level of the individual atoms.

---

[1] For the definition of the coulomb (C) consult any elementary physics textbook.

- The negative electric charge resides on the atomic particle called the *electron*, and the positive charge on the *proton*.
- In the Rutherford model of the atom the massive protons reside in the atomic nucleus and the much lighter electrons in a large space around the nucleus. The electrons are much more weakly bound to the atom than are the protons.

7.2 ELECTRIC CURRENT AND CONDUCTORS

Did you know? The terms "positive" and "negative" for the two types of charge were first applied by Benjamin Franklin (1706-1790).

As a result of the last item in the previous section, most electric currents (and all that we will deal with) are a result of the movement of electrons (i.e., negative electric charge) from one place to another. The binding of the electrons to the atoms does, however, vary over a wide range. In some atoms the electrons are very tightly bound and so do not move easily through the material; these materials are called *insulators*. Other materials, particularly the metals, have a few outer-orbit electrons so weakly bound that, in the bulk material, they are virtually free and so can move to make an electric current; these materials are called *conductors*.

It is one of the great unfortunate circumstances in physics that the mobility of the negative charges was realized long after the definitions and conventions had been firmly established. Unfortunately these conventions were established assuming that it was the positive charges that moved. As a result we maintain the fiction that the *electric current* is the flow of positive charge in some direction in a conductor when we know very well that in most cases it is really a flow of electrons in the opposite direction. This fiction will plague students to the end of time!

7.3 ELECTRIC POTENTIAL ENERGY AND POTENTIAL

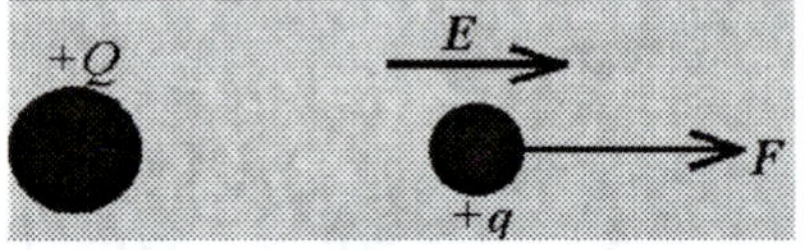

Figure 7-1 Electric Field and Force

Since an electric charge is able to exert a force on another electric charge, the first charge (say +Q in Fig. 7-1) is said to be surrounded by an *electric field* [2] (***E***) which acts on the second charge (+q) to produce the force ***F***. The field and the force are in the same direction for two positive charges. The electric field of Q is defined by

$$\mathbf{E} = \frac{\mathbf{F}}{q} \qquad [7\text{-}2]$$

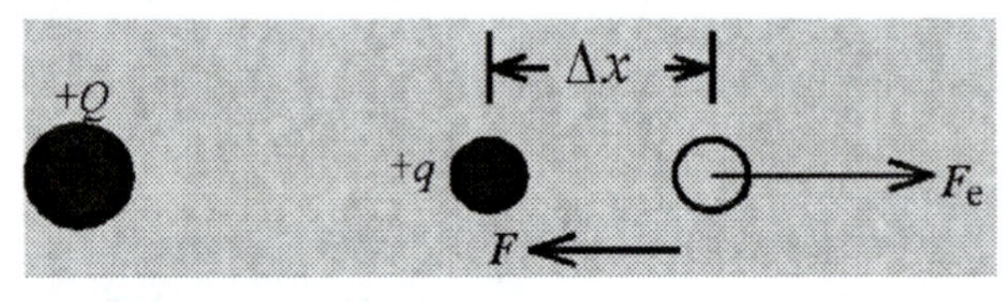

Figure 7-2 Electric Potential Energy

Notice that the units of the field are "newtons per coulomb," (N/C).

[2] When boldface type is used, as in ***E*** and ***F*** the symbol stands for the vector quantity i.e. the magnitude as well as the direction. *E* and *F* in ordinary type stand for the magnitude of the vector alone without the direction.

In Fig. 7-2 the charge q is acted on by the electric field to produce the electric force $\boldsymbol{F}_e$. Imagine that an external force $\boldsymbol{F}$, essentially equal in magnitude to $\boldsymbol{F}_e$, is applied in the opposite direction and the charge is moved a distance Δx. The amount of work done by this force on the charge is ΔW where

$$\Delta W = F\,\Delta x = qE\,\Delta x \qquad \textbf{[7-3]}$$

Since an amount of work ΔW has been done on the charge q its *electric potential energy* (U) has been increased by the same amount ΔU, i.e., $\Delta W = \Delta U$. [3]

A useful new quantity, which is related to the potential energy, is called the *electric potential* or *potential difference* (ΔV); it is the *electric potential energy per unit charge* or

$$\Delta V = \frac{\Delta U}{q} \qquad \textbf{[7-4]}$$

The units of this new quantity are joules per coulomb, termed *volt* (V). [4] If q is eliminated between Eq. [7-3] and [7-4] (noting that $\Delta W = \Delta U$) then

$$E = \frac{\Delta V}{\Delta x} \qquad \textbf{[7-5]}$$

Notice that the units of the electric field can be volts per metre (V/m).

In Fig. 7-3 a charge $+Q$ creates an electric field $\boldsymbol{E}$ around it; the direction of an electric field is always away from a positive charge. The small test charge $+q$ experiences a force in the direction of the field. Left to itself it would move from the point A to point B. We say, therefore, that "the potential at A is higher than at point B", that is, it would move from the point of higher potential to the point of lower potential. The situation is analogous to the case of a mass in a gravitational field.

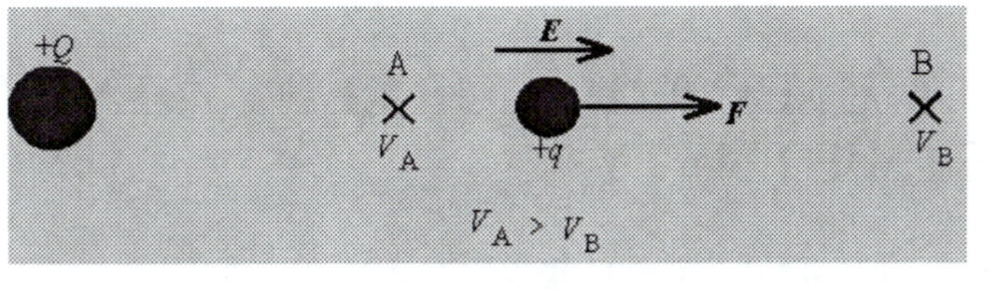

Figure 7-3 Electric Potential for the Case $V_A > V_B$.

The analogy (between gravitational field and electric field) breaks down when one of the charges has a negative sign. For example if Q is negative then the direction of the field would be reversed and the potential of A would be lower than that of B. In that case the charge $+q$ would move from B to A, which is as expected since "unlike charges attract." On the other hand negative charges move from positions of low potential to positions of high potential. Specific examples are left to the exercises at the end of the chapter.

[3] Recall in mechanics: If a mass m is lifted vertically a distance h at constant velocity by a force F in Earth's gravitational field, the force exerted is $F = mg$ and the work done by the upward applied force is $Fh = mgh$. The increase in *gravitational potential energy* is also mgh.

[4] Unfortunately in this case the symbol for the quantity (electric potential) and the unit (volt) is the same i.e., V.

EXAMPLE 7-1

Two metal conductors are charged such that the potential of the first is 100 V higher than that of the second, i.e., $V_1 = V_2 + 100$. A small object carrying a charge $q = 5\times10^{-19}$ C is released at conductor 1 and moves to conductor 2. What is the change in KE of the object?

SOLUTION

$\Delta U = q\Delta V$, $\therefore \Delta U = (5\times10^{-19}\text{ C})(-100\text{ J/C}) = -5\times10^{-17}\text{ J}$
$\Delta KE = -\Delta U$ $\therefore \Delta KE = 5\times10^{-17}\text{ J}$

EXAMPLE 7-2 (The "Electron-Volt")

The *electron volt* or "eV" is a unit of energy which is very useful for systems on the atomic scale. It is the energy expended when a positive charge equal in magnitude to that of one electron is moved through a potential of one volt. What is one eV in joules?

SOLUTION

Using Eq. [7-4]; $\Delta U = q\Delta V = (1.602\times10^{-19}\text{ C})(1\text{ V}) = 1.602\times10^{-19}\text{ J}$

7.4 ELECTRIC CURRENT, RESISTANCE, AND RESISTIVITY

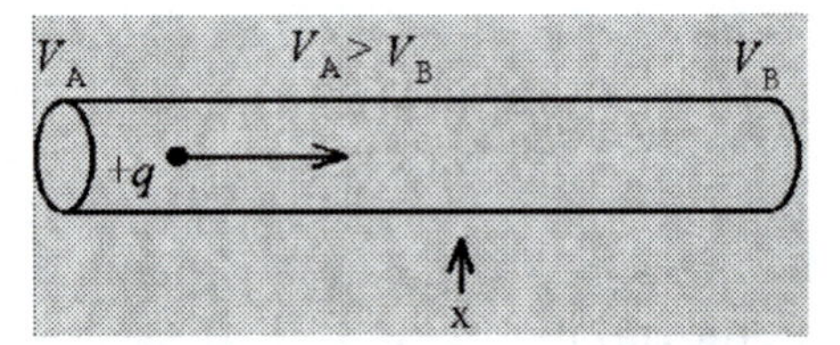

Figure 7-4 Electric Current.

It is important to remember the comment about the *conventional current* made in Sec. 7.2; even though an electric current in a metallic conductor is, in fact, a flow of electrons we imagine that it is a flow of positive charge in the opposite direction. If an electric conductor such as a copper wire as shown in Fig. 7-4 has one end at a positive potential with respect to the other end, that is, $V_A > V_B$, the charges will flow from A to B creating a current of charge in the conductor for as long as the potential difference is maintained. We define the *electric current* (I) at the point x as the total amount of charge (Q) that passes the point per unit time.

$$I = \frac{Q}{t} \qquad \textbf{[7-6]}$$

The units of I are coulombs per second or *amperes* (A).

Of course, the situation shown in Fig. 7-4 cannot last for long; eventually enough charge will flow from A to B to eliminate the potential difference and the current will stop. Useful currents must flow in continuous circuits with some device that has the ability to remove the charge at B and re-insert it at A thus maintaining the potential difference, much as a pump removes water from the low pressure end of a closed pipe system and inserts it again at the high pressure end maintaining the current of liquid. There are several devices that perform this function of providing a source of potential difference; the most common is the dry battery found in flashlights and other devices, the automobile battery, and a large number of rotating machines generally called "dynamos."

It seems logical that in a given conductor a higher potential difference should produce a larger current but the mathematical relation between I and V is not obvious. This relation was investigated experimentally by Georg Simon Ohm (1789-1854) in 1826. He found that for most conductors the current was proportional to the potential difference, i.e., $I \propto V$. Introducing a proportionality constant R this can be written as

$$V = I\,R \qquad \textbf{[7-7]}$$

This simple, but important, equation is known as *Ohm's Law*. The quantity R is called the *resistance* and has the units volts/amperes (V/A) which we rename the *ohm* with the symbol Ω.

EXAMPLE 7-3

A battery with potential difference 1.5 V causes a current of 0.30 mA to flow through a conductor. What is the resistance of the conductor? (mA means "milli-ampere" or 10^{-3} A)

SOLUTION

$V=IR$, therefore
$R=V/I=1.5\text{V}/(0.30\times10^{-3}\text{A}) = 5.0\times10^{3}\ \Omega$

Most common conductors, like metals, obey Ohm's Law and are called *Ohmic conductors*. Some materials, however, do not have a linear relationship between potential difference and current, but a more complicated one; they are called *non-ohmic* conductors. Some of these materials, like semiconductors, are very important and will be discussed later in connection with solar cells; for now our attention will be restricted to ohmic conductors.

Resistivity

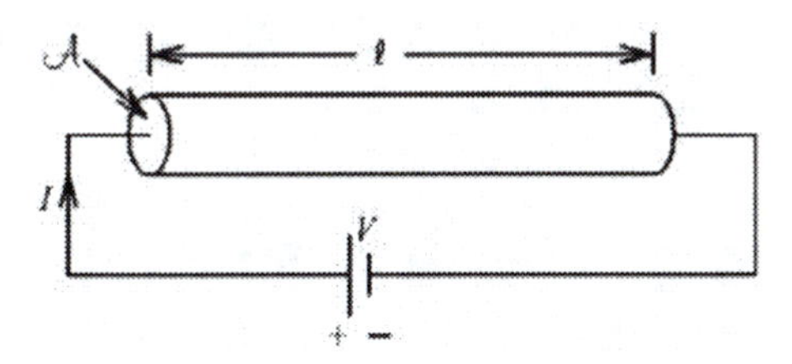

Figure 7-5 Resistivity

Figure 7-5 shows a conductor of length l and uniform cross sectional area A connected to a battery of voltage V producing a current I; what does the current depend on? Clearly, from Ohm's Law, it depends on V. It is reasonable that it varies inversely with the length l; if the length were doubled the resistance would be doubled and the current would be halved. It is also reasonable that it depends on A; if the area were doubled the resistance would surely be halved (see Sec. 7.5 below). Finally it will depend on some intrinsic resistive property of the material itself; if copper were changed to iron and then to silver, etc., the current would differ in each case. This property of the material is called the *resistivity* ρ. From these facts we can write

$$I = \frac{\mathcal{A}V}{\rho\, l} \qquad [7\text{-}8]$$

The resistance is

$$R = \frac{\rho\, l}{\mathcal{A}} \qquad [7\text{-}9]$$

The units of resistivity are ohm-metres (Ω m), and values for some common materials are given in Table 7-1.

Table 7-1: Resistivities of Materials at Room Temperature

Material	ρ (Ωm)
Copper	1.72×10^{-8}
Silver	1.47×10^{-8}
Steel (10% Ni)	29×10^{-8}
Tungsten	5.51×10^{-8}
Carbon	3.5×10^{-5}
Silicon	2.6×10^{3}
Wood	4×10^{11}
Fused Quartz	5×10^{13}

EXAMPLE 7-4

What is the resistance of a conductor made of copper 1.0 cm in diameter and 1.0 km in length?

SOLUTION

Using Eq. [7-9],
$R = (1.72 \times 10^{-8}\ \Omega\ \text{m}) \times (1000\ \text{m})/[\pi \times (0.50\times10^{-2})^2\ \text{m}^2] = 0.22\ \Omega$

7.5 SIMPLE ELECTRIC CIRCUITS AND RESISTANCE COMBINATION RULES

The simplest possible electric circuit is shown in Fig. 7-6 where a source of potential difference V (perhaps a battery) is connected by conductors of negligible resistance (wires) to some device whose resistance is R. A current I flows from the positive terminal of the battery through the resistance to the negative terminal. The figure

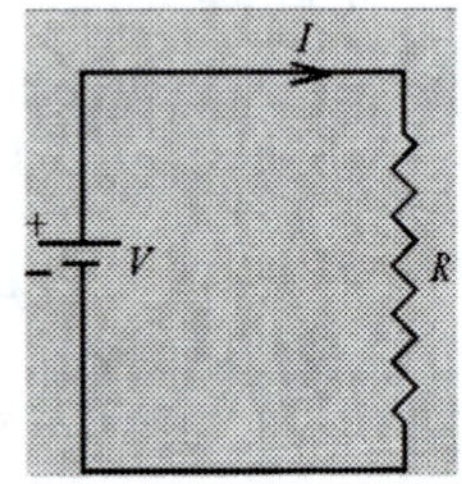

Figure 7-6
Simple Electric Circuit.

also shows the standard symbols used for batteries and resistors.

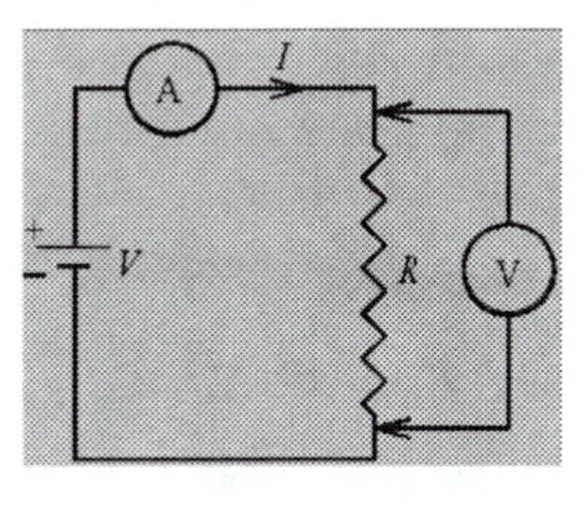

Figure 7-7 Electric Circuit with Meters.

There are devices capable of measuring both voltage and current which are called *voltmeters* and *ammeters*. The former are assumed to have infinite resistance and so can be connected across any part of a circuit without affecting it in any way; today's modern instruments approach that ideal very closely. Ammeters are assumed to have zero resistance so they can be placed in the circuit to measure the current, again without disturbing the circuit. In Fig. 7-7 is shown the circuit of Fig. 7-6 with a voltmeter and ammeter added. We say that the ammeter is measuring the current through, and the voltmeter is measuring the voltage across, the resistor.

Now let us suppose, as shown in Fig. 7-8 that the resistor R is in reality two resistors R_1 and R_2 connected in series; what is the relationship between R, R_1 and R_2? In other words, "What is the equivalent resistance of the two resistors in series?" Clearly the current I goes through both resistors 1 and 2 and the voltages across them V_1 and V_2 must add up to the original voltage V. Therefore we have, using Eq. [7-7]

$$V = IR = V_1 + V_2 = IR_1 + IR_2$$

From this it follows that

$$R = R_1 + R_2$$

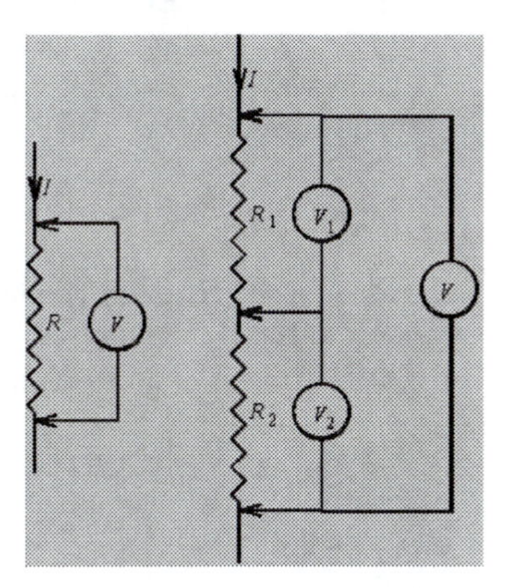

Figure 7-8 Resistors in series.

This can be generalized to any number of resistors in series:

> **When resistors are connected in series the total resistance is the sum of the individual resistances.**

Alternatively, the resistor R might be made up of two resistances connected in parallel as shown in Fig. 7-9. Now the voltage across each resistor is V but a current I_1 goes through R_1 and I_2 through R_2. The two currents must, however, add up to the original current. Thus

$$I = \frac{V}{R} = I_1 + I_2 = \frac{V}{R_1} + \frac{V}{R_2}$$

from which it follows that

$$\frac{1}{R} = \frac{1}{R_1} + \frac{1}{R_2}$$

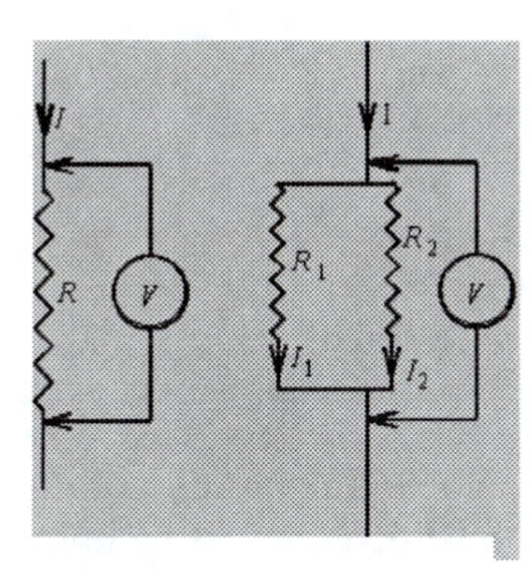

Figure 7-9 Resistances in Parallel

Again this can be generalized to any number of resistors in parallel:

When resistors are connected in parallel the reciprocal of the total resistance is equal to the sum of the reciprocals of the resistances taken individually.

EXAMPLE 7-5

Figure 7-10 shows a circuit consisting of three resistors connected in series and parallel across a battery. What is the current through each resistor and what is the voltage across each resistor?

SOLUTION

If the resistance of R_2 and R_3 in parallel is R' then

$1/R' = 1/R_2 + 1/R_3 = 1/(10\ \Omega) + 1/(30\ \Omega) = 4/30\ \Omega$
$\therefore\ R' = (30\ \Omega)/4 = 7.5\ \Omega$
(The two resistances in parallel are equivalent to a single resistance of 7.5 Ω)

The total resistance $R = R_1 + R' = 2.5\ \Omega + 7.5\ \Omega = 10\ \Omega$

The current through the battery and through R is $I = V/R = (20\ \text{V})/(10\ \Omega) = 2.0\ \text{A}$.
The voltage V_1 across R_1 is: $V_1 = IR_1 = 2.0\ \text{A} \times 2.5\ \Omega = 5.0\ \text{V}$.

The voltage V' across the two parallel resistors can be obtained in two ways:
i) $V' = IR' = 2.0\ \text{A} \times 7.5\ \Omega = 15\ \text{V}$,

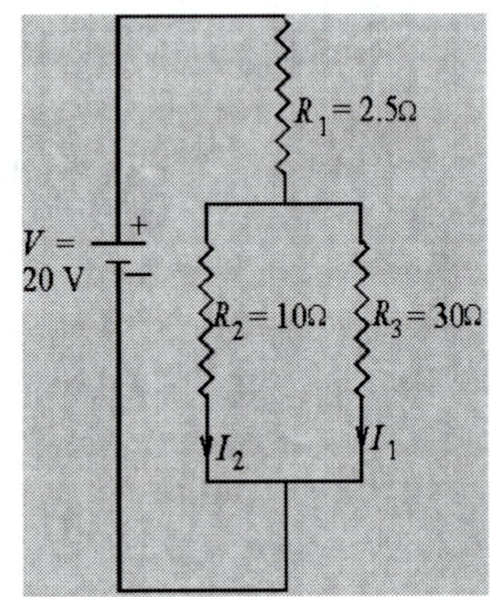

Figure 7-10 Example 7-5

OR

ii) The total voltage is 20 V; 5 V appears across R_1 so the rest, or 15 V must appear across the other resistors.

The current through R_2 is $I_2 = V'/R_2 = (15\ \text{V})/(10\ \Omega) = 1.5\ \text{A}$.
The current I_3 through R_3 can now be obtained in two ways:

i) $I_3 = V'/R_3 = (15\ \text{V})/(30\ \Omega) = 0.50\ \text{A}$,

OR

ii) The total current is 2.0 A; 1.5 A goes through R_2 so the remainder, or 0.5 A, must go through R_3.

You will perhaps note from Example 7-5 two useful points that might almost be called "circuit rules" that help in solving problems:

1. Current is not lost but is conserved; if a current I enters a device the same current must leave.

2. The voltage of the source (e.g. a battery) is equal to the sum of the "voltage drops" in the various elements of the external circuit.

7.6 ELECTRIC POWER

Just as water flowing in a continuous loop, driven by a pump, encounters resistance to the flow due to friction, so electric charges flowing in a closed loop driven by a battery also encounter resistance to the flow. The energy lost due to friction in the flowing water appears as heat and that energy must be provided by the pump. In exactly the same way the resistance to an electric current creates heat from energy provided by the battery. If a potential difference V causes a total charge Q to flow in a conductor during some time interval t then, from Eq. [7-4], we can write that the work done W is given by

$$\Delta U = \Delta W = VQ$$

Divide both sides of the equation by the time t to give

$$\frac{\Delta W}{t} = \frac{VQ}{t}$$

But Q/t is, from Eq. [7-6], just the current I, and W/t is the power P so

$$P = IV \qquad \textbf{[7-10]}$$

It is left as an exercise to show that the units of the product VI are indeed watts as they must be.

EXAMPLE 7-6

What power is provided by the battery of Example 7-3?

SOLUTION

$P=VI \quad \therefore P=1.5 \text{ V} \times 0.3\times10^{-3} \text{ A} = 0.45\times10^{-3} \text{ W} = 0.45 \text{ mW}$

Using Eq. [7-7] and [7-10] the power can be expressed in other ways for ohmic conductors:

$$P = \frac{V^2}{R} = I^2 R \qquad \textbf{[7-11]}$$

What is happening at the atomic level is that the electrons are accelerated by the electric force, but before they can travel very far they collide with atoms setting them into vibratory motion; such motion of the atoms of a solid is the basis of heat. The slowed-down electrons are again accelerated by the electric force and again suffer an energy loss by collisions. Thus the electrical energy which is the kinetic energy of the moving electrons can be converted to heat by letting a current flow through a material that offers resistance to the passage of current.

This is what happens in a toaster or in any electric heating element. The process is very efficient; all the energy is converted into heat. Obviously the 100% efficient conversion of electricity into heat is extremely useful. We should remember though that the initial generation was accomplished from fossil fuel or uranium at only 30-40% efficiency.

EXAMPLE 7-7

The power used by a toaster is typically 1200 W; if the voltage applied is 120 V, then what is the current drawn by the toaster and how much energy does it require to make a piece of toast if the toasting time is 2.0 minutes?

SOLUTION

From Ohm's Law, Eq. [7-7], the current drawn is 1200 W/120 V = 10 A.
If it takes 2 minutes to make toast then the energy used is 1200 J/s × 120 s
$= 1.44\times10^{5}$ J.

In toasters, irons, stoves, kettles and other devices that operate through resistive heating the wire is often not bare since at high temperatures it can react chemically with the air. It is supported or embedded in a material (often a ceramic) that is solid and chemically inert at the temperature encountered.

Sometimes the heating effect can cause problems and is wasteful. Electrical wiring in buildings and electrical transmission lines across the country are made of copper because its low resistivity affords a low resistance (see Example 7-4). If the electrical energy is to be used in a device one does not want it being wasted as heat in the wiring. Although the resistivity of copper is very small, the length of transmission lines is very great between generating sites and cities. The resistance of transmission lines can contribute significantly to the further loss of efficiency in electrical distribution systems as shown in Example 7-8.

EXAMPLE 7-8

A transmission line is made from copper of diameter 1.0 cm. The distance from the generating station to the user is 160 km. The station develops 10 MW of power at a voltage of 40 kV as shown in Fig. 7-11. What is the current in the transmission line and what power is lost in heating the line?

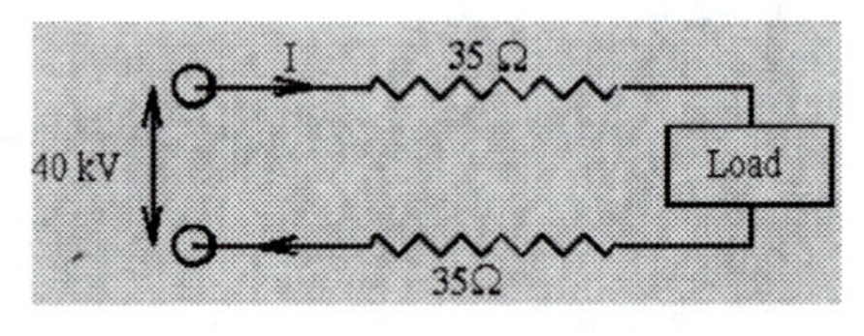

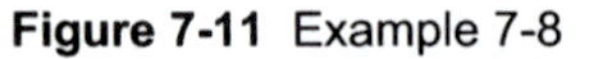
Figure 7-11 Example 7-8

SOLUTION

For a complete circuit there must be a transmission line going in both directions between generator and load. (In reality this return is via the ground but for simplicity here we assume that it is a similar returning line.) Using Example 7-4 the resistance of 160 km of line is

$R = 0.22\ \Omega/\text{km} \times 160\ \text{km} = 35\ \Omega$

Since there must be a complete circuit, two such lines are needed, therefore the resistance of the total transmission line is 70 Ω.
Since we know the power developed by the station and the voltage across it we can calculate the current through the station and thus through the whole circuit.
From Eq. [7-10], the current is $I = P/V = (10\times10^6 \text{ W})/(4\times10^4 \text{ V}) = 250 \text{ A}$

Using Eq. [7-11], the power loss in the lines is

$I^2R = 250^2 \text{ A}^2 \times 70\ \Omega = 4.3\times10^6 \text{ W}$ (4.3 MW) or 43% of the output!

How could the performance of the transmission lines in Example 7-8 be improved without making them thicker and heavier? Since the power loss varies with the square of the current, increasing the voltage and thereby decreasing the current will reduce the power loss markedly. If the voltage is increased to 400 kV the loss falls to 0.043 MW; you should verify this result for yourself. (This shows why electricity is transmitted at high voltage using tall unsightly towers.)

7.7 BATTERIES AND FUEL CELLS.

There are many uses of direct current (DC) electricity on a small scale, for example, electronic devices are replete with DC power supplies or small batteries. Most large scale energy systems rely on alternating current (AC) which is discussed in Chapter 8. DC is used on a large scale in a few applications, the most important of which is in electric traction. Most electric trains, trolleys and subways use DC generators and motors. There are two other classes of DC electric supply which are important in medium scale power: One is the chemical cell or battery, familiar to all of us in the flashlight cell and the automobile storage battery. The other is the Fuel Cell which is increasingly important in transport and industry.

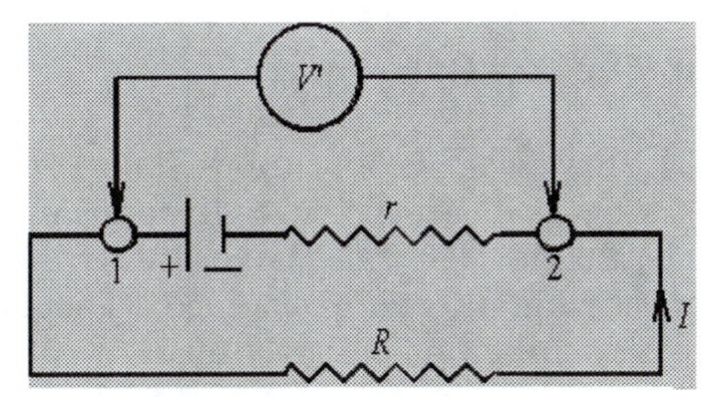

Figure 7-12 Internal resistance.

All sources of voltage, whether they are generators or batteries, are not perfect but have an *internal resistance.* This is illustrated in Fig. 7-12 where a battery of voltage V and internal resistance r is connected to an external resistance R passing a current I. The actual terminals of the battery are labelled 1 and 2 and a voltmeter could be connected across them. The circuit current flows through the internal resistance producing a voltage drop Ir so that the voltmeter does not read V but $V'=V-Ir$. The more current that is drawn from the battery the greater is Ir and the less is the effective voltage V' of the battery. If a battery is to deliver a useful current and maintain an almost constant effective voltage then the internal resistance must be kept as small as possible. When batteries go "dead" what has happened is that the internal resistance has increased to the point where a small current produces a large voltage drop and so large currents are impossible.

Batteries and storage cells are of interest to us since they are a means of storing chemical energy which they liberate in the form of electrical energy. When two dissimilar metals are immersed in a common, conducting medium called the "electrolyte", chemical processes produce a potential difference between them.

1. The Simple Voltaic Cell

This cell shown in Fig. 7-13 consists of copper and zinc plates immersed in dilute sulphuric acid. In the electrolyte, hydrogen ions H^+ and sulphate ions SO_4^{2-} exist separately. Zinc dissolves in the acid much more easily than does the copper; the zinc goes into solution as Zn^{2+} ions which react immediately with the SO_4^{2-} ions to form neutral zinc sulphate $ZnSO_4$; hydrogen ions in solution thus find themselves without negative partners. Meanwhile the constant loss of Zn^{2+} ions leaves the zinc plate with a net negative charge on it. The same process happens at the copper plate but much more slowly so that the zinc plate becomes negatively charged with respect to the copper. When the battery is not in use, it comes to an equilibrium where the further dissolving of zinc is prevented because the negative charge prevents the loss of further positive zinc ions. The voltage developed by this cell is 1.1 V.

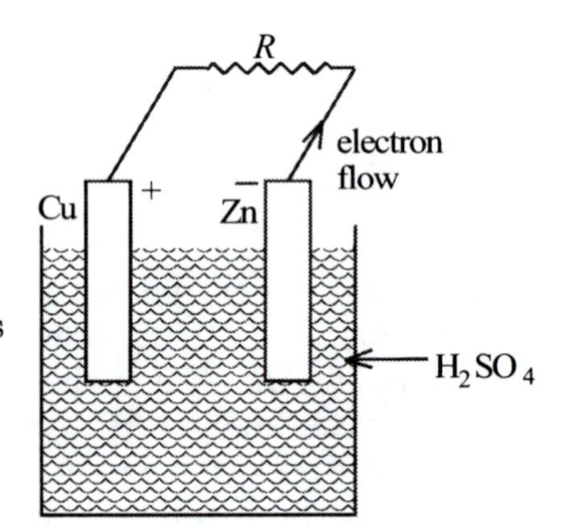

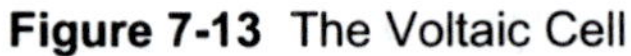

Figure 7-13 The Voltaic Cell

If the two terminals are connected by a wire, the excess electrons will flow to the positive copper plate through the wire (conventional current from copper to zinc) where they neutralize H^+ ions causing hydrogen gas to bubble up. The removal of electrons from the zinc plate allows more ions to be dissolved replacing the hydrogen ions that have escaped in the gas. Overall, zinc is replacing hydrogen in the solution. Eventually the battery becomes dead when the sulphuric acid is saturated with zinc sulphate.

2. The Dry Battery

The dry cells, such as are used in transistor radios and flashlights have zinc (the outer casing) and carbon (the central terminal) electrodes and an ammonium chloride paste electrolyte. They also have a finite lifetime as a result of chemical saturation and increasing internal resistance. (There are also combinations of electrodes that operate with an alkaline electrolyte.)

3. The Lead-Acid Storage Cell

A storage cell differs from a battery in that, when it runs down, it can be returned to its initial state by passing an electric current through it in the opposite direction from an external source and reversing the chemical reactions; it is rechargeable. The best known example is the automobile battery.

The lead-acid cell consists of a negative lead and a positive lead oxide electrode in dilute sulphuric acid. The electric potential of this cell is 2 V; three in series provide a 6 V "battery" and produce 12 V. Lead from the lead plate forms a lead sulphate coating right on the plate. The electrons released in this reaction travel through the external circuit and combine with SO_4^{2-} ions to coat the positive plate with lead sulphate. When both plates are covered with lead sulphate so that no further reactions can take place then the cell is "dead." Passing a current through the cell in the other direction re-dissolves the sulphate and "recharges" the battery. Clearly if the cell is to provide a lot of charge and last a long time the plates must have a large surface area.

The lead-acid cell is not the only rechargeable cell; there are many others of which the nickel-iron and the nickel-cadmium or NiCad are the most common.

4. Fuel cells

A fuel cell is similar to a storage cell in that its function is to convert chemical energy directly into electrical energy. This is done at a fixed temperature so the device is not a heat engine and thus is not subject to the thermodynamic limitation on efficiency; as with a storage cell the efficiency of energy conversion can be very high. The fuel cell differs from the storage cell in that it is supplied with a constant flow of two chemicals, one being the fuel and the other oxygen or air. Instead of burning the fuel with the oxygen to produce heat, the two chemicals react so as to release electrons directly which flow as a current around an external circuit.

An example is the hydrogen-oxygen fuel cell illustrated in Figure 7-14. The H_2 and O_2 flow at high pressure into two porous electrodes immersed in a liquid electrolyte solution such as potassium hydroxide (KOH). The H_2 is fed to the anode and the O_2 to the cathode. The anode has its surface treated with platinum which acts as a catalyst to convert the hydrogen into H^+ ions, thereby releasing electrons to the external circuit. The H^+ ions combine with the OH^- ions in the electrolyte to form water. The reaction at the anode is

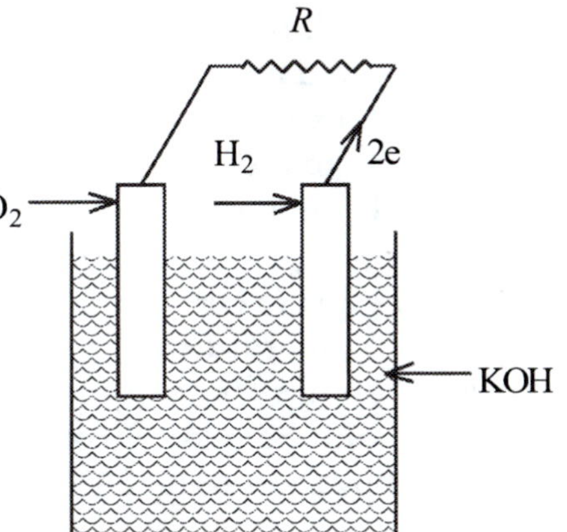
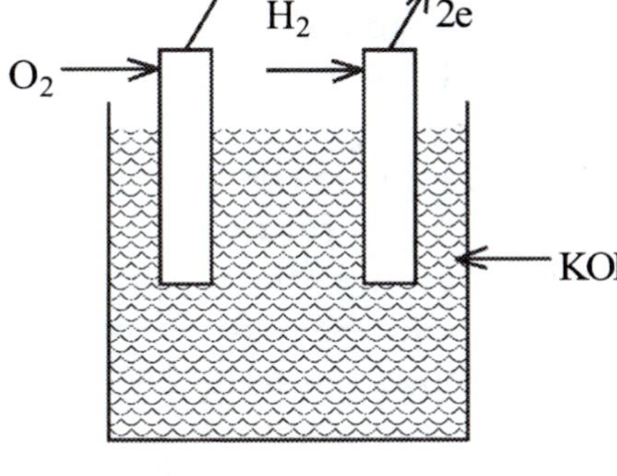

Figure 7-14 The Hydrogen-Oxygen Fuel Cell

$$H_2 + 2OH^- \Rightarrow 2H_2O + 2e^- \qquad \textbf{[7-12]}$$

At the cathode the electrons combine with oxygen and water in the reaction

$$2e^- + O_2 + H_2O \Rightarrow 2OH^- + O \qquad \textbf{[7-13]}$$

Thus OH^- ions are continuously formed at one electrode and destroyed at the other, maintaining equilibrium in the electrolyte. Combining Eq. [7-12] and [7-13] the total reaction is

$$2H + O \Rightarrow H_2O \qquad \textbf{[7-14]}$$

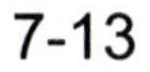

The KOH is not affected and the waste product is water which must be continuously removed. This constant flow method of operation eliminates the necessity of recharging as in the case of storage cells.

The voltage output is determined by the two chemicals used; for H_2–O_2 this is 1.229 V, which reflects the molecular binding energy released when the reaction of Eq. [7-12] proceeds. This cell can produce 100-200 mA of current per cm^2 of electrode surface. There is no aquatic thermal pollution since no condensers are used, and no gaseous or particulate air pollution since there is no burning. Efficiencies are of the order of 50%.[5] Hydrogen is not the only possible fuel nor is oxygen the only oxidant. Of course a fuel-cell that uses the oxygen directly from air is very attractive from the point of view of cost. So far, the fuel-cell has had limited use most notably in the space programme where it uses the rocket fuels, hydrogen and oxygen, that are already on board and its waste product, water, is essential as well. Pure hydrogen is neither a convenient nor an easily stored fuel (except on space ships).

In terrestrial applications hydrogen-rich fuels like gasoline or natural gas can be processed using high temperature and catalysts in a *fuel-processor* or *reformer* to make hydrogen as it is needed in the fuel cells. Most current applications envisage this method.

More recent emphasis has been on the development of the *proton exchange membrane* (PEM) fuel cell in which the fluid electrolyte is replaced by a polymer membrane on which is coated a suitable catalyst to break the hydrogen into protons. The power density of these fuel-cells is higher than the fluid electrolyte ones and their rigid construction makes them suitable for use in vehicles as discussed in Chapter 17. These cells have been developed by Ballard Power Systems of Vancouver, Canada.

Solid oxide fuel (SOFC) cells use a ceramic material as the electrolyte and operate at very high temperatures (~1000 C). A typical SOFC uses a matrix of tubes about 1 metre long and can use H_2 or CO as a fuel. Tubular SOFCs are produced by Siemens in Germany and are close to commercialization for stationary applications such as auxiliary power in industry or for power in remote locations.

Fuel cells are already finding uses in stationary applications. Ballard produces a 250 kW fuel cell power plant for industrial and utility use, and a 1 kW version for single dwelling use that has been applied to residences in Japan.

EXERCISES

7-1. An electric charge of 2.0×10^{-15} C is placed in the vicinity of another electric charge where it experiences a force of 5.0×10^{-13} N. What is the electric field strength at the site of the 1st charge?

---

[5] The efficiency is the chemical potential divided by the operating voltage while supplying current. In a typical cell this is about 0.6V so the efficiency for producing electric current is 0.6/1.23 = 0.5 or 50%.

7-2. The charge in Exercise 7-1 moves a distance of 1.0 mm in the direction of the field. Assuming that the field is constant over that distance how much work is done on the charge?

7-3. If the charge in Exercises 7-1 and 7-2 was initially at rest and was carried by a dust mote of mass = 5.0 nanograms, what is its speed after moving the 1.0 mm?

7-4. Through what potential difference has the charge of Exercise 7-2 moved?

7-5. Show that the units of electric field can be either N/C or V/m.

7-6. In a chemical process an average current of 25 A was maintained for 3 hours. What electric charge was transferred?

7-7. If the chemical process in Exercise 7-6 was an electroplating process and a singly charged metal ion was involved, how many moles of the metal were plated?

7-8. A battery of voltage 1.1 V is connected across a resistor of 2.2 Ω; what current is drawn?

7-9. What power is dissipated in the resistor of Exercise 7-8?

7-10. A copper bar 0.50 cm. in diameter is 5.0 m long; what is its resistance?

7-11. Two resistors each of 0.50 Ω are connected in series; what is the total resistance?

7-12. The resistors of Exercise 7-11 are connected in parallel; what is the resultant resistance?

PROBLEMS

7-13. A battery of internal resistance 2.0 Ω is used to drive current through a device which is equivalent to a 20 Ω resistive load. What fraction of the battery's power is dissipated in the load?

7-14. A 100 V power supply contains a 10 A fuse. What is the maximum power rating of any device run from this supply?

7-15. The resistors in the network in Fig. 7-16 are:
$R_1 = 6.0\ \Omega$ $R_4 = 7.5\ \Omega$
$R_2 = 3.0\ \Omega$ $R_5 = 5.0\ \Omega$
$R_3 = 15\ \Omega$ $R_6 = 4.0\ \Omega$
If the current I is to be 14 A, what is the voltage of the battery that must be connected across a and b? What is the voltage across R_1 and R_2? What is the voltage across the three resistor section?

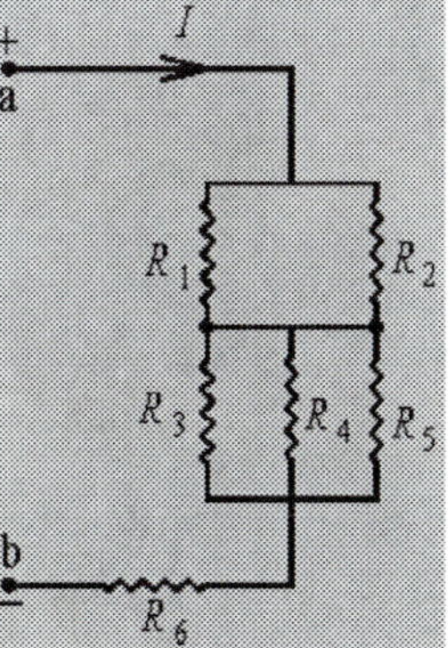

Figure 7-15
Problem 7-15

7-16. An electric kettle of 7.0 Ω resistance containing 500 cm^3 of water at 10 °C is connected to the 110 V supply. If 10% of the electrical energy is wasted by imperfect insulation, calculate how long it takes to convert all the water into steam. (Latent heat of vaporization of water = 2.26×10^6J/kg and its specific heat is 4189 J/kg/K).

7-17. Many cheap electric stoves have three settings on their switch (besides OFF) labelled LOW, MED and HIGH. This is actually achieved by making the heating element with three resistive coils each of equal resistance R. The switch then arranges these resistors

i) in series, ii) with two in parallel connected in series to the remaining one, and iii) all three in parallel.
a) Which combination gives LOW, MED, and HIGH?
b) What is the total element resistance in each case?
c) If the HIGH setting is 1000 W what is the power of LOW and MED? Is it really a good set of values for LOW, MED, and HIGH?

7-18. A factory is supplied with 100 kW of power from a generating station some distance away. The supply and return lines each have 5.0 Ω resistance. The factory receives the power at 4000 V. What power does the generator actually have to supply? If the factory took its power at 2000 V, what power would have to be generated?

7-19. A copper transmission line, 100 miles long and 1.0 cm. in diameter is replaced by one of twice the diameter. At the same time new users are connected and the current drawn increases by 50%. What is the fractional change in the heating losses in the transmission line?

7-20. A transmission line is made of a core of copper surrounded by a concentric circular sheath of steel to give it strength. The copper core is 1.0 cm in diameter and the tightly fitting sheath has an outer diameter of 2.0 cm. What is the resistance of 1.0 km of this transmission line? What is the ratio of the current carried by the core to that carried by the sheath? What is the ratio of the power lost in the core to that lost in the sheath?

7-21. 100 MW of electrical power is delivered at 200 kV to a distribution centre over a transmission line which consumes 10% of the total power developed by the generating station in heat losses.
a) What is the value of the voltage fed to the transmission line at the generator end?
b) How much power would be delivered to the same load if the resistance of the transmission line was halved but the generator voltage was unaltered?

7-22. Consider a 1.0 kWe fuel-cell assembly operating at 50% efficiency. a) How many litres of water does it produce per hour? b) If its fuel is natural gas (CH_4), what fraction of a cubic metre does it consume each hour and what is the cost if natural gas costs 25¢ per m^3?

ANSWERS

7-1. 250 N/C
7-2. 5.0×10^{-16} J
7-3. 0.4 mm/s
7-4. 0.25 V
7-6. 27×10^{4} C
7-7. 2.8 moles
7-8. 0.50 A
7-9. 0.55 W
7-10. 4.4×10^{-3} Ω
7-11. 1.0 Ω
7-12. 0.25 Ω
7-13. 91%
7-14. 1 kW
7-15. 119 V, 28 V, 35 V
7-16. 850 s
7-17. **b.** i) $3R$, ii) $3R/2$, iii) $R/3$, **c.** P_H = 1000 W, P_M = 222 W, P_L = 111 W
7-18. 106 kW, 125 kW
7-19. 44% decrease
7-20. 0.19 Ω, 5.6, 5.6
7-21. **a.** 222 kV, **b.** 111 MW
7-22. **a.** 0.14 L/hr, **b.** 0.17 m³, 4¢

CHAPTER 8

MAGNETISM AND A.C. ELECTRICITY

In Chapter 7 some properties of direct current (DC) electricity were examined and several important relationships were derived. Most of commercial electricity, however, is distributed not as direct current but as an alternating current (AC). That is, the current alternates in direction, flowing first one way and then in the reverse direction many times each second.

The widespread use of AC (rather than DC) for consumer use is for a technological reason; electric current is most easily generated, manipulated, and distributed on the large scale as an alternating rather than a direct current. Since AC electrical machinery utilizes the magnetic effects of an electric current, in this chapter it seems logical to explore magnetism first.

Did you know? The technique of rendering magnetic field lines visible by means of iron filings was described by William Gilbert (1544-1603) and "magnetic figures" were apparently known to the Roman philosopher Lucretius (95BC-55BC)

8.1 BASIC MAGNETISM

The magnetic effects of natural magnets such as lodestone have been known for centuries. It is the effect which causes a compass needle or filings of iron to align themselves up along invisible *lines of force* which seem to exist in the *magnetic field* surrounding a natural magnet.[1] Figure 8-1 is a sketch of iron filings sprinkled on a sheet of paper placed over a bar magnet. The lines of force as delineated by the filings seem to point toward two locations (or at least small regions) near the ends of the magnets called the *poles*. They are designated "north" (N) and "south" (S) according to which pole of the Earth they point to if the magnet is suspended and allowed to swing freely about its centre.

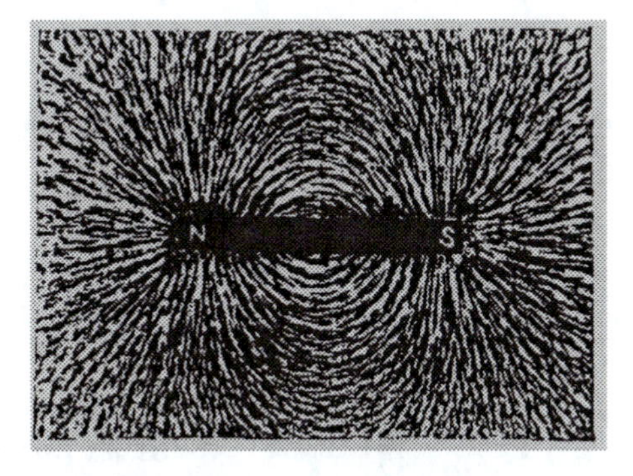

Figure 8-1 The Magnetic Field of a Bar Magnet

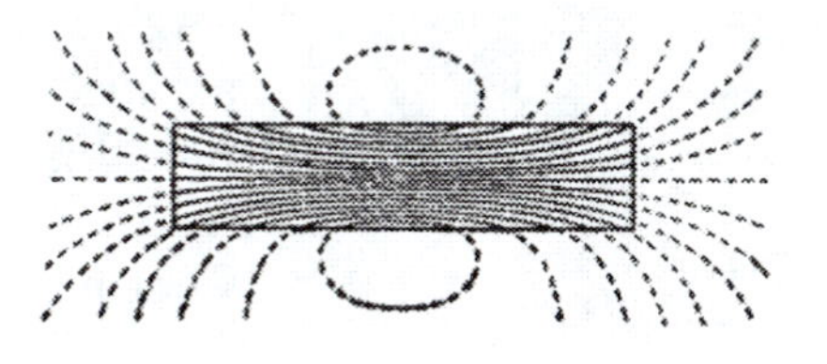

Figure 8-2 The Magnetic Field Inside a Bar Magnet.

A further observation about the poles is that unlike (N-S) poles attract and like (N-N, S-S) poles repel each other, much like positive and negative electric charges. A basic difference between electric and

[1] Much investigation of the magnetic phenomena occurred before these terms were adopted. They were used extensively, if not actually coined, by Michael Faraday (1791–1867).

magnetic fields is that whereas electric field lines originate on positive charges and terminate on negative charges the magnetic field lines occur only in closed loops. In Fig. 8-1 the lines must be imagined to extend from the N to the S pole and return to close the loop within the magnet itself as shown in Fig. 8-2.

8.2 ELECTROMAGNETISM

For a long time electricity and magnetism were thought to be unrelated phenomena until, in 1820, H. C. Oersted (1777-1851) discovered that a compass needle placed near a current-carrying wire was deflected. With this discovery of *electromagnetism* it was realized that magnetic fields are a result of moving charges. In the case of the wire the moving charges are the flowing electrons; in the case of a permanent magnet they are the electrons orbiting in their individual atoms. (It is important not to confuse electric and magnetic fields: electric fields exist between electric charges whether stationary or moving; magnetic fields accompany only moving charges.)

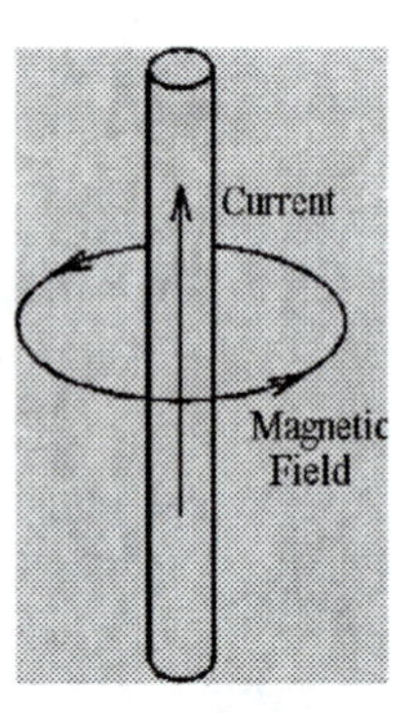

Figure 8-3 The Magnetic Field About a Current Carrying Wire.

Oersted's investigations showed that the magnetic field which accompanied a current-carrying wire was in the form of circular lines about the wire as shown in Fig. 8-3. The direction of the field can be determined from the following *right-hand thread rule:*

To advance a right-hand threaded screw (the ordinary kind) in the direction of the conventional current, the screw driver would have to be rotated in the direction of the magnetic field.

Examine Fig. 8-3 in the light of this rule.

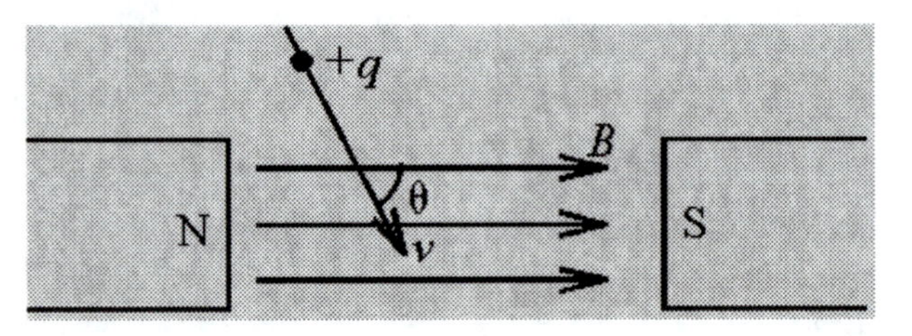

Figure 8-5 The Interaction of a Moving Charge With a Magnetic Field.

Figure 8-4 shows a uniform magnetic field established between the poles of two large magnets. If a charge $+q$ is injected at a velocity v, at angle θ, into this field of strength B (defined below) it will experience a force which is at right angles to both the magnetic field lines and the direction of the velocity. This means that in Fig. 8-4 the force is directed either into, or out of, the page. Whether it is into or out of the page is determined by another *right-hand thread rule* shown in Fig 8-5.

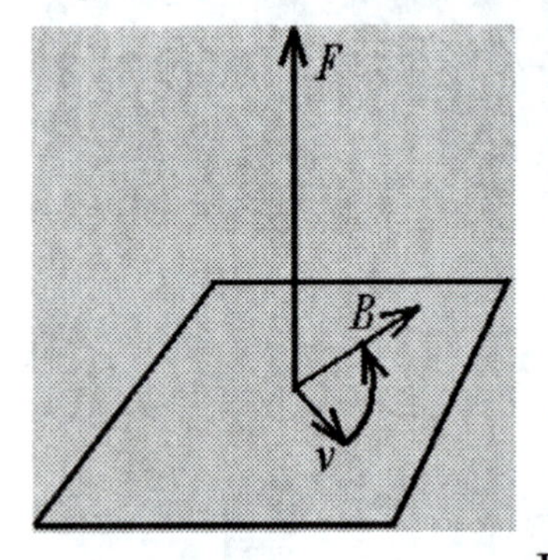

Figure 8-4 The Direction of the Electromagnetic Force.

A screwdriver turned in the direction of vector v rotated into vector B would advance a right-hand threaded screw in the direction of the force.

In the situation shown in Fig. 8-4 the force would be out of the page.

This rule, like all rules in electricity, has been given for a positive charge. Of course, if the charge is negative the force will be in the opposite direction.

Experiments show that the force experienced by the charge is proportional to the magnitude of the charge, the speed, and the sine of the angle θ. From these observations *magnetic field strength B* is given by

$$F = qvB\sin\theta \qquad \textbf{[8-1]}$$

In Eq. [8-1] when the force is in newtons (N), the charge in coulombs (C), and the speed in m/s, the magnetic field strength has units N s C^{-1} m^{-1}; this unit is called the "tesla" [2] (T). A magnetic field has a strength of 1 T if it exerts a force of 1 N on a 1 C charge that is moving at 1 m/s at right angles to it.

EXAMPLE 8-1

A He^{2+} ion travels at 45° to a magnetic field of 0.80 T with a speed of 4.0×10^5 m/s. Find the magnitude of the force on the ion.

SOLUTION

The charge on a He^{2+} ion is twice the elementary charge or $2\times1.602\times10^{-19}$ C

From Eq. [8-1], $F = qvB\sin\theta = 2(1.602\times10^{-19}\text{ C})(4.0\times10^5\text{ m/s})(0.80\text{ T})\sin45^\circ = 7.2\times10^{-14}$ N.

8.3 ELECTROMAGNETIC INDUCTION

Oersted's discovery that an electric current produced a magnetic field raised the question as to whether or not a magnetic field would produce an electric current or, more precisely, an electric potential which could, in turn, drive a current in a circuit. This question was first seriously investigated by Michael Faraday who, after lengthy investigation, discovered *electromagnetic induction.*

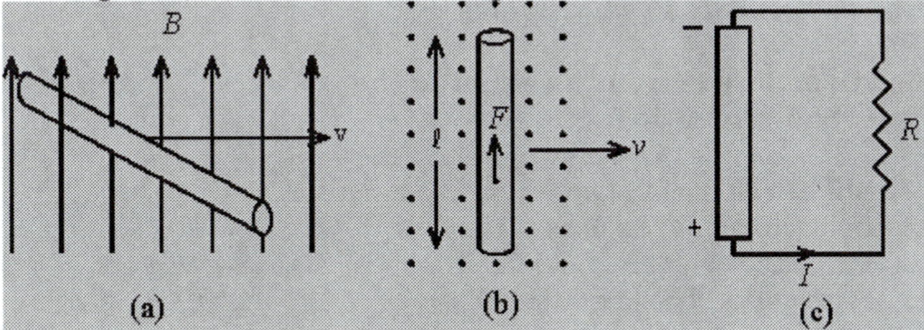

Figure 8-6 A Conductor Moving in a Magnetic Field.

[2] Named after Nicola Tesla (1875–1943), a brilliant but eccentric Croatian electrical engineer who invented the induction motor and the transformer.

In Fig. 8-6(a) a conductor of length l, such as a wire, is being pulled by an externally applied force, at a constant speed υ through a magnetic field of strength B; for simplicity we choose υ, l, and $\boldsymbol{B}$ to be at right angles to each other.

In Fig. 8-6(b) the situation is viewed from above where the dots represent the points of the arrows representing $\boldsymbol{B}$ which are pointing out of the page. Imagine a free electron in the wire being carried along with the wire. From the right-hand thread rule it will experience a force $\boldsymbol{F}$ in the direction shown. Since the electron, like all the others, is free to move in the wire, it will move toward one end making it negatively charged with respect to the other end. When sufficient charge has been separated the electric force will counterbalance the magnetic force and the process will come into equilibrium so long as the wire continues to move in the magnetic field. A voltage has been developed along the wire called the *induced voltage.* Balancing the electric and magnetic forces of Eq. [8-2] and [8-1] ($\theta=90^\circ$) we have $eE = e\upsilon B$ or $E = \upsilon B$.

But we also have $E = V/l$ (See Eq. [8-5]) so

$$V = \upsilon B l \qquad \textbf{[8-2]}$$

Thus the motion of the wire through the field induces a voltage across it. The wire can be considered as a voltage source for some external circuit containing a resistance, and so the moving wire would act like the battery in a circuit, driving current, I, through the external resistance, R as shown in Fig. 8-6(c). The current would continue to flow so long as the wire continued to move. The current is given by

$$I = V/R = \upsilon B\, l/R \qquad \textbf{[8-3]}$$

In a time dt the wire moves a distance υdt tracing out an area $d\mathcal{A} = \upsilon l dt$ as shown in Fig. 8-7, and thus, $d\mathcal{A}/dt = \upsilon l$

Using Eq. [8-2]

$$V = B(d\mathcal{A}/\,dt) = (d/dt)(B\mathcal{A}) = (d/dt)\varphi \qquad \textbf{[8-4]}$$

The quantity $\varphi = B\mathcal{A}$ is called the *magnetic flux* [3]. Eq. [8-4] is Faraday's *law of electromagnetic induction* which states that

> **The voltage induced in a closed circuit is equal to the time rate of change of the magnetic flux.**

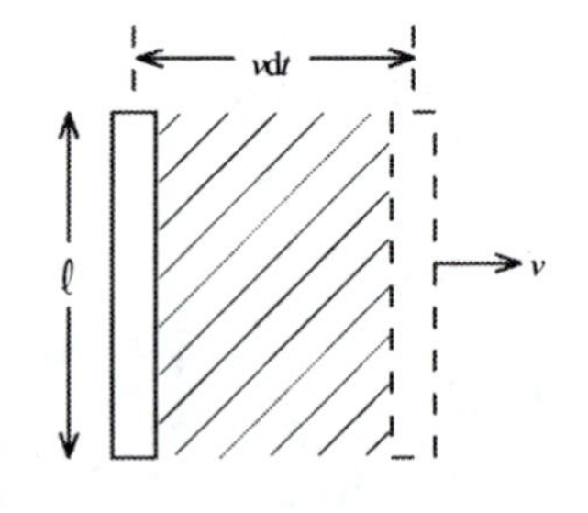

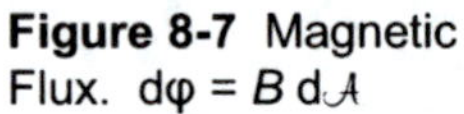
Figure 8-7 Magnetic Flux. $d\varphi = B\, d\mathcal{A}$

Obviously the flux in a circuit can change in two ways: The circuit can move in the magnetic field, changing the area through which the magnetic field acts or the magnetic field itself could change in a circuit of fixed size. The operation of a generator is an example of the former and a transformer is an

[3] The word "flux" is often encountered in science where some physical quantity flows or passes through some area. For example, a "sound flux" would be the total sound energy falling on some area; a "fluid flux" is the flow from a pipe of a given cross–sectional area.

example of the latter. Of course in practical cases both processes might occur simultaneously.

If the closed circuit is a coil of N turns then the magnetic flux through the coil is just N times the flux through one coil or

$$\varphi = NB\mathcal{A}. \qquad [8\text{-}5]$$

EXAMPLE 8-2

A circular loop of wire 5.0 cm in diameter is placed between the poles of a magnet perpendicular to the lines of a uniform magnetic field of strength 0.20 T. The leads from the loop are fed through a tube and are connected to a resistor of 393 Ω. A steady pull on the leads causes the loop to shrink until its area is halved after a time of 2.0 s. What is the average current in the resistor during this process?

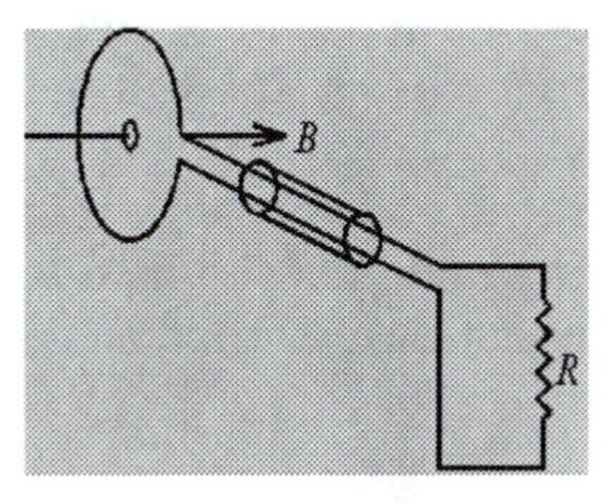

Figure 8-8 Example 8–2

SOLUTION

The initial flux through the loop is, from Eq. [8-5],
$\varphi = NB\mathcal{A} = 1(0.2\text{ T})\pi(2.5\times10^{-2}\text{ m})^2 = 3.93\times10^{-4}\text{ T m}^2$

The final flux is half this amount since the area is reduced to one half. Therefore the change in the flux is $\Delta\varphi = 3.93\times10^{-4}/2 = 1.96\times10^{-4}\text{ T·m}^2$.

From Eq. [8-4], $V = d\varphi/dt = 1.96\times10^{-4}\text{ T m}^2/2\text{ s} = 9.81\times10^{-5}\text{ V}$

From Ohm's Law $I = V/R = 9.81\times10^{-5}\text{ V}/393\ \Omega = 2.5\times10^{-7}\text{ A} = 25\ \mu\text{A}$

8.4 THE GENERATION OF AC POWER

Figure 8-9 shows a single-loop coil of area $\mathcal{A}$ rotating with an angular frequency [4] ω in a magnetic field of strength B. As the coil rotates, the flux through the coil varies from 0, when the coil sides lay in the direction of the field to a maximum value of $\varphi_{max} = BA$ when the plane of the coil is perpendicular to the field. At some arbitrary angle $\theta=\omega t$, as shown in the figure, the area presented by the coils to the field is $\mathcal{A}\sin\omega t$ and so, from Eq. [8-4] the induced voltage is

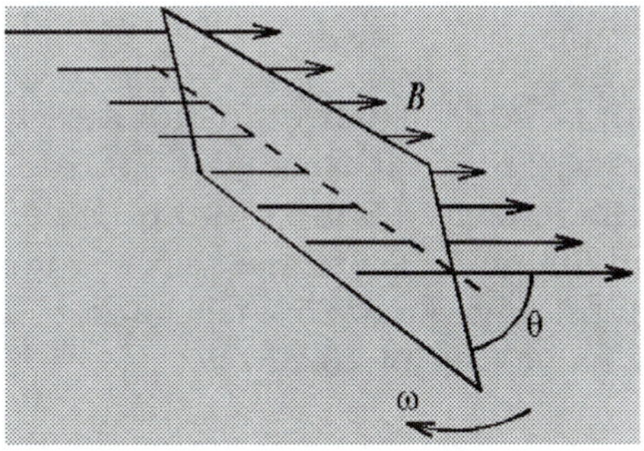

Figure 8-9 Rotating Coil in a Magnetic Field.

$$V = (d/dt)(\mathcal{A}B\sin\omega t)$$

$$V = \omega\mathcal{A}B\cos\omega t \qquad [8\text{-}6]$$

---

[4] The angular frequency is the angle swept out per unit time measured in “radians per second”.

If the coil has N turns then the induced voltage appears N times so

$$V = \omega \mathcal{A} BN\cos\omega t = V_p\cos\omega t \qquad [8\text{-}7]$$

where $V_p = \omega \mathcal{A} BN$ is the peak value of the induced voltage. This is an alternating voltage and is entirely different from the direct voltage produced by chemical cells. In one complete rotation of the coil, the voltage acts in one direction round the loop for one half of the cycle and in the reverse direction for the other half as shown in Fig. 8-10. There are $f = \omega/2\pi$ complete oscillations of the voltage per second, where f is the *frequency* of the voltage measured in *hertz* (Hz).

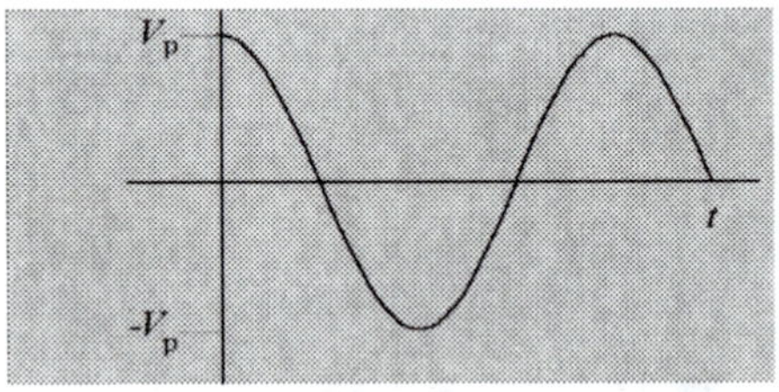

Figure 8-10 Alternating Voltage From a Rotating Coil.

If the ends of the wire making the coil are led out of the region of the field to an external circuit of resistance R, then a current will flow such that

$$I = I_p\cos\omega t \qquad [8\text{-}8]$$

where $I_p = V_p/R$. The current therefore, flows first in one direction and then in the reverse direction, following the voltage.

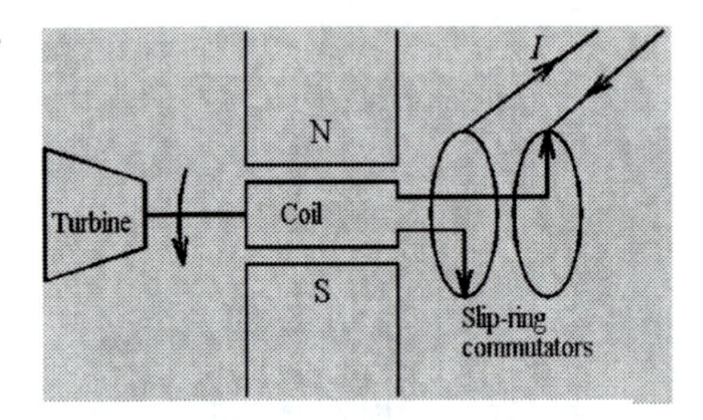

Figure 8-11 Basic AC Generator.

The basic elements of an AC generator are shown in Figure 8-11. The coil is rotated by a turbine (or any other engine) in the magnetic field of a pair of magnets. In practice, in a generator of high power output, the magnets would be electro-magnets . The leads from the coil are connected to sliding contacts (brushes) on a *slip-ring commutator* which applies the AC voltage to the external circuit.

Electrical generators are very efficient, converting well over 90% of the turbine's rotational energy into electrical energy. The limitation on the overall efficiency of a fuel-heat-turbine-generator system is the thermodynamic limitation of the engine explained in Chapter 4. (This reduces the overall efficiency for the conversion of chemical energy into electric energy to 30 - 40%.)

In large scale power installations this machinery is very large. At the Robert Moses power plant at Niagara Falls, each turbine has a mass of 200 tonnes and is rotated by the impact of falling water. Each generator contains a rotor 10 m in diameter on a shaft of diameter 1.2 m and mass 55 tonnes. One complete generator has a mass over 10,000 tonnes of which 65% is rotating. The electrical power output of one generator is 150 kW at a voltage of 13,800 V.

8.5 POWER IN AC CIRCUITS

From Eq. [7-10] the power at any instant t is given by

$$P(t) = V(t)I(t) = V_p I_p \cos^2\omega t$$

What is of more interest to us is the average value of this power P_{av}. Common AC power in North America has a frequency of 60 Hz,[5] and most electrical applications involve times much longer than 1/60 of a second. Since the average of cos² or sin² is ½ we have

$$P_{av} = ½V_p I_p \quad \textbf{[8-9]}$$

A useful voltage (or current) is termed the *root mean square (rms)* voltage (or current). To determine this quantity, first square the voltage (or current), next calculate its average (mean) value during one complete oscillation, and then take the square root. The square of the voltage is

$$[V(t)]^2 = V_p^2 \cos^2\omega t$$

Since the average of cos² is ½ then

$$[V(t)]^2_{av} = (1/2)V_p \quad \textbf{[8-10]}$$

and taking the square root we have

$$V_{rms} = V_p/\sqrt{2} \quad \textbf{[8-11]}$$

Similarly

$$I_{rms} = I_p/\sqrt{2} \quad \textbf{[8-12]}$$

Rewriting Eq. [8-9] using [8-11] and [8-12] we have

$$P_{av} = V_{rms}I_{rms} \quad \textbf{[8-13]}$$

which is identical in form to Eq. [7-10]. Thus the equations developed for DC electricity can be applied to the AC situation if the rms values of the voltage and current are used. When AC voltmeters and ammeters are used their scales are usually calibrated in rms values.

8.6 TRANSMISSION OF AC POWER

The same considerations of transmission-line power-loss as discussed for DC power in Chapter 7 apply as well to AC power. Because the power-loss heating of the transmission lines depends on the square of the current it is important that the current

[5] In Europe and many other parts of the world it is 50 Hz. In Ontario in Canada and parts of New York State it was 25 Hz until the end of WWII; you could see the flicker in the lights!

be kept as low as possible. The only way this can be done is to raise the transmission-line voltage as high as possible. Modern high-power transmission lines can be operated at voltages as high as 750 kV. The generators that produce the electric power, however, have much lower output voltages of the order of 10 to 30 kV because of size and magnetic field limitations. The higher transmission voltages are achieved by means of the voltage *transformer*.

A simple transformer is depicted in Fig. 8-12; it consists of an iron yoke in the form of a closed loop on which two coils of N_1 and N_2 turns have been wound. If an AC source of voltage $V_1(t)$ is connected to coil 1, a field $B(t)$ is created which has a flux $\varphi(t)$. This flux is almost entirely inside the iron yoke and is related to the voltage V_1 by Eq. [8-4]:

$$V_1 = N_1(d\varphi/dt)$$

Virtually the entire flux is trapped inside the iron and so is carried through the second coil inducing a voltage in it given also by Eq. [8-4]:

$$V_2 = N_2(d\varphi/dt)$$

Dividing V_1 by V_2 we have:

$$V_1/V_2 = N_1/N_2 \qquad \textbf{[8-14]}$$

The ratio of the *primary* to *secondary* voltage on the transformer coils is the ratio of the number of turns on the primary and secondary coils.

EXAMPLE 8-3

A turbine generator produces AC power at a voltage of 20 kV. It is desired to transmit this at a voltage of 200 kV. If the transformer has 500 turns on its primary coil how many turns must there be on the secondary coil?

SOLUTION

From Eq. [8-14], $20/200 = 500/N_2$

$N_2 = 10 \times 500 = 5000$ turns

The transformer described in Example 8-3 is a *step-up* transformer. At a power sub-station the transmission line voltage is reduced to a few thousand volts with a *step-down* transformer and further reduced by the pole transformer near your home to the familiar 120 V_{rms}. Resistive heating losses in the transformer require that they be cooled efficiently; larger units are usually filled with oil for this purpose.

Figure 8-12 A Step-Down Pole Transformer

A typical pole transformer that serves a street of homes is

shown in Fig. 8-12. The 8000 V high voltage connection to the primary is from the well-insulated line at the top of the pole; the other side of the primary is grounded. The secondary consists of two 120 V high-current leads and a grounded line which emerge from the side of the container and are connected to the local distribution.

EXAMPLE 8-4

A 2400 V to 120 V suburban step-down transformer, of output 10 kW operates with 90% efficiency. What is the turns-ratio, the power in the primary, and the primary and secondary currents?

SOLUTION

From Eq. [8-14] $N_2/N_1 = 120/2400 = 1/20$

Secondary power = 10 kW

Primary power = 10 kW/.9 = 11.1 kW

Primary current = $I_1 = P_1/V_1$ = 11100 W/2400 V = 4.6 A

Secondary current = $I_2 = P_2/V_2$ = 10000 W/120 V = 83 A

A small amount of current delivered to the transformer is stepped up to supply enough current for several homes.

8.7 POLY-PHASE POWER GENERATION

1. Single-, or Split-phase Power
Let's return to Example 7-8 that posed a simple problem of a generator delivering power to a load through two wires. This example is so simple that it is in fact unrealistic. Power companies and their customers are always mindful of the economics of the system and its shock hazard.

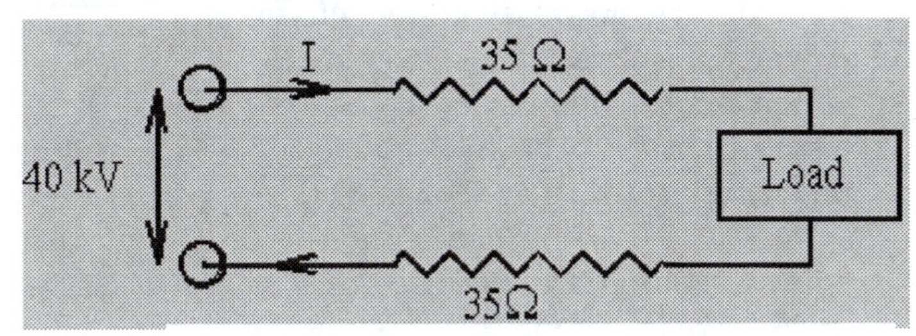

Figure 8-13 Repeat of Fig. 7-11

No user would be able to cope with power delivered at 40 kV. At some point the voltage would be lowered (at the cost of higher current) and the customer would receive it at a few hundreds of volts. In fact there is a way to lower the voltage for both the

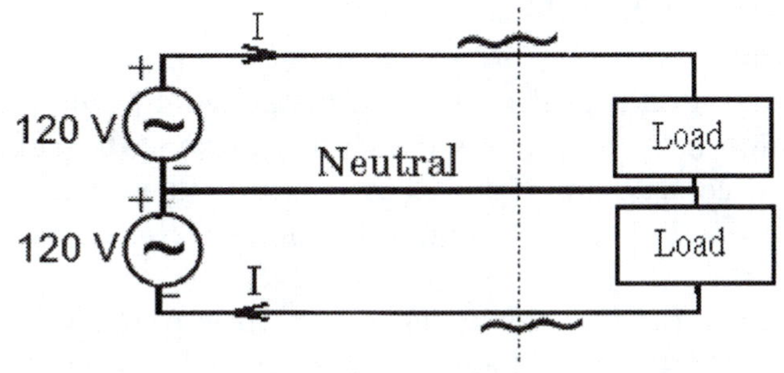

Figure 8–14 Split–Phase Power Distribution.

power company and the customer. This is the split-phase system. Imagine that the power company connects two 120 V transformers in series and delivers the output to two separate customers (or two separate major services for one customer) as shown in Fig. 8-14.

Notice:
1. The AC power in each circuit is exactly out of phase so although power is being delivered at 240 V the generators (windings of the transformers) operate at 120 V.

2. Since the voltages are exactly 180° out of phase the neutral line is very near zero potential and if the loads are "balanced" (drawing exactly the same current and having exactly the same capacitance and inductance) then it carries no current.

3. Within each load voltages are single-phase and 120 V, but if we did something stupid like connecting a single phase appliance between the two loads (across the two major circuits in the house) then voltages of 240 V would appear, possibly creating a hazard. Of course the 240 v service is available for appliances made for it such as electric stoves.

4. Almost all low power motors used in appliances, small tools, etc. are designed to run on split-phase power. Such motors will not start by themselves because the two windings on the motor armature, which are the two loads, are always exactly out of phase. Special circuitry has to be included, such as a special starting winding or the brute-force method of putting a large capacitor across one motor winding to momentarily shift its phase. When the motor is up to speed than the capacitor must be taken out or the motor will overheat and burn out.

2. Three-phase power.

Real electric power systems deliver power via a three-phase system which is sketched in Fig. 8-15. Generalizing on the split-phase case, now three sources are connected symmetrically. The connection shown in the figure is called the "Y" or "Star" connection but it is not the only one possible; it is just the simplest to visualize. The three sources, which could be three windings on a transformer or three coils in a generator, each produce a voltage of 120 V, each of which is shifted 120° in phase from its neighbours. This is delivered via four conductors to three loads which, if exactly balanced, require no current to flow in the neutral conductor. If one of the circuits is switched off the system still operates for the other two since the neutral wire then carries current.

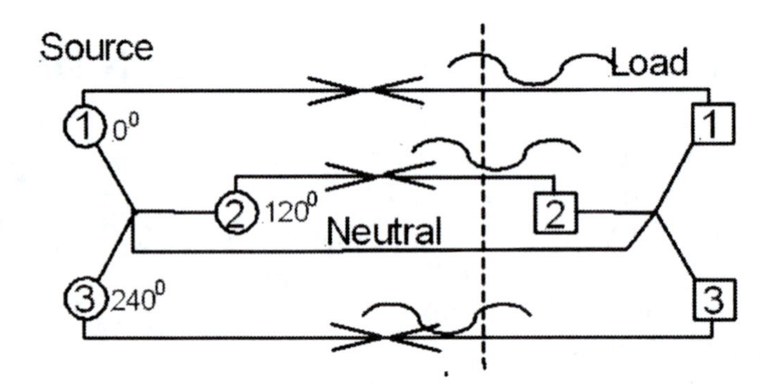

Figure 8-15 Three-Phase Power Systems.

One of the great advantages of the three phase system at low voltages is with electric motors. If the three loads are three windings on the armature of an electric motor then, because of the phase difference, there is a rotating magnetic field set up in the motor and the motor will begin to rotate by itself. Three-phase induction motors require no special starting circuits. In addition, for a given power and voltage, three-phase motors are smaller and lighter.

At the high voltage supply end the power companies have the advantage that the voltage of any one generator required to deliver a certain power or current is reduced compared to the simple circuit discussed in Chapter 7 and embodied in Fig. 7-11 above. The saving in copper metal alone in distribution wires justifies the method.

EXERCISES

8-1. If a horizontal wire carries a DC current from east to west, and is located in a magnetic field that is vertically downward, what is the direction of the magnetic force on the (moving charges in the) wire?

8-2. In Figure 8-16 a positive charge is moving with a velocity v in a magnetic field B. The non-zero components of v and B are given by the subscripts. Find the direction of the resulting force.

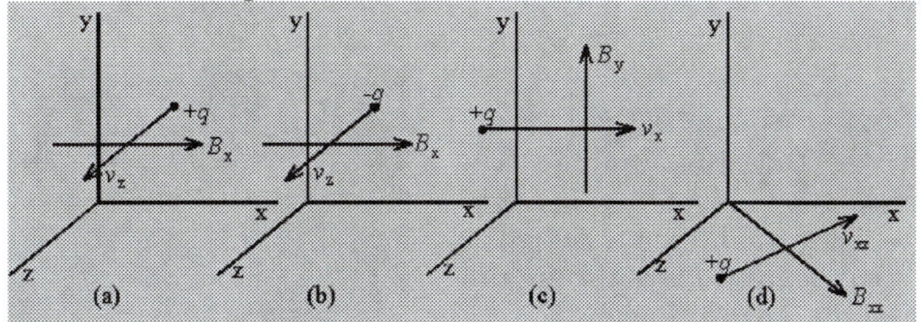

Figure 8-16 Exercise 8-2

8-3. A wire of length 0.20 m moves at a speed of 15 m/s at right angles to a magnetic field of strength 0.0040 T. The wire is also at right angles to the field. What is the induced voltage in the wire?

8-4. Consider a rotating rectangular generator coil as shown in Fig. 8-17. When the area is perpendicular to the field $\boldsymbol{B}$, as in Fig. 8-17(a), consider the direction of the force on a negative charge in each of the four sides of the coil by answering the following questions:

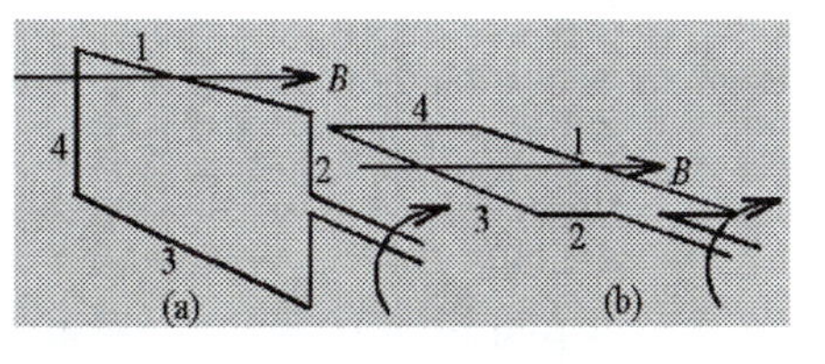

Figure 8-17 Exercise 8–4.

a) What is the value of the force on a charge in any portion of the coil in Fig. 8-17(a)?
b) What is the instantaneous current in the wire?
c) What is the direction of the force on a negative charge in portions 1 and 3 in Fig. 8-17(b)? Does this force contribute to a current in the coil?
d) What is the direction of the force on a negative charge in portions 2 and 4 in Fig. 8-17(b)? Do they contribute to the current flowing in the coil?
e) In what direction is the (conventional) current in the coil in Fig. 8-17(b)?

8-5. A 900 W (2 sig. digits) toaster is plugged into a standard 120 V_{rms} outlet.
a) What is the rms current in the toaster?
b) What peak current does this correspond to?
c) What is the resistance of the heating element in the toaster?

8-6. An electric stove element operates at 220 V and has a power of 2000 W. Which of the following fuses will blow if put in the protective circuit of this element: 1 A, 5A, 10 A, 15 A, 20 A?

8-7. What is the resistance of the element in Exercise 8-6?

8-8. If V_{rms} is 120 V, what is the peak voltage?

8-9. A step-down transformer in a TV set reduces the 120 V_{rms} line supply to 6.3 V_{rms} for the tube filaments. If the primary has 400 turns, how many has the secondary?

PROBLEMS

8-10. A horizontal wire 0.75 m long is falling at a speed of 4 m/s perpendicular to a magnetic field of strength 2.0 T. The field is directed from north to south. What is the induced voltage of the wire? Which end of the wire is positive?

8-11. A copper bar 1.0 cm long is dropped from rest at a height of 3.0 m and falls in a horizontal orientation through the gap of a magnet which produces a uniform field of 0.2 T. What voltage appears across the ends of the bar when it is in the gap?

8-12. In Fig. 8-18 a metal rod rests on two conducting rails completing a circuit which includes the resistor of 3 Ω. The circuit is perpendicular to a magnetic field, of strength 0.15 T, which acts into the figure as shown. The rod is pulled at a constant speed of 2.0 m/s. What current flows through the resistor? In what direction does the current flow?

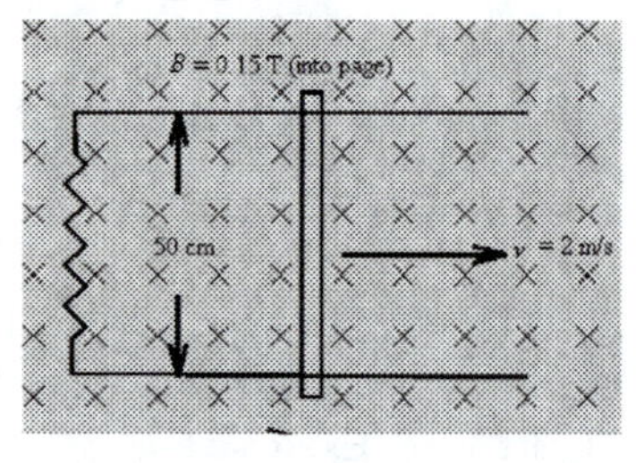

Figure 8-18 Problem 8–12

8-13. An electric generator has a 20-turn coil of area 0.04 m^2 rotating uniformly at 60 rev/s in a magnetic field of 1.0 T. Calculate the peak value of the induced voltage and the power dissipated in a 100 Ω resistive load.

8-14. A square coil 5.0 cm on a side and containing 828 turns spins at 3600 rpm in a uniform field of 0.20 T. a) What is the peak power developed if this coil is connected to a resistor of 55 Ω? b) What is the rms voltage, current and power?

8-15. A neighbourhood pole transformer reduces the voltage from 44,000 V (2 sig. digits) to 120 V for household distribution. If the current flowing in the primary coil is 5.5 A, what is the current flowing in the secondary coil?

ANSWERS

8-1. south

8-2. a) +y, b) –y, c) +z, d) –y

8-3. 0.012 V

8-4. **a.** zero **b.** zero **c.** 1) toward 4, 3) toward 2, Yes **d.** Perpendicular to the length of the wire, No **e.** 1 ➔ 2 ➔ 3 ➔ 4

8-5. **a.** 7.5 A **b.** 10.6 A **c.** 16 Ω

8-6. 1 and 5 A

8-7. 24 Ω

8-8 170 V

8-9. 21

8-10. 6.0 V, west

8-11. 15 mV

8-12. 0.050 A, counter- clockwise

8-13. 302 V, 455 W

8-14. 440 W, 110 V, 2.0 A, 220 W

8-15. 2.0×10^3 A

CHAPTER 9

COMMERCIAL ELECTRICITY

We are all accustomed to flicking a switch and having a light turn on, essentially instantaneously. The electrical energy used by the light has not been stored anywhere — it is being generated at the moment. How do electrical utilities ensure that there is always enough energy, and not too much, for everyone?

This chapter is concerned with various aspects of the commercial generation and distribution of electricity. What are the advantages and disadvantages of hydroelectric energy? How do utility companies deal with changing electrical demand? Are electromagnetic fields from power lines and appliances dangerous to health?

9.1 MEETING ELECTRICAL DEMAND

Power companies normally have a number of electrical generators that can be used, some of them virtually all the time, and some only during periods of peak demand. Electrical demand is continuously monitored and the number of generators in operation is adjusted to meet the demand. Figure 9-1 illustrates how a typical company might use various generators to meet demand on a daily basis. There is a baseload that must be generated no matter what the time of day. This load is met by large reliable generators that use the least expensive fuel — hydroelectric generators, nuclear plants, and perhaps some coal-fired generators. On a short timescale it is difficult to alter the output of these plants or to shut them down. The hydroelectric stations used to provide baseload power might be a mix of run-of-river stations and dam installations. (As the name implies, a run-of-river station uses water that is actively flowing down a river; a hydroelectric dam creates a lake from which water is extracted when needed.)

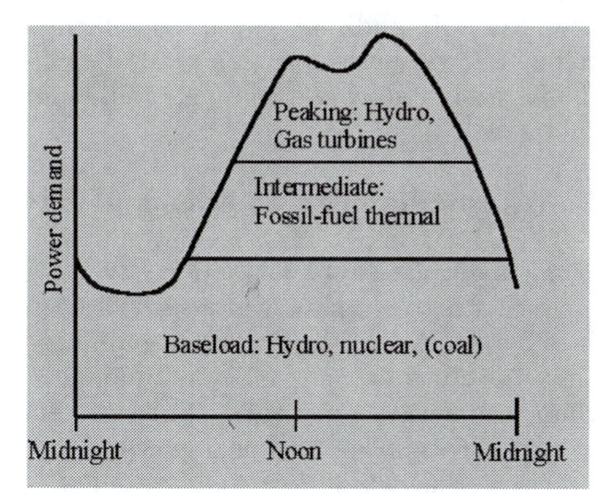

Figure 9-1 Typical Generation-stacking-order for an Electrical Utility.

The intermediate load is met by smaller generators that can be varied easily in power output; steam turbines using coal, oil, or gas as a fuel are often used for this purpose. The peak demand is met by bringing more turbines into service at hydroelectric dams, and by using gas turbines (sometimes called combustion turbine units). In a gas turbine, natural gas is burned and the hot exhaust gases themselves turn a turbine without the use of steam. Gas turbines have the virtue of being able to be turned on and off quickly.

Of course, it takes some minutes to put another generator into service or to increase the output of a running generator (for example, by burning fuel at a faster rate). How are fluctuations in demand on a smaller time scale handled? Suppose that there is an sudden increase in power demand. There is a large amount of energy stored as rotational kinetic energy in the rotors of the generators themselves, and short-time surges in demand can be taken from this energy, resulting in a slightly slower rate of rotation of the generators. The generators are designed to rotate with a frequency of 60 Hz,[1] but this can decrease to as low as 59.5 Hz as energy is withdrawn, or increase to as high as 60.5 Hz if power demand suddenly drops. Changes in frequency are kept no larger than this because of the use of synchronous electrical motors in industry. These motors run exactly in step with the applied current, and their efficiency decreases drastically if the frequency changes very much. In particular, the efficiency of the synchronous motors that lubricate and cool the generators themselves would be adversely affected by appreciable changes in frequency.

Utilities need to be prepared for shutdowns of generators for maintenance or repair, and must have spare generating capacity in reserve. Most utilities can accommodate a shutdown of about 20% of the generating capacity. However, occasionally a large number of generators can be out of commission because of mechanical or other problems, and a utility is then forced to import energy from neighbouring utilities (at high cost). In addition, power can be cut off to large industrial users who have elected to pay a lower price for using interruptible electricity. In extreme cases, a utility can deliberately reduce the voltage, producing a "brownout," or curtail power to users in a "blackout."

Long-term Planning and Demand-side Management

It takes ten to twenty years for a new power plant to be approved and constructed; electric utility companies are therefore continuously projecting demand for many years in the future and planning for new plants. In recent years, many utilities have realized that it is cheaper, more efficient, and more environmentally sound to save a megawatt than to generate a megawatt, and have embarked on measures to reduce both total energy consumption and peak power demand. This aspect of a utility's operation is referred to as *demand-side management.*

To reduce total consumption of energy, utilities can offer cash rebates to consumers who purchase more efficient end-use devices such as fluorescent lights and high-efficiency industrial motors. Utilities can also offer advice to large industrial or commercial clients on ways to use less energy without negatively affecting production or business.

Did You Know? In Nova Scotia in 1989, about 40% of the estimated 250 active osprey nests were on transmission-line structures.
(Source: *The State of Canada's Environment*, Government of Canada, Ottawa, 1992, p. 12-26)

It is also useful to decrease the peak power demand, which determines how many generators a power company needs to have available. Reducing future peak demand

[1] Although 60 Hz is the frequency used in North America, most of the rest of the world uses 50 Hz.

means that construction of new generators can be postponed. To accomplish this, a power company can shift load from peak times to off-peak times by changing the pricing structure to offer lower rates, especially to large industrial users, during the night when total demand is low. A utility can also offer preferential rates to customers willing to have power interrupted during peak periods.

9.2 TRANSMISSION TECHNOLOGIES

Once electricity has been generated, it must be transmitted to the consumers. As both the number of consumers and the per-capita consumption of electricity increase, utilities are presented with the problem of delivering an increasing amount of electrical power.

The basic physics of this topic has already been covered in Chapters 7 and 8. Recall that the power P produced by a generator is the product of the voltage V and current I: $P = VI$ (Eq. [7-10] and [8-13]). Some power is lost as heat in the transmission lines themselves, and is usually expressed as I^2R, where R is the resistance of the lines: $P_{LOST} = I^2R$. These ohmic losses can be decreased by increasing the voltage fed to the lines by the step-up transformer at the generator, thus decreasing the current. Typical voltages for long-distance transmission of electricity are now in the range of several hundred kilovolts, up to about 1000 kV. Most overhead and underground transmission is accomplished in the AC mode. Overhead lines are held high in the air by unsightly towers to avoid arcing (sparking) from the high-voltage lines to the ground. Underground transmission lines tend to be considerably shorter and operate in the DC mode.

Suppose, for example, that an urban area doubles its demands for power; if the power is supplied at the same voltage, then the current must double. The I^2R heating loss in the cables therefore quadruples. What options present themselves if this loss is to be avoided?

(a) Duplicate the entire transmission line using the same voltage.

In terms of land use, this is neither an attractive nor even, in many cases, a feasible solution. In scenic areas the impact on the landscape is adverse, while in many urban areas the necessary land is simply not available or is extremely expensive. Even when land is available, many people do not want a transmission line located near their property. (This is one example of the NIMBY syndrome: “not in my back yard.”)

(b) Increase the voltage applied to the line.

If the voltage is doubled, then the doubled power can be transmitted with the same current and thus the same I^2R losses. Increasing the voltage introduces limitations from both technology and land use. The higher the voltage on an object, the more efficiently it must be insulated to prevent arcing. Higher transmission voltages require improvements in the insulating materials used to support the cables at the transmission towers. In addition, the main insulator between a cable and the ground is air, and the only way to increase the insulating power of air is to provide more of it by using much higher and larger towers, thus increasing the land required for right-of-way. There is increasing opposition to demands for land for the reasons mentioned in (a). However, the land required for one ultra-high-voltage line is less than that needed for two lines at half the voltage.

(c) Reduce the resistance of the line.

Since copper is already the least resistive conductor at reasonable cost, this requires a new strategy. One method that might prove useful would be simply to cool the line, since resistance falls with temperature (Problem 9-2). However, another approach is beginning to become available through the phenomenon of *superconductivity*, which is the complete loss of electrical resistance in certain materials when cooled to extremely low temperatures. Superconducting metals at liquid-helium temperature (4 K) have been used successfully in other applications, such as in the magnets of the huge atom-smashing machines used in nuclear research. In 1986, Alex Müller and Georg Bednorz, working at the IBM Research Laboratory in Zurich, Switzerland, discovered that an oxide of barium, lanthanum, and copper exhibited superconductivity at a much higher temperature, about 35 K. Continued research has produced superconductors that operate at temperatures greater than 120 K, well above the temperature of liquid nitrogen (77 K), which is readily producible.

These high-temperature superconductors (HTSs) can now be manufactured in the form of wires capable of carrying a large enough current to be of use to the electrical industry. Projects demonstrating the transmission of AC electricity over HTS wires of lengths 100-500 m have been successfully carried out in China, Denmark, Japan, Mexico, and the US.[2] As of September 2006, HTS cables are just beginning to be used in actual electrical grids. A 200-m cable is distributing electricity to customers in Groveport, Ohio, and a cable almost a kilometre in length is being planned for use by the Long Island Power Authority in New York.[3] HTS cable systems are still quite expensive, and these two US projects are funded partly by the US Department of Energy. A superconducting line must be kept cold by liquid nitrogen, and as HTS technology continues to improve, the cost of this cooling and the necessary heat insulation will eventually be more than compensated for by the ability to pass immense quantities of power with very small losses. In addition to the small I^2R losses, an advantage of HTS cables is that they can carry three to five times the power than a conventional cable of the same size.

It should be pointed out that for superconducting lines, the resistance to direct current is zero, but, unfortunately, for alternating current there is a small problem. Alternating current gives rise to a continually varying magnetic field, which exerts a force on electrons in the material around the cable; the electrons move as a result of this force, heating the material. To remove this heat, additional refrigeration is necessary.

There have been tremendous strides forward in developing HTS materials that are superconducting at higher and higher temperatures. If it becomes possible in the future to develop a *room-temperature* superconductor, it would have tremendous applications for long-distance transmission of electricity.

---

[2] A.L. Malozemoff et al, *High-Temperature Cuprate Superconductors Get to Work*, Physics Today, April 2005, p. 41-47

[3] American Superconductor Co. website, www.amsuper.com

(d) Use underground cables.

Land demands can be greatly reduced by laying cables underground, but there are disadvantages. Insulating the wires is more difficult, and the soil is not as good as air in carrying away the I^2R heat that is generated, thus producing the possibility of overheated cables (which degrade the insulation).[4] One improvement being investigated is to cool the cables with oil or gas circulated in a pipe surrounding the cables. An additional problem is that for the same capacity of power transmission, underground lines cost much more than overhead lines; long underground lines require additional equipment which further increases the cost differential.

(e) Use underwater cables.

It is likely that in the future substantial amounts of electricity may be generated at locations that are separated from the users by hundreds of kilometres of sea. In this scenario, DC underwater cables appear to be an appropriate solution that is both economic and reliable.

(g) Hydrogen

It seems not unreasonable to envisage that in the future, the best means to transport power generated at certain remote sites (more than 2000 km from the users) might be to use electricity to produce hydrogen by the electrolysis of water. In this process the electrical energy is converted into chemical potential energy of the hydrogen gas. The gas could then be pumped through pipelines — a well-established transportation technology. At the receiving end, the hydrogen could be used directly as a fuel or used to generate electricity.

9.3 HEALTH EFFECTS OF ELECTRIC AND MAGNETIC FIELDS

There has been much concern expressed in the news media in recent years about the possible health effects of the electric and magnetic fields produced by transmission lines, house wiring, and common electrical appliances. These fields oscillate with a frequency of 60 Hz in North America (50 Hz in Europe and some other parts of the world). Because the frequency is so low, the energy carried in 60-Hz fields is extremely small. The energy, E, of a photon of electromagnetic radiation of frequency f is given by $E = hf$, where h is Planck's constant. For a frequency of 60 Hz, the photon energy is 4×20^{-32} J, or 2×10^{-13} eV. This energy is much too low to break molecular bonds — the energy of a carbon-carbon single covalent bond is about 3.6 eV, and that of a hydrogen bond is about 0.1 eV. Hence, if 60-Hz fields do produce health effects, then new interaction mechanisms would have to be discovered to explain the results.

[4] One of the major power cables into central London goes under the towpath of the 18th century Prince Regent's Canal. It is cooled by the water in the canal.

A "Resource Letter"[5] in the American Journal of Physics provides an extensive compilation of articles, studies and papers on this topic. Although most research has shown no adverse health effects due to 60-Hz radiation, some experiments have suggested that these fields can have effects. Three general types of studies have been carried out: laboratory studies that expose single cells, groups of cells, or organs to fields; laboratory studies that expose animals or humans to fields; and epidemiological studies of various human populations to investigate possible connections between exposure to fields and various diseases. Some experiments have shown that under certain circumstances, fields can interact with cell membranes to produce changes in the rate at which the cell makes hormones, enzymes, and other proteins. The details of the processes that are occurring are not understood. Many of the results are rather unusual — sometimes weak fields seem to produce larger effects than strong fields, and other effects appear only for pulsed fields with special pulse shapes. Some people exposed to strong fields under laboratory conditions can experience changes in heart rate and reaction time, and studies of people who use electric blankets report changes in the level of the hormone melatonin, which is important in our circadian rhythm (daily biological cycle). However, it is not at all clear what (if any) the health effects of these changes might be.

The two epidemiological studies [6, 7] that have received the most attention involve childhood leukemia in Denver, Colorado. These studies indicate that the incidence of leukemia is greater by about a factor of two in children who live near major electrical transmission lines. (For comparison, smoking increases the risk of lung cancer[8] by about a factor of 10.) However, because of the small number of people in the Denver studies, the statistical uncertainty is very large, and the effect could be much smaller or even nonexistent. Even if the statistical uncertainty were smaller, the studies do not necessarily mean that the fields around the lines are a cause of leukemia. Major transmission lines are usually built near busy roads, and such areas typically have more air pollution and noise. Other factors might also be important, such as diet and various socioeconomic considerations.

Because of public concerns about this issue, the American Physical Society undertook an extensive review of more than 1000 research papers written about the health effects of electric and magnetic fields. In 1995 this Society published its findings and issued a public announcement, which stated (in part)[9]:

> "The scientific literature and the reports of reviews by other panels show no consistent, significant link between cancer and power line fields. This literature includes epidemiological studies, research on biological systems, and analyses of theoretical interaction mechanisms. No plausible biophysical mechanisms for the systematic initiation or promotion of cancer by these power line fields have been identified. Furthermore, the preponderance of the epidemiological and biophysical-biological research findings have failed to substantiate those studies which have reported specific adverse health effects from

---

[5] D. Hafemeister, Amer. J. Phys. **64** No. 8, Aug. (1996), p. 974.
[6] N. Werthheimer, E. Leeper, Am. J. Epidemiol. **109**, p. 273 (1979).
[7] D.A. Savitz, H. Wachtel, F.A. Barnes, E.M. John and J.G. Tvrdih, Am. J. Epidemiol. **128**, p. 21 (1988)
[8] M. Shepard, EPRI Journal, Oct./Nov. (1987) p. 7
[9] *Power Line Fields and Public Health*, Statement by the Council of the American Physical Society, April 22, (1995).

exposure to such fields. While it is impossible to prove that no deleterious health effects occur from exposure to any environmental factor, it is necessary to demonstrate a consistent, significant, and causal relationship before one can conclude that such effects do occur. From this standpoint, the conjectures relating cancer to power line fields have not been scientifically substantiated."

The more recently published report of the UK Childhood Cancer Study [10] reached the same negative conclusion.

9.4 HYDROELECTRICITY AND THE ENVIRONMENT

The three major sources of electrical energy are fossil fuels, nuclear energy, and hydraulic (water) energy. The problems associated with the use of fossil fuels have been discussed in Chapters 5 and 6, and nuclear energy is addressed in Chapters 10 to 13. The present section shifts the focus to hydroelectricity; a typical large plant is shown in Fig. 9-2.

Figure 9-2 A Hydroelectric Power Station.
Photo courtesy of Ontario Power Generation.

Hydroelectricity has many positive features: it is renewable, highly efficient, and has no fuel costs, no combustion products, and no radioactive waste. Once constructed, a hydroelectric plant has low operating costs, and hydroelectric generators can be started quickly to meet peak demands for electrical power.

However, there are a number of disadvantages. To generate large amounts of reliable hydroelectric energy, dams must be constructed at great expense and huge areas of land flooded. To take one example, the five reservoirs of the La Grande Phase I portion of the James Bay hydroelectric project in Quebec cover 11 400 km^2, of which 9 700 km^2 are flooded land.[11] There are many obvious environmental problems associated with a dam and flooding: sedimentation, erosion, spawning difficulties for fish, and large changes to ecosystems. In addition, the toxic metal mercury is released from soil and vegetation in areas flooded by new dams, and makes its way, for example, into fish, building up to extremely high concentrations. Another example is the dramatic alteration of the Mekong basin in S.E. Asia, where Thailand, Laos, Vietnam and Cambodia have 15 hydroelectric dams on tributaries of the Mekong River, and China has built one dam on the main river and is planning 14 more. The annual Mekong floods fertilize the agricultural area around the giant Tonle Sap lake and also bring in vast quantities of fish. This is the region of the fabled city of Angkor, where space shuttle earth-imaging radar has revealed the remains of a vast ancient urban complex surrounded by 1000

[10] *Exposure to power-frequency magnetic fields and the risk of childhood cancer*, UK Childhood Cancer Study Investigators, Lancet **354**, p. 1925, (1999).
[11] *The State of Canada's Environment*, Government of Canada, Ottawa, (1992), p. 12-26

square kilometres of irrigation systems and reservoirs. Archaeologists now suspect that Angkor's demise was caused by sedimentation clogging up the waterways and irrigation channels. Now ecologists are increasingly concerned that history might repeat itself.

Concern has recently been expressed about methane (a greenhouse gas) released by rotting vegetation that is flooded after construction of a dam. Flooding also means that some people will probably have to be moved to new locations; such displacement has produced serious distress for native peoples affected by the James Bay project. Finally, with a dam there is always the spectre of a catastrophic failure, with loss of human life.

Remaining Hydroelectric Resources

Most of the best sources of hydroelectric energy have already been tapped. Worldwide, it has been estimated that about one third of the economically feasible potential has been exploited.[12] With increasing awareness of the negative aspects of hydroelectric power, many new sites that could have been developed without difficulty 20 or 30 years ago would meet stiff opposition now. It is likely that much of the increase in hydroelectric energy production in the near future will come from improving current hydroelectric installations by adding or upgrading generators, converting non-hydro dams to power-producing dams, and developing small run-of-river projects such as the 200 MWe Murray Hydroelectric Station on the Mississippi River.[13]

In Canada, the generating capacity of hydroelectric plants in the year 2002 was 67 100 MW.[14] The total remaining potential is approximately 180 000 MW, but when future sites are removed from consideration because of technical, economic, or environmental constraints, the remaining potential is only about 34 000 MW. Development of some of these sites would likely encounter resistance from various groups of people.

PROBLEMS

9-1. Suppose that electrical power is distributed from a power plant through a 350-kV transmission line; the power lost to heat in the line is 14% of the total power supplied. The plant is going to be doubled in size, and has the choice of duplicating the 350-kV line, or using a single 700-kV line. This single line would have the same resistance, and would pass the same current, as one of the 350-kV systems. What percentage of the total power supplied would be lost in the lines for (a) the two 350-kV line system? (b) the single 700-kV line system?

9-2. The resistivity of copper at 77 K (the boiling temperature of liquid nitrogen) is 13% of the value at 295 K (22°C). Compare two current-carrying copper transmission lines of the same length and diameter, one at 77 K and the other at 295 K. If the thermal losses are the same in the two lines, what is the ratio of the current in the cold line to that in the warmer line?

ANSWERS

9-1. (a) 14% (b) 7% **9-2.** 2.8

[12] World Energy Council website, www.worldenergy.org
[13] *Profiles in Renewable Energy: Case Studies of Successful Utility Sector Projects.* National Renewable Energy Laboratory, US Department of Energy, DOE/CH10093-206 (1993).
[14] Natural Resources Canada website, www.canren.gc.ca

CHAPTER 10

NUCLEAR PHYSICS

To understand nuclear fission as an energy source, we first need to understand a little of the structure and masses of atomic nuclei. And because the principal waste product of nuclear power is intensely radioactive material, we need to understand the mechanisms by which this is created, and the rearrangements of the nuclear structure which give rise to the phenomenon of *radioactivity*.

10.1 STRUCTURE OF NUCLEI

The Rutherford model of the atom pictures a very small dense atomic nucleus which contains virtually all the mass of the atom and all of its positive electrical charge. Orbiting around this nucleus in a volume about 10^{12} times greater are the negatively charged electrons, equal in number to the number of positive charges in the nucleus, so that the atom as a whole is electrically neutral. These orbiting electrons determine the chemical nature of the atom (a neutral atom with 6 electrons is always carbon for example) and in our discussion we have almost no further interest in them; what interests us here is the nucleus.

The nucleus is composed of two types of particles of almost equal mass: the proton (p) and the neutron (n). The total number of neutrons and protons (collectively nucleons) in the nucleus is the *mass number* (A) of the nucleus. The proton has a single positive electrical charge and the neutron has no charge. Some basic properties of these particles are given in Table 10-1.

Table 10-1: Properties of Proton, Neutron, Electron

Particle	Mass	Charge
Proton	1.6726×10^{-27} kg	e
Neutron	1.6750×10^{-27} kg	0
Electron	9.1×10^{-31} kg	-e

(e = 1.602×10^{-19} coulomb)

A chemical element can exist, however, with different nuclear masses; for this to be the case it must be the number of neutrons which varies because the number of protons is fixed. For example, the element carbon exists with mass numbers 10, 11, 12, 13 and 14. Since all carbons have six protons, then these forms of carbon must have 4, 5, 6, 7, and 8 neutrons in their nuclei. These forms are called the *isotopes* of carbon.

The notation we use to specify a particular isotope of element X is

$$^{A}{}_{Z}X_{N} \quad [10\text{-}1]$$

where Z is the *Atomic Number*, i.e., the number of protons (also the number of orbiting electrons in the neutral atom), N is the number of neutrons in the nucleus, and A as defined previously is the *mass number*.[1] Clearly $A = Z + N$ and so one of the numbers is redundant; usually the number N is omitted and the symbol becomes,

$$^{A}{}_{Z}X \quad [10\text{-}2]$$

The isotopes of carbon mentioned above are $^{10}{}_{6}C$, $^{11}{}_{6}C$, $^{12}{}_{6}C$, $^{13}{}_{6}C$, and $^{14}{}_{6}C$.

10.2 MASS AND ENERGY UNITS

In dealing with the masses of individual nuclei we must use very small numbers. The atomic mass scale is based on the definition that the atomic mass of ^{12}C is exactly 12 atomic mass units (u). Since 12×10^{-3} kg of ^{12}C contains exactly Avogadro's number ($N_a = 6 \times 10^{23}$) of atoms then it follows that 1 atomic mass unit is given by

$$1 \text{ u} = 12\times10^{-3} \text{ kg}/12 \times N_a = 1.66\times10^{-27} \text{ kg} \quad [10\text{-}3]$$

Einstein's famous equation,

$$E = mc^2 \quad [10\text{-}4]$$

where c is the speed of light (2.998×10^{8} m/s), tells us that mass and energy are inter-convertible and so the mass in Eq. [10-3] could just as well be expressed in energy units. Then

$$1 \text{ u} = 1.66\times10^{-27} \text{ kg} \times (2.998\times10^{8} \text{ m/s})^2 = 1.49\times10^{-10} \text{ J} \quad [10\text{-}5]$$

Usually, however, a different and better suited energy unit is used in nuclear physics. Recalling that the electron volt (eV) is 1.602×10^{-19} J, we find that

$$1 \text{ u} = 931.5\times10^{6} \text{ eV} = 931.5 \text{ MeV} \quad [10\text{-}6]$$

The MeV unit is the usual unit in which nuclear masses and energies are measured.

---

[1] Notice that to 2 or 3 significant figures the mass number, A, is the same as the molar mass. The mass number of ^{16}O is 16; its molar mass is 15.99491 g/mol.

10.3 THE STABILITY OF NUCLEI

There are about 100 different elements but many elements have more than one stable isotope; there are about 300 stable isotopes. (For example, beryllium has only one stable isotope $^{9}_{4}Be$ whereas tin has ten.) Not every imaginable mixture of protons and neutrons will form a stable nucleus. Almost all stable nuclei have a number of protons (*Z*) which is less than the number of neutrons (*N*). This stability can be understood on the basis of simple electrostatics; if there are too many protons the mutual Coulomb electrical repulsion of the positive charges overcomes the forces holding the nucleus together. In a very few cases a nucleus is stable with a number of neutrons equal to, or even less than the number of protons. These few cases occur only for the very lightest elements: $^{3}_{2}He$ has N (1) less than Z (2); for $^{4}_{2}He$ the numbers are equal.

For all heavier nuclei N is slightly greater than Z but not by a large amount; nuclei also are unstable if they have too many neutrons. Figure 10-1 is a plot of N vs. Z showing the narrow band of nuclear stability. The band lies almost entirely above the line of $N = Z$.

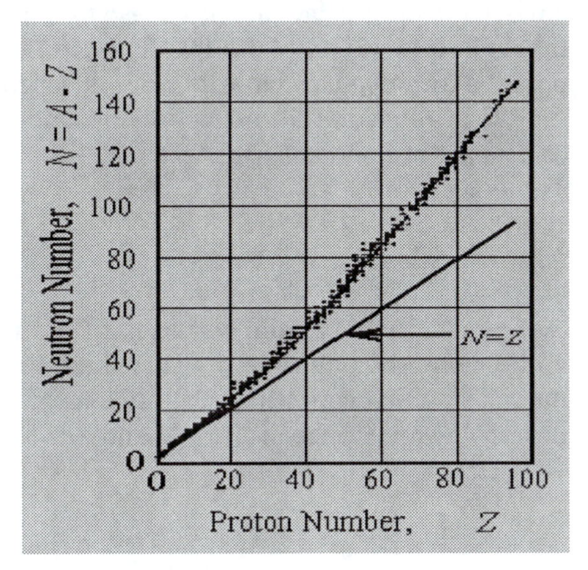

Figure 10-1 N-Z plot for Stable Nuclei

Theories of the structure of the nucleus are much more complicated than for the atom. In the latter case the structure is dominated by the Coulomb force between the orbiting electrons and the central nucleus, which on the atomic scale is just a point. In the nucleus itself there is no such simplicity; fortunately the details of nuclear structure are not required for a discussion of energetics. It is sufficient to know that any nucleus which has too few or too many neutrons relative to protons will be unstable and that the nucleus will change in some self-initiated way to redress the imbalance. The method by which it does this is to emit various particles in a process called *radioactivity*.

10.4 RADIOACTIVITY

It might be expected that an unstable nucleus (the *parent)* with an excess of some particle might simply emit the requisite number of those particles and so produce a new stable nucleus called the *daughter* nucleus.

Table 10-2: Radioactive Emissions

Particle	Identity
alpha(α)	Nucleus of helium atom $^{4}_{2}He$
beta(β)	β^-, Ordinary electron $^{0}_{-1}e$
	β^+, Positive electron or positron $^{0}_{1}e$
gamma (γ)	Electromagnetic wave of very short wavelength

Because of the internal structure of the nucleus this almost never happens. In practice, the unstable nucleus achieves stability by emitting two other types of particles. In the early days of nuclear physics these two particles were unidentified and were simply labeled alpha (α) and beta (β). The alpha had positive electric charge but betas of both positive and negative charge were encountered. It was also recognized that there was another radiation that often accompanied α and β and it was labeled gamma (γ). In the early part of last century these radiations were identified; their properties are given in Table 10-2.

Alpha Emission

Some nuclei – particularly those at the high-mass end of the periodic table - achieve stability by the emission (at first sight surprising) of a tightly bound cluster of particles which constitute the nucleus of normal helium ($^{4}_{2}He$). Two examples of practical importance are $^{238}_{92}U$ and $^{239}_{94}Pu$. The decay scheme of the former is

$$^{238}_{92}U \rightarrow {}^{234}_{90}Th + {}^{4}_{2}He \qquad \textbf{[10-7]}$$

These transformations are subject to two conservation laws which determine the balance of the two sides of the equation. First, electrical charge must be conserved; this is the same as the atomic number (subscript). It follows that the atomic numbers on both sides of the equation must add up to the same value (92 = 90 + 2). Secondly, the mass number on the nuclear scale is conserved, this means that the superscripts on each side of the equation must add up to the same value (238 = 234 + 4). [2]

The α-particles are emitted with well defined energies, typically a few MeV, and because of their large mass and charge, they interact strongly with matter. As a result they are easily shielded, being effectively stopped by as little as a sheet of paper. Alpha emitters tend to have long lifetimes.

[2] This is actually a loose way of formulating the correct rule which is called the "law of conservation of baryon number".

Beta emission

The predominant method of radioactive adjustment for unstable nuclei is by β-emission. For example, tritium, ^{3}H, undergoes radioactive decay by emitting a β-particle. The transformation equation describing this process is written as

$$^{3}_{1}H \rightarrow ^{3}_{2}He + ^{0}_{-1}\beta \quad \textbf{[10-8]}$$

Note that the rules for the conservation of atomic number (charge) and atomic mass still hold. It must be noted that the mass of the electron is negligible on the scale of nuclear particle masses. Accordingly it is assigned a mass number of zero. In fact the mass of the electron is 1/1840 of the proton's mass.

The transformation equation for ^{14}C is

$$^{14}_{6}C \rightarrow ^{14}_{7}N + ^{0}_{-1}\beta \quad \textbf{[10-9]}$$

In this process a nucleus of carbon has been transformed into one of nitrogen; no further transformations will take place in these cases as both of the daughters $^{3}_{2}He$ and $^{14}_{7}N$ are stable.

An example of a nucleus that emits a positron is $^{11}_{6}C$; its decay is given by

$$^{11}_{6}C \rightarrow ^{11}_{5}B + ^{0}_{1}\beta \quad \textbf{[10-10]}$$

Again the daughter boron nucleus is stable. On occasions when the daughter of a β-decay is not stable one or more subsequent decays will occur until stability is achieved.

Beta particles from radioactive nuclei have speeds close to that of light, and kinetic energies of the order of one MeV. They travel for about one metre in air or a few mm in water or human tissue before coming to rest. In the process of coming to rest in tissue they can do damage as will be seen in Chapter 12. It is rather easy to shield a β-emitter; a plastic sheet one cm thick affords protection. If, however, β-emitting materials are ingested via food, air or water, the betas can cause considerable damage.

A given radioisotope emits betas of varying energies. What is observed is a continuous spectrum of energies from zero to some maximum. A useful approximation is that the average energy E_{av} and the maximum energy E_{max} are related by

$$E_{av} \approx E_{max}/3 \quad \textbf{[10-11]}$$

Because the initial and final nuclear masses are fixed, the energy released in a β-decay should also be fixed. This requirement can be met if there is another particle released along with the β to share the energy. For this reason physicists assumed that such a particle existed and it was indeed subsequently found. It is a particle without charge, and if it has a mass it is very small. The particle, called the *neutrino*, has a velocity essentially equal to that of light and is very difficult to stop or detect; a neutrino will go right through the planet Earth unaffected. We will not consider neutrinos further as they are irrelevant to terrestrial problems of energy generation.

Gamma Emission

Gamma rays are very short wavelength electromagnetic waves. They are essentially high energy X-rays. After the emission of an α- or β-particle, the daughter nucleus, in most cases, is left with excess energy in an *excited state.* This excess energy is emitted, as a γ-ray, very shortly (10^{-14}s) after the primary event, and this shedding of energy permits the nuclear particles to re-adjust into their lowest energy (*ground*) state. This is very similar to the re-adjustment of orbital electrons in excited atoms, where low energy electromagnetic waves are emitted as X-rays or light. Gamma rays are very penetrating, having energies around one MeV. Typically several centimetres of lead are required to attenuate them to an acceptable level and form an effective shield.

Natural and human-made radionuclides

Only a very few radioactive isotopes occur naturally in substantial quantities. The reason is that their lifetime must be very long for them to have survived since their formation in whatever cosmological event was involved, e.g., the formation of the universe itself (20×10^9 yr) or the formation of the solar system (5×10^9 yr). Examples are $^{235}_{92}U$ (α-emitter) and $^{40}_{19}K$ (β-emitter). A few unstable nuclei are produced continuously by the action of cosmic rays in the atmosphere but the quantities are minuscule. Examples are the production of $^{14}_{6}C$ (so important in *carbon dating* in archaeology), and $^{3}_{1}H$ or *tritium*, the radioactive form of hydrogen found in trace quantities in water. Most of the exposure to radiation that we experience comes from a small number of natural radioactive isotopes (see Chapter 12).

Some very heavy nuclei are so very far from nuclear stability that they require many radioactive events to occur before they achieve stability. This results in a *radioactive series.* Such series usually begin with a long lived parent whose slow rate of decay determines how many of each nuclide of the subsequent species are found downstream in the various daughter nuclei. An example of such a series is the one which begins with $^{238}_{92}U$ and ends with $^{206}_{82}Pb$; fifteen transformations occur before the stable end-product is reached. Three other series are $^{235}_{92}U$ - $^{207}_{82}Pb$, $^{232}_{90}Th$ - $^{208}_{82}Pb$, and $^{241}_{94}Pu$ - $^{209}_{83}Bi$.

With nuclear reactors and high energy particle accelerators we have the technology to transmute stable nuclei into radioactive ones by adding or removing neutrons or protons. For example, the isotope ^{60}Co, widely used in cancer treatment, is manufactured by exposing the natural stable ^{59}Co to neutrons in a nuclear reactor. The resulting ^{60}Co is long lived; each radioactive nucleus decays by emitting a β-particle followed by two γ-rays. The highly penetrating nature of these gammas enables them to reach and destroy deep-seated tumors.

10.5 RADIOACTIVE DECAY CALCULATIONS

Suppose that we start with a number N_0 of nuclei at time zero. The time at which a given nucleus decays is entirely random so we can only look at the average behavior of a large number of nuclei. Let λ be the probability that in unit time a given nucleus will decay; this is called the *decay constant*. If after a time t the number of nuclei remaining is N, then in the next short time dt the number decaying will be proportional to both N and dt, therefore

$$dN = -\lambda N\, dt \qquad \textbf{[10-12]}$$

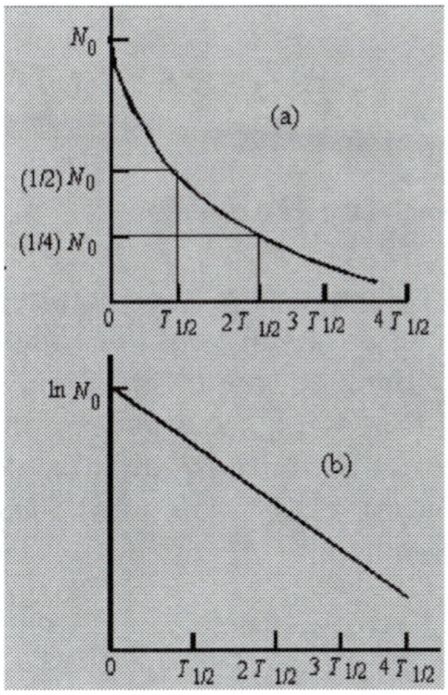

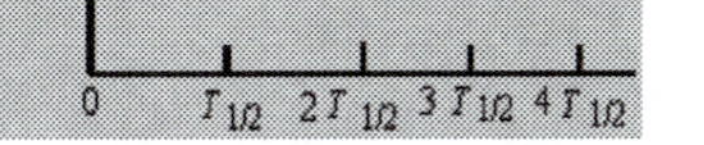

Figure 10-2 Exponential Decay

The minus sign expresses the fact that the number N can only decrease as t increases. Eq. [10-12] in the form,

$$dN/N = -\lambda\, dt \qquad \textbf{[10-13]}$$

has the well known solution

$$N = N_0\, e^{-\lambda t} \qquad \textbf{[10-14a]}$$

or, alternatively

$$\ell n(N/N_0) = -\lambda t \qquad \textbf{[10-14b]}$$

The decrease takes place in an exponential manner with time. The behaviour described by Eq. [10-14a] is illustrated in Fig. 10-2a and that of Eq. [10-14b] in Fig. 10-2b. A characteristic of exponential decay is that it can be characterized by a unique time called the *half-life* ($T_{1/2}$), that is, the time for any given starting number N_0 to decrease to $N_0/2$. Substituting $N = N_0/2$ into Eq. [10-14] gives

$$T_{1/2} = 0.693/\lambda \qquad \textbf{[10-15]}$$

The half-life is clearly illustrated in Fig. 10-2a. Half-lives are usually specified for radioisotopes in preference to decay constants since it immediately conveys the important information of how long the isotope will survive. For example, ^{239}Pu, which merits concern because of its cancer-inducing properties, has a half-life of 24,400 years.

EXAMPLE 10-1

How long will it take for a stored radioactive ^{239}Pu waste to decay to 1% of its present level?

SOLUTION

$$\lambda = 0.693/24{,}000\ \text{yr}^{-1} = 2.84 \times 10^{-5}\ \text{yr}^{-1}$$

$$N = 0.010\, N_0$$
$$0.010 = e^{-\lambda t}$$
$$t = -(\ell n\, 0.010)/\lambda = -(\ell n\, 0.010)/2.84 \times 10^{-5}\ \text{yr}^{-1} = 162{,}000\ \text{yr}$$

Effective Half-life

If a radioactive species is ingested by a living organism the effective half-life of the species in the organism can be significantly altered by the biological activities of the organism. Although the isotope is decaying with a physical half-life of $T_{1/2,\,p}$ (decay constant = λ_p), the organism may be eliminating the isotope in some manner; animals excrete, perspire, would also decay exponentially with a biological half-life $T_{1/2,b}$ (decay constant = λ_b). [3] The total decay factor is given by the product of the two decay exponentials

$$e^{-\lambda_p t} \cdot e^{-\lambda_b t} = e^{-(\lambda_p + \lambda_b)t} = e^{-\lambda_e t}$$

where λ_e is the *effective decay constant* and

$$N = N_0\, e^{-(\lambda_p + \lambda_b)t} = N_0\, e^{-\lambda_e t}$$

Therefore

$$\lambda_p + \lambda_b = \lambda_e \qquad \textbf{[10-16]}$$

From Eq. [10-14] it follows that

$$\frac{1}{T_{1/2,p}} + \frac{1}{T_{1/2,b}} = \frac{1}{T_{1/2,e}} \qquad \textbf{[10-17]}$$

EXAMPLE 10-2

When iodine is ingested by humans they eliminate it such that one half of the body's iodine content is excreted every 4.0 days. Radioactive ^{131}I with a physical half-life of 8.1 days is administered to a patient. When will only 1% of the isotope remain in the patient's body?

SOLUTION

$1/T_{1/2,\,e} = 1/T_{1/2\,,p} + 1/T_{1/2\,,b} = 1/8.1\ \text{d} + 1/4\ \text{d} = 0.37\ \text{d}^{-1}$
$\therefore\ T_{1/2\,,e} = 2.7\ \text{d}$
$\lambda_e = 0.693/T_{1/2,e} = 0.693/2.7\ \text{d} = 0.26\ \text{d}^{-1}$
$N/N_0 = e^{-0.26t} = 0.010$
$\ell n(0.010) = -0.26t$
$t = 18\ \text{d}$

---

[3]Note that this half life has nothing to do exhale etc. The rate of elimination is often proportional to the amount present, so the amount present in the organism with radioactivity; it is purely biological.

Activity

Activity is a term that refers to the number of radioactive nuclei that disintegrate per second and might be considered a measure of the *strength* of the sample. The activity, A, of a sample of radioactive material will depend on both the number, N, of nuclei present and the half-life; the shorter the half-life the faster the nuclei decay and the greater the strength. Using Eq. [10-13]

$$A = |dN/dt| = \lambda N \qquad \textbf{[10-18]}$$

Since N decays exponentially then so also will A. It follows that

$$A = A_0 e^{-\lambda t} \qquad \textbf{[10-19]}$$

Thus the amount of radiation, α, β or γ, emitted per second falls off exponentially in the same way and at the same rate as N. The unit of activity is called the B*ecquerel* and is defined as one disintegration per second; the abbreviation is Bq. An older unit which is gradually losing currency is the Curie (Ci) which is 3.7×10^{10} Bq.

EXAMPLE 10-3

A 4.3×10^{14} Bq source of ^{60}Co is used for cancer treatment. Each disintegrating nucleus emits two γ-rays, one of energy 1.17 MeV and one of 1.33 MeV. What is the mass of ^{60}Co present in the source and how much energy is emitted per second in the form of γ-rays? The half-life of ^{60}Co is 5.3 years.

SOLUTION

$\lambda = 0.693/(5.3 \text{ yr} \times 365 \text{ day/yr} \times 24 \text{ hr/day} \times 3600 \text{ s/hr}) = 4.15 \times 10^{-9} \text{ s}^{-1}$
Since $A = \lambda N$ (Eq. [10-18]) the number of radioactive nuclei is

$N = A/\lambda = 4.3 \times 10^{14} \text{ s}^{-1} / 4.15 \times 10^{-9} \text{ s}^{-1} = 1.04 \times 10^{23}$ atoms

Since 6.02×10^{23} atoms of ^{60}Co have a mass of 60 g or 0.060 kg,
1.04×10^{23} atoms have a mass of 1.04(0.06/6.02) = 0.010 kg.

The energy per second = $4.3 \times 10^{14} \text{ s}^{-1}(1.17+1.33)$ MeV
= 10.75×10^{14} MeV/s = 10.75×10^{14} MeV/s $\times 1.6 \times 10^{-13}$ J/MeV
= 1.7×10^{2} J/s = 1.7×10^{2} W

Some Important Radioactive Isotopes

Although hundreds of radioactive isotopes have been produced artificially, there are a few which, for various environmental reasons, merit special consideration:

- Cobalt-60, ^{60}Co. This isotope which has been referred to earlier has a half-life of 5.3 years. Each disintegrating nucleus emits a β-particle, whose kinetic energy is in the range 0 – 0.3 MeV, followed by two γ-rays of 1.17 and 1.33

MeV (see Example 10-3 above). The time delay between these three successive emissions are each about 10^{-12} s, i.e., negligible on our time scale so we regard them as simultaneous. The isotope is produced by bombarding natural cobalt with neutrons in a nuclear reactor. It finds widespread use in the radiation treatment of deep tumors.

- Strontium-90, ^{90}Sr, is a principal component of fallout from atmospheric tests of nuclear bombs. It has a half-life of 28 years and its betas have a mean energy of 0.2 MeV. The daughter nucleus ^{90}Y has a 64 hour half-life, emitting betas of mean energy 0.8 MeV. Strontium is particularly important as its chemistry is similar to that of calcium and so it becomes incorporated in the skeleton of animals which ingest it. Concern about it has been decreasing due to the worldwide ban on atmospheric nuclear-bomb testing.

- Cesium-137, ^{137}Cs, is an important waste product of fission reactors which has some chemical similarity to potassium. It decays mainly by betas of mean energy 0.2 MeV, with a half-life of 30 years. The beta is followed by a 0.66 MeV gamma.

- Plutonium-239, ^{239}Pu, is an α-emitter with a 24,400 year half-life; it is formed by the bombardment of uranium in nuclear reactors. Of more importance than its radioactivity is the extreme chemical toxicity of plutonium.

- Carbon-14, ^{14}C, is a β-emitter of very low energy (0.05 MeV) and half-life 5700 years. It is formed by the interaction of cosmic ray neutrons with ^{14}N in the atmosphere. The atmospheric ^{14}C then enters the plant-animal cycle so it is present in all biological material. When the organism dies, no more ^{14}C enters and the amount already present declines through radioactive decay as time progresses. By measuring the remaining ^{14}C content the age of the specimen can be determined.

10.6 NUCLEAR REACTIONS

It was mentioned earlier that reactors and accelerators can transmute stable nuclides into radioactive ones, of which the production of ^{60}Co is an example; these transmutations are called *nuclear reactions*. Reactors, as we shall see shortly, produce copious amounts of neutrons; being uncharged these can easily penetrate within a stable nucleus, and if captured there they increase the mass number A by one unit without affecting Z. Our example was

$$^{59}{}_{27}\mathrm{Co} + {}^{1}{}_{0}\mathrm{n} \rightarrow {}^{60}{}_{27}\mathrm{Co} \qquad \textbf{[10-20]}$$

Notice that the rules for balancing these equations are still obeyed.

In the case of bombardment of a stable nuclide by charged particles, e.g., protons, the bombarding particle must acquire sufficient kinetic energy to enable it to overcome the electrostatic repulsion of the positively charged target nucleus. This can be done by accelerating it to a high speed in a cyclotron or similar accelerating machine. In charged particle reactions a new nuclide is again produced but frequently an additional particle is also emitted. For example, when a ^{26}Mg target is bombarded with protons of high energy, neutrons are observed coming off; the reaction is

$$^{26}_{12}Mg + ^{1}_{1}p \rightarrow ^{26}_{13}Al + ^{1}_{0}n \qquad \textbf{[10-21]}$$

The particular interest in specific types of nuclear reactions will become clear as nuclear fission and fusion are discussed. For example, neutron-induced reactions play a major role in creating the hazardous radioactivity that builds up in the structures of nuclear reactors.

An important concept in the physics of nuclear reactions is that of the *cross section*. You might imagine that a reaction would automatically take place if a particle (p, n, etc.) collided with a target nucleus. If this were automatically so, then we could calculate the number of reactions. Consider a flux of N particles per second per square meter striking a thin target at right angles as in Fig. 10-3. The target material has a mass of m grams. It contains, in its 1 m^2 of area, a number of nuclei n given by

$$n = N_a \, m/A \qquad \textbf{[10-22]}$$

where A is the molar mass of the target material and N_a is Avogadro's number. Each of these nuclei has a cross sectional area σ m^2 and so the total area presented to the in-coming beam is $n\sigma$ m^2. It follows that the number of particles, ΔN, intercepted from the beam is given by [4]

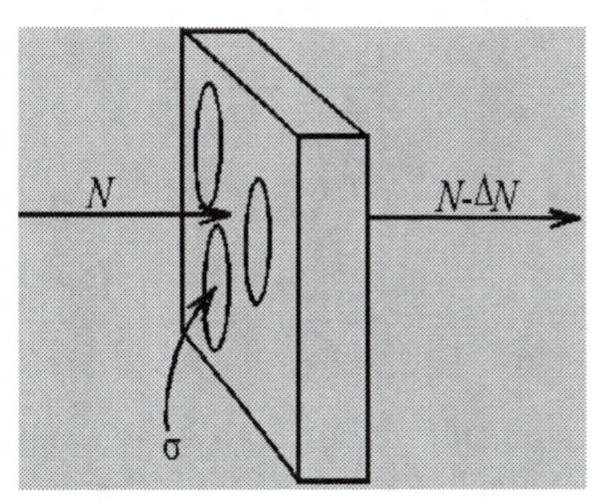

Figure 10-3 Absorption Cross Section

$$\Delta N/N = n\sigma/1 = N_a \, m\sigma/A \qquad \textbf{[10-23]}$$

The number of reactions that occur per second is

$$\Delta N = n\sigma N = N_a \, m\sigma N/A \qquad \textbf{[10-24]}$$

When a reaction is studied experimentally it is found that a different number of reactions occur than is predicted by Eq. [10-24]. Obviously a collision does not automatically ensure that a reaction will take place. The probability of occurrence of a reaction is dependent on details of the internal structure of the target nucleus and on the energy of the incident particle. It is still possible to describe nuclear reactions by the simple form of Eq. [10-24] if we use a more flexible definition for the quantity σ. Rather than being the *geometrical cross section* of the nucleus, we think of it as a *reaction cross section*. Thus the nucleus presents an "effective" area to the beam rather

[4] The assumption is made that the sample is sufficiently thin that no nucleus is hidden behind another. That is, all the nuclei are exposed to the particle beam.

than its real area; σ has a large value if the reaction is a highly probable one and a low value if it is less probable.

Values of σ must be determined experimentally by bombarding thin sections of material with particle beams at various energies. Equation [10-24] permits a calculation of the cross section. Cross sections are measured in units of area; in this case m^2 is very inconvenient since the areas are so very small. A unit called the *barn* is used.[5]

$$1 \text{ barn} = 10^{-28} \text{ m}^2 \qquad \textbf{[10-25]}$$

Values for a few neutron reaction cross sections are given in Table 10-3.

Table 10-3: Neutron Absorption Cross Sections

Isotope	Cross section (barns)	Isotope	Cross section (barns)
$^{1}_{1}H$	0.332	$^{135}_{54}Xe$	2.64×10^{6}
$^{2}_{1}H$	5.2×10^{-4}	$^{157}_{64}Gd$	2.54×10^{5}
$^{59}_{27}Co$	37	$^{197}_{79}Au$	99
$^{113}_{48}Cd$	2×10^{4}	$^{235}_{92}U$	582
$^{115}_{49}In$	198	$^{238}_{92}U$	2720

Most of the entries in this table have been chosen because of their relevance in later sections, but it is useful to note a few things at this stage:

- Heavy hydrogen $^{2}_{1}H$ is a much poorer absorber of neutrons than is ordinary hydrogen $^{1}_{1}H$ by a factor of over 600. This is very relevant to the design of the Canadian CANDU power reactor which uses heavy water instead of ordinary water (see Chapter 11).

- The relatively light metal cadmium (Cd) is often used as a shield against neutrons as its isotope $^{113}_{48}Cd$, which constitutes 12% of the natural material, has such a very high reaction cross section for neutrons.

- To make measurements of the strength of neutron radiations, the soft metal indium is often used in the form of thin foils. $^{115}_{49}In$ constitutes 96% of the natural material and has a rather high reaction cross section for neutrons. The resulting isotope $^{116}_{49}In$ is radioactive with a half-life of 54 min. It transforms according to

[5] The word "barn" was used as a code-word in the atomic bomb project during World War II. It is thought to have been suggested by the common phrase that something is "as big as a barn" since a cross section of 10^{-28} m^2 is indeed large relative to the real cross-sectional area of a nucleus.

$$^{116}_{49}\text{In} \rightarrow {}^{116}_{50}\text{Sn} + {}^{0}_{-1}\text{e}$$

A measure of the induced radioactive activity of the indium can be used as a measure of the neutron flux to which it was exposed.

- The isotope $^{135}_{54}$Xe is produced continuously by fission of uranium in copious quantities in nuclear reactors; it is radioactive with a half-life of 9.2 hr. This isotope is of great importance to the operation of nuclear reactors because of its enormous cross section for neutrons. (It becomes a major source for the loss of neutrons in the reactor and thus is an issue for reactor control.)

- Natural uranium consists of two isotopes ^{238}U (99.3%) and ^{235}U (0.7%). The absorption of a neutron by ^{235}U leads to fission as discussed in the next chapter. The isotope ^{238}U, which is the most common, absorbs neutrons without fission and with a rather large cross section.

$$^{238}_{92}\text{U} + {}^{1}_{0}\text{n} \rightarrow {}^{239}_{92}\text{U}$$

The further fate of ^{239}U will be deferred till later in the chapter.

EXAMPLE 10-4

A 1 g piece of cobalt is bombarded by neutrons in a reactor for one hour. The neutron intensity is 1.0×10^{17} per s per m^2. How many nuclei are transmuted to ^{60}Co? Express the resulting radioactivity in Bq units. Ignore the fact that some of the radioactive nuclei decay during the first hour since the half-life of ^{60}Co is 5.3 years.

SOLUTION

The number of nuclei, n = (1/59) mole × 6.02 × 10^{23} atoms/mole = 1.02×10^{22}
The value of σ for ^{59}Co, from Table 10-3 is 37 barns
$\therefore$ The area presented to the neutron beam = $n\sigma = 1.02\times10^{22} \times 37 \times 10^{-28}$ m^2
= 3.78×10^{-5} m^2
The number of neutrons intercepted is given by Eq. [10-24]
$\Delta N = n\sigma N = 3.78 \times 10^{-5}\ \text{m}^2 \times 1.0\times10^{17}\text{m}^{-2}\ \text{s}^{-1} = 3.78 \times 10^{12}\ \text{s}^{-1}$
The number activated per hour = 3600s × 3.78×10^{12} s^{-1} = 1.36×10^{16}
From Eq. [10-18], the activity = λ × Number of nuclei
$\lambda = 0.693/(5.3 \times 365 \times 24 \times 3600)\ \text{s} = 4.1 \times 10^{-9}\ \text{s}^{-1}$
Activity = $4.1 \times 10^{-9}\ \text{s}^{-1} \times 1.36 \times 10^{16} = 5.6 \times 10^{7}$ Bq

---

Calculations such as those in Example 10-4 can also be used to calculate the buildup of radioactivity in the structural materials of a nuclear reactor. In practice the calculation is more complicated as several other factors have been ignored, in particular, the fact that as the radioactive species build up they in turn start to decay; this omission is more serious for short half-lives.

10.7 NUCLEAR BINDING ENERGY

Since we know the masses of the proton (m_p = 1.0078252u) and of the neutron (m_n = 1.0086652u) very accurately, and we know the number of each of these in every nucleus, we ought to be able to predict the mass of every nucleus. For example the mass of ^4_2He should be the mass of 2 protons and 2 neutrons or,

$$2 \times 1.0078252 + 2 \times 1.0086652 = 4.0329808\text{u} \qquad \textbf{[10-26]}$$

In fact, the mass that is measured for the ^4_2He nucleus is 4.002603u, a number that is slightly smaller than expected. Expressed in energy units, Eq. [10-26] is

$$2 \times 938.79 + 2 \times 939.57 = 3756.7 \text{ MeV} \qquad \textbf{[10-27]}$$

whereas the actual mass of ^4_2He is 4.002603u × 931.5MeV/u = 3728.4 MeV. This is a difference of 28.3 MeV.

In general for a nucleus with Z protons and N neutrons the predicted mass (or energy) would be $Zm_p + Nm_n$. However in every case experiment has demonstrated that the actual nuclear mass is less, i.e.,

$$Zm_p + Nm_n > m_{nucleus} \qquad \textbf{[10-28]}$$

This extra mass (or energy) is a result of the *nuclear binding energy*. It is the extra energy we would have to provide to the nucleus to overcome the nuclear forces and break it up into its constituent parts. For ^4_2He there is a total binding energy of 28.3 MeV. It is usual to express this as the binding energy per nucleon; since ^4_2He has 4 nucleons (2 p + 2 n), its binding energy per nucleon is 28.3/4 = 7.07 MeV.

As we will see below, except for the lightest nuclei, the binding energy per nucleon is approximately constant throughout the periodic table. This immediately tells us something about the force that binds the nucleons together. Suppose that the nucleon- nucleon force had a long range, so that a given nucleon bonded to every other nucleon in the nucleus. In this case its binding energy per nucleon would simply increase proportionately to the nucleon number A, but this is not what we observe. The approximate constancy of the binding energy per nucleon indicates that a given nucleon bonds to only a very small number of close neighbours; the addition of further nucleons to make a heavier nucleus has little further effect on the nucleon we are considering. This immediately suggests that the nuclear

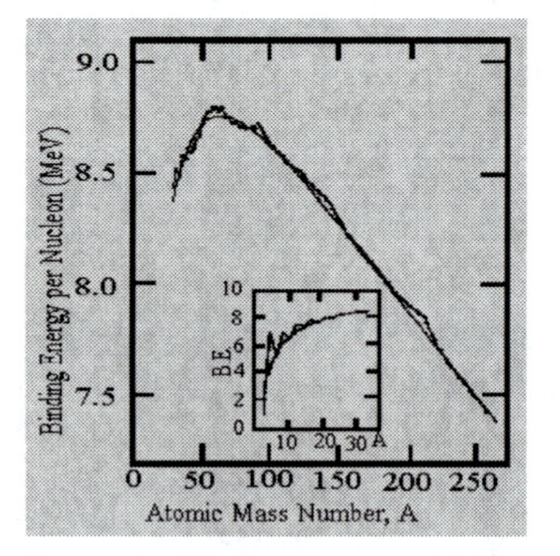

Figure 10-4 Binding Energy per Nucleon

forces which bind the nucleus together are of short range. The molecules in a water drop provide an analogy; each molecule binds only to its closest neighbours. Consider a drop of mass m containing N molecules. The binding energy is the *heat of vaporization*, Q, which is the energy needed to dissociate (evaporate) the drop into its constituent molecules. The average binding energy per molecule is Q/N. For a drop which has twice the volume - that is, double the number of water molecules - the energy (heat) needed to dissociate the drop also doubles. The binding energy per molecule has not changed.

This simple comparison gives rise to the *Liquid Drop Model* of the nucleus from which we are able to predict and understand many nuclear properties including fission and fusion.

Table 10-4: Binding Energy per Nucleon

Nucleus	B.E. (MeV)	Nucleus	B.E. (MeV)	Nucleus	B.E. (MeV)
$^{2}_{1}H$	1.112	$^{10}_{5}B$	6.475	$^{140}_{54}Xe$	8.295
$^{3}_{1}H$	2.827	$^{12}_{6}C$	7.68	$^{235}_{92}U$	7.59
$^{3}_{2}He$	2.572	$^{14}_{7}N$	7.475	$^{236}_{92}U$	7.586
$^{4}_{2}He$	7.074	$^{56}_{26}Fe$	8.791	$^{238}_{92}U$	7.57
$^{6}_{3}Li$	5.332	$^{95}_{38}Sr$	8.552	$^{239}_{92}U$	7.558
$^{7}_{3}Li$	5.606	$^{100}_{40}Zr$	8.531	$^{239}_{94}Pu$	7.56
$^{7}_{4}Be$	5.371	$^{139}_{54}Xe$	8.314	$^{240}_{94}Pu$	7.556

In Fig. 10-4 the average binding energy per nucleon is plotted vs. A (mass number) for all the stable elements of the periodic table. Except for some large variations for the light elements, the curve has low values for light and heavy elements with a maximum in between; the maximum actually occurs for ^{56}Fe. The average value of the binding energy per nucleon for a few important nuclei is given in Table 10-4.

EXAMPLE 10-5

The mass of $^{56}_{26}Fe$ is 55.934934 u. Calculate the binding energy per nucleon for this nucleus, i.e., derive the entry for $^{56}_{26}Fe$ in Table 10-4.

SOLUTION

The mass of 26 protons and 30 neutrons is
$26 \times 1.0078252 \text{ u} + 30 \times 1.0086652 \text{ u} = 56.4634112 \text{ u}$.
The binding energy $= 56.4634112 \text{ u} - 55.934934 \text{ u} = 0.5284772 \text{ u}$
$= 0.5284772 \text{ u} \times 931.5 \text{ MeV/u} = 492.277 \text{ MeV}$
The binding energy per nucleon $= 492.277/56 = 8.791$ MeV

Figure 10-4 and Table 10-4 show that there are, in principle, two ways to extract energy from the nucleus. If a very heavy nucleus like uranium were to split into two almost equal pieces, (each with Z ~ 50) the resulting two lighter nuclei would have more binding energy per nucleon and so energy would be released; this process is called *nuclear fission.* In the second case if two very light nuclei were to join together to form one heavier nucleus, again the binding energy per nucleon would be increased and energy would be released; this process is called *nuclear fusion.*

10.8 NUCLEAR FISSION

Spontaneous Fission

Let us examine a possible case of nuclear fission of ^{235}U into a nucleus of ^{140}Xe and ^{95}Sr; the equation for the reaction is

$$^{235}_{92}U \rightarrow \, ^{140}_{54}Xe + \, ^{95}_{38}Sr \qquad \textbf{[10-29]}$$

Note that the rules given in Section 10.4 for balancing nuclear equations are obeyed. Using Table 10-4, the energy released in this fission is:

$$140 \times 8.295 + 95 \times 8.552 - 235 \times 7.590 = 190 \text{ MeV}$$

If a ^{235}U nucleus were to split in this way, 190 MeV of energy would be released, which would appear mostly as kinetic energy of the two resulting *fission fragments* as they fly away from each other at high speed. When these fragments collide with other atoms the energy quickly becomes transformed into vibratory motions of the surrounding atoms, in other words as heat.

This type of *spontaneous fission* actually does occur but it is rather rare; it was only observed several years after the observation of *induced fission*, which is discussed next.

Induced Fission

It is the induced fission that is technologically important, and as will be seen, it is induced fission in the isotope ^{235}U (which only constitutes 0.72% of natural uranium) rather than in the more plentiful ^{238}U that can provide useful energy. In induced fission a stray neutron encounters the nucleus and in the process of becoming bound to the nucleus causes it to fission.

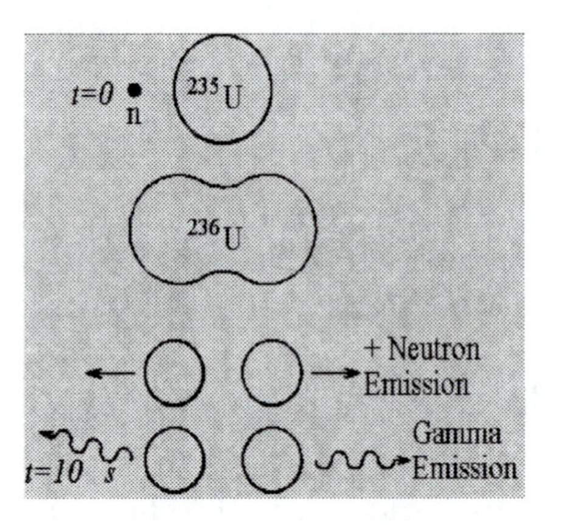

Figure 10-5 The "Liquid Drop" Nucleus Undergoing Induced Fission

It is possible to understand the process of induced fission on the basis of what is known about binding energy and by using the Liquid Drop Model. It is important to realize that the stability of the nucleus is a result of a balance between forces. The short-range nuclear forces are acting to attract their nearby neighbours, but the electrical (Coulomb) forces of the protons are acting to push the nucleus apart. If the heavy nucleus happens to capture a neutron, that neutron becomes bound with an energy of about 7 MeV. This amount of energy is therefore released into the nucleus and sets it into violent vibration. The vibrations may result in the nucleus splitting into two fragments as illustrated in Fig. 10-5.

Let us examine the neutron-induced fission of ^{235}U. Calculations show that 6.6 MeV of energy is required to cause uranium to fission. If a neutron is absorbed by ^{235}U, a short-lived nucleus of ^{236}U will be created. From Table 10-4 we calculate that the total binding energy of ^{236}U is 1791 MeV, whereas for ^{235}U it is 1784 MeV. The act of binding the neutron has released 7 MeV which is more than enough to cause the nucleus to fission. In this case the neutron need not even bring in any kinetic energy of its own; the release of its binding energy is more than sufficient.

When neutron-induced fission takes place it is most effective if the neutrons are moving slowly, that is, they should have a small kinetic energy of the order of 1/40 eV; such neutrons are said to be *thermalized* because their kinetic energy is roughly the same as the kinetic energy of room-temperature air molecules. A neutron has no charge so it is not affected by the charge on the protons. If it moves slowly it will spend more time in the vicinity of the other nuclear particles, increasing its chance of being captured by the short-range nuclear force. A rapidly moving neutron spends only a short time near the nucleons and its probability of capture is greatly reduced. This fact is very important in the operation of nuclear reactors.

Chain Reaction

Did you know? The first nuclear chain reaction we know of took place 2 billion years ago in a very rich uranium deposit in the present Gabon in Africa.

For neutron-induced fission to take place there must be a source of neutrons; this is provided by the fission reaction itself. When the fission takes place, the fission fragments, which are discussed in more detail below, are so neutron-rich that they quickly emit two or three neutrons. These neutrons have large velocities but if they can be slowed down, or thermalized, by some "*moderation*" process they can go on to induce further fissions.

The fact that more than one neutron is produced, on average, for each neutron absorbed means that the process can, under the right circumstances, grow in intensity; this is the so-called *chain reaction*. For this to happen it is necessary to slow the neutrons down and to ensure that they are not lost by other processes. Losses can take place by the neutrons passing out of the reaction volume and by being absorbed by nuclei like cadmium which have a large cross section for neutron absorption but do not fission. This absorption process is one way of controlling the chain reaction.

Fission fragments

Returning to the neutron-induced fission of ^{235}U we can write the reaction as

$$^{235}_{92}\mathrm{U} + \mathrm{n} \rightarrow \, ^{236}_{92}\mathrm{U} \rightarrow \mathrm{X} + \mathrm{Y} + q\mathrm{n} \qquad \textbf{[10-30]}$$

where X and Y are the two nuclei (*fission fragments*) produced by the fission, and q is a small integer (0,1,2,etc). Many different X,Y pairs are possible, grouped around the region where the mass is one half of the mass of uranium. The distribution of abundance of fission fragments is shown in Fig. 10-6. There is actually a minimum at half the mass; the nucleus is more likely to fission into unequal fragments of about 40% and 60% of the mass. Some of the important fission fragments which will be examined later have their high abundance as a result of this fact: ^{90}Sr, ^{137}Cs, ^{135}Xe, etc.

After the initial fission there is still more energy available. The neutron-rich fragments emit several neutrons and gamma rays almost instantaneously (~10^{-12} s) as they rearrange their internal structure to reach some temporary degree of stability. They still have a neutron excess, however, which makes them highly radioactive and over a long period of time will undergo a series of β-decays. These decays are usually accompanied by γ-emission until eventually two stable nuclei are reached. The intermediate nuclei in this process constitute most of the radioactivity hazard that results if a nuclear reactor should happen to leak.

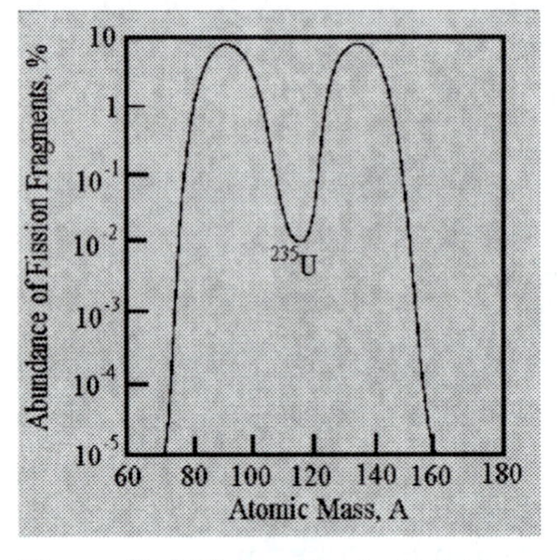

Figure 10-6 Fission Product Yields for Uranium-235

Fission Energy

Table 10-5 summarizes the various forms of energy released during the entire process of fission and subsequent decay of the fragments of ^{235}U. The total energy, 207 MeV, agrees very well with the 190 MeV predicted earlier. Of this energy, that of the antineutrinos that accompany the β decay (see Section 10.4) is lost, but all the rest will be captured and converted to heat. On the average then, about 200 MeV is available in practice from the fission of one uranium nucleus.

Table 10-5: Typical Energy Released in ^{235}UFission

Fragment Kinetic Energies	170 MeV
KE of 2 or 3 neutrons emitted	5
Prompt γ-rays (~5 gammas)	6
~7 β-particles	8
Fission Fragment Decay - Antineutrinos	12
-Gammas	6
Total	207 MeV

EXAMPLE 10-6.

How many fission events of uranium nuclei must occur to produce one joule of energy?

SOLUTION

$1\ \text{J} = 1\ \text{J}/(1.6 \times 10^{-13}\ \text{J}\cdot\text{MeV}^{-1}) = 6.25 \times 10^{12}\ \text{MeV}$
$= 6.25 \times 10^{12}\ \text{MeV}/(200\ \text{MeV per fission}) = 3 \times 10^{10}$ fissions

---

The number of fissions in Example 10-6 might appear to be a very large number but remember that there are a very large number of nuclei in even a very modestly sized piece of uranium.

EXAMPLE 10-7.

a) How many joules of energy are produced by the fissioning of all of the ^{235}U nuclei in 1 cm^3 of natural uranium (the size of a sugar cube)? The density of U is 19×10^3 kg/m^3, and ^{235}U constitutes 0.72% (by mass) of natural uranium.

SOLUTION

Mass of U = $1\times10^{-6}\ \text{m}^3 \times 19\times10^3\ \text{kg/m}^3 = 19\times10^{-3}$ kg.

Mass of ^{235}U = $0.0072 \times 19\times10^{-3}\ \text{kg} = 1.37\times10^{-4}\ \text{kg} = 0.137$ g.
Number of ^{235}U atoms = $0.137\ \text{g} \times (1\ \text{mole}/235\ \text{g}) \times (6.02\times10^{23}\ \text{atoms}/1\ \text{mole})$
$= 3.51\times10^{20}$
The energy produced is $3.51\times10^{20} \times 200\ \text{MeV} \times 1.6\times10^{-13}\ \text{J/MeV} = 1.1\times10^{10}$ J

b) How much water would this energy boil starting at room temperature?

$1.1\times10^{10}\ \text{J} = CM\Delta T + ML_v = [4186\ \text{J}\cdot\text{kg}^{-1}\cdot{}^\circ\text{C}^{-1}(100\text{–}20)^\circ\text{C} + 2.26\times10^6\ \text{J}\cdot\text{kg}^{-1}]M$
$M = 4.2\times10^3$ kg; over 4 tonnes!

Controlled Fission

It is interesting to compare the energies released by 1 kg of uranium and of coal. The heat of combustion of coal is 5600 Cal/kg so one kg of coal produces 2.3×10^7 J of energy. One kg of ^{235}U produces:
$[(1\ \text{kg}/0.235\ \text{kg}\cdot\text{mole}^{-1}) \times 6.02\times10^{23}\text{atoms}\cdot\text{mole}^{-1}] \times 200\ \text{MeV} \times 1.6\times10^{-13}\ \text{J/MeV} = 8\times10^{13}$ J. Because only 0.72% of uranium is ^{235}U then one kg of natural uranium produces $0.0072 \times 8\times10^{13} = 5.9\times10^{11}$ J. Weight for weight, uranium produces 25000 times more energy than coal. This number is not just a curiosity; among other environmental factors, it is an indication of how much less strain is placed on the transportation infrastructure by nuclear fuels compared with coal. In addition, since the mass of waste (solid and gaseous) produced by either coal-burning or nuclear fission is of the same order of magnitude as the mass of the fuel, this number indicates that the quantity of waste from fission also is much less than that from coal.

It can be seen in Table 10-5 that the energy released in the radioactive decay of the fission fragments is responsible for about 9% of the total energy. This is important in the case where the cooling of a reactor may fail. The safety systems will shut down the fission reactions but the fragments keep on producing energy by means of their radioactivity. In a 300 MWe reactor operating at 33% efficiency there is 1000 MW of heat being produced in the reactor core. If the reactor is shut down there will still be about 90 MW of fission-fragment heat. This is the most important reason for back-up cooling systems in nuclear reactors.

Uranium-238

In the case of ^{238}U, the very abundant isotope, the situation is quite different. Using Table 10-4 we see that if ^{238}U captures a neutron to form ^{239}U the total binding energy of ^{239}U is 1759 MeV whereas for ^{238}U it is 1754 MeV, a difference of only 5 MeV; this is not enough to cause fission. Nevertheless ^{238}U is not just a passive spectator in the ^{235}U fission process. It has a nuisance value in that it absorbs neutrons, removing them from availability to fission ^{235}U. This fact makes it very difficult (but not impossible) to have controlled fission in natural uranium. As a result, a large industry has grown up in some countries (particularly those with nuclear weapons programs) to enrich the concentration of ^{235}U in the uranium to be used in reactors.

There are several techniques used for this enrichment. The oldest technique, originally developed for weapons production, is to convert the uranium to its hexafluoride (UF_6) which is a gas. This is allowed to diffuse through a large number of small apertures. The $^{238}UF_6$ component diffuses a little bit more slowly than the lighter one, and after many such treatments an enrichment can be achieved. The process is very expensive, requiring immense plants to process the tons of ore necessary for the operation of reactors. The second enrichment option is high speed centrifugation of UF_6. About 60% of enrichment is accomplished using diffusion and about 40% by the centrifuge method. Research is proceeding to develop other methods that require less energy such as selective optical excitation in laser beams, but none are yet in the production stage.

When the ^{238}U nucleus absorbs a neutron, it is transformed into ^{239}U, which β-decays within minutes ($T_{1/2}$ = 23.5 min) to ^{239}Np which in turn β-decays ($T_{1/2}$ = 2.35 days) to ^{239}Pu. ^{239}Pu is a fissionable nucleus and so this process transforms the useless ^{238}U into useful fuel. This process is called *breeding*; it can be used in a reactor to produce more fuel than is actually consumed. (Breeding will be discussed in more detail in the next chapter). Some of the ^{239}Pu produced in a reactor undergoes fission and contributes to the energy produced by the reactor.

EXERCISES

10-1. Write nuclear reaction equations for the following processes:

a) The production of ^{14}C from ^{14}N in the atmosphere by interaction with cosmic-ray neutrons.

b) The decay of ^{226}Ra to ^{222}Rn by α emission.

c) The decay of ^{14}C by β-emission.

d) The production of tritium ($^{3}_{1}H$) in the atmosphere from normal hydrogen by cosmic rays (high energy protons from the Sun).

e) The decay of tritium to ${}^{3}_{2}He$.
f) The β-decay of ${}^{40}_{19}K$.

10-2. Alpha-particles emitted by nuclei generally have energies in the range 4 - 9 MeV. Calculate the corresponding speeds in m/s.

10-3. A simplified sketch of a deuterium atom is shown at the right. The open circle is a proton; filled circle is a neutrons; the dot is an electron in its orbit. Make similar sketches for: ${}^{3}_{2}He$, ${}^{4}_{2}He$, ${}^{6}_{3}Li$.

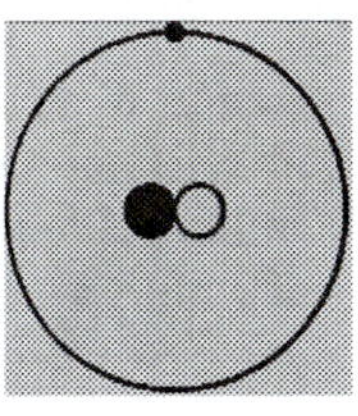

Figure 10-7

Exercise 10-3

10-4. How many neutrons are there in each of the nuclei ${}^{11}_{5}B$, ${}^{234}_{90}Ac$, ${}^{235}_{92}U$, ${}^{238}_{92}U$?

10-5. The decay constant of ${}^{232}Th$ is 5.0×10^{-11} yr^{-1}. What is its half-life? On the basis of this value for $T_{1/2}$ do you expect to find ${}^{232}Th$ in the Earth's rocks? Why?

10-6. The radioactive isotope ${}^{131}I$ has a half-life of 8 days. A sample is prepared on July 1 and placed in front of a Geiger counter where 20,000 counts are recorded in each second. It is left in place until July 25; how many counts per second are recorded on that day? (Think about it; it can be solved in your head!)

10-7. The stable nucleus ${}^{115}_{49}In$ absorbs a neutron to form a radioactive nucleus. Write the equation of the reaction.

PROBLEMS

10-8. The radioactive isotope ${}^{22}Na$ has a half-life of 2.60 y and ${}^{35}S$ has a half-life of 87.0 days. They each emit positive β-particles or positrons. A sample of each is prepared on June 1 and when placed on the window of a Geiger counter the ${}^{22}Na$ gives 5000 counts per second and the ${}^{35}S$ 20,000 counts per second. When will the two samples have the same activity, and what will the count rate be at that time?

10-9. Assume that when the earth was created there was the same amount of ${}^{235}U$ and ${}^{238}U$; the present ratio of ${}^{235}U$ to ${}^{238}U$ is 0.7%. The half-life of ${}^{235}U$ is 8.8×10^{8} yr. and for ${}^{238}U$ is 4.5×10^{9} yr. What is the age of the earth?

10-10. The biological activity of some organs can be investigated using radioactive isotopes. For example, iodine is important in the action of the thyroid. A subject is injected with a small amount of radioactive ${}^{131}I$ and after a few days the following measurements are made with a Geiger counter:
1. The counter is placed in contact with the abdomen and 2750 counts per second are observed. Six days later the counter records 600 counts per second.
2. With the counter over the thyroid, 5500 counts per second are observed and 6 days later 3290 counts per second are observed.
a) If the physical half life of ${}^{131}I$ is 194 hr, what is the biological half life of iodine in the bulk of the human body?
b) What can you say about iodine in the thyroid?

10-11. One estimate of stored radioactive waste in the USA in the year 2000 is 1.2×10^{21} Bq.[6] One problem is the heat generated, which may be sufficient to damage containers. Calculate the total heat generated per second by this waste if, on average, each disintegration yields about 1.0 MeV in the total kinetic energies of the emitted α-, β- and γ-rays? How does this compare with the power output of a large power station?

10-12. Radon is a radioactive, α-emitting gas which constitutes a hazard in uranium mines. Its half-life is 3.8 days, but it is continuously generated by the decay of a long lived radioactive parent, radium. How many α-particles are emitted in 1 minute from 5.0 cm^3 of radon at room temperature and standard atmospheric pressure? For radon the density of the gas at 20 °C and 1 atm. is 9.7 kg/m^3; Mass number = 222.

10-13. Suppose that a sample of radioactive waste consists of equal activities of ^{89}Sr, ^{137}Cs, and ^{106}Rh. Each of these isotopes decays into a stable daughter nucleus. After one year, what fraction of the original activity of the total sample remains, and which isotope(s) would be primarily responsible for this residual activity? The half-lives are: ^{89}Sr - 54 days, ^{137}Cs - 30 yr, ^{106}Rh - 30 s.

10-14. ^{90}Sr emits β-particles with a maximum energy of 0.546 MeV with a half-life of 28.1 yr. Its daughter is ^{90}Y which also emits β-particles with a maximum energy of 2.280 MeV and a half-life of 64.2 hr. After a few days equilibrium is established and there are the same number of Y and Sr nuclei decaying each second; it is just as if the Sr were emitting both β-particles. About 60% of the energy is taken away by the neutrinos and is lost leaving 40 % in the β-particles.
a) What is the activity of 100 mg of ^{90}Sr after it has achieved equilibrium?
b) What is the power developed by this source?

10-15. The old unit of radioactive activity, the curie (Ci) was defined as the activity (number of disintegrations per second) from one gram of radium ^{226}Ra. The half-life of ^{226}Ra is 1628 years. How many disintegrations per second are there in 1.0 Ci of radioactive material?

10-16. The neutron flux in a reactor is 1.0×10^{17} neutrons/($s\cdot m^2$) in the moderator and cooling water.
a) If the reactor uses ordinary water as coolant and moderator how many heavy hydrogen nuclei are produced in 1.0 year by neutron capture on $^1_1$H in each kilogram of water?
b) If the reactor uses heavy water (D_2O), how much tritium ($^3_1$H) is produced in one kilogram of water in one year? (Since the half-life of tritium is 12.3 yr., the decay of the tritium can be neglected.)
c) Tritium is radioactive with a half-life of 12.3 years. What is the activity of one kilogram of the reactor water after one year?

10-17. Consider the fission of ^{239}Pu according to the equation

$$^{239}_{94}\text{Pu} \rightarrow {}^{100}_{40}\text{Zr} + {}^{139}_{54}\text{Xe}$$

Calculate the fission energy released in this reaction. Use Table 10-4.

10-18. Using the nuclear masses given below, calculate the binding energy per nucleon for the nuclei of $^4_2$He, $^{12}_6$C, and $^{235}_{92}$U as given in Table 10-6.

---

[6] J.F. Ahearne, Physics Today, **50**, June (1997), p. 24.

Table 10-6

Nucleus	Mass u
$^{4}_{2}He$	4.002603
$^{12}_{6}C$	12
$^{235}_{92}U$	235.04394

10-19. If the nucleus $^{239}_{94}Pu$ absorbs a neutron what is the excess binding energy? Use Table 10-4.

10-20. A common neutron detector is made from a tube filled with boron trifluoride (BF_3). The isotope $^{10}_{5}B$, which constitutes 20% of natural boron, reacts with a neutron according to

$$n + {}^{10}_{5}B \rightarrow {}^{7}_{3}Li + {}^{4}_{2}He$$

What is the energy released in this reaction?

10-21. When a ^{235}U nucleus fissions, about 200 MeV of energy is released. What fraction of the mass of the nucleus is converted into energy?

10-22. The heat of combustion of carbon is 94 Cal/mole. What mass is converted to energy when one mole of carbon is burned? Express the mass loss as a percent of the initial mass.

10-23. The mass of the proton is 1.6726×10^{-27} kg and that of the neutron is 1.6750×10^{-27} kg. The mass of the $^{206}_{82}Pb$ nucleus is 341.92×10^{-27} kg. What is the binding energy per nucleon of the $^{206}_{82}Pb$ nucleus?

ANSWERS

10-2. 1.4×10^{7} m/s at 4 MeV
10-4. 6, 144, 143, 146
10-5. 1.4×10^{10} yr; yes
10-6. 2500
10-8. Dec. 9, 4350 counts per second
10-9. 8×10^{9} yr, 2 × too large
10-10. a. 100 hr, **b.** I is not eliminated from the thyroid
10-11. 192 MW, less
10-12. 1.7×10^{16}
10-13. 1/3, Cs
10-14. a. 1.05×10^{12} Bq, **b.** 0.095 W
10-15. 3.6×10^{10} s^{-1}
10-16. a. 7×10^{21} nuclei, **b.** 10^{19} nuclei, **c.** 1.8×10^{10} Bq
10-17. 2×10^{2} MeV
10-19. 6.6 MeV
10-20. 2.8 MeV
10-21. 0.09%
10-22. 4×10^{-8} %
10-23. 7.99 MeV

CHAPTER 11

NUCLEAR ENERGY: 20TH CENTURY

11.1 NUCLEAR POWER REACTORS

Nuclear reactors are devices for establishing and maintaining a chain reaction in a fissionable nuclear fuel, usually ^{235}U. While some are designed for research purposes, the ones that we are interested in here are those built for generating large amounts of thermal energy, which can then be converted to electrical energy.

Did you know? The first attempt to construct a nuclear reactor was made in Canada by Dr. G. Laurence in 1940. It failed because of a lack of pure materials.

The essential features of a nuclear reactor are shown in Fig. 11-1. The reactor core consists of a chamber containing a critical mass of the *fuel elements* surrounded by a *moderator*. Through the core a cooling fluid or *coolant* (liquid or gas) is passed to remove the heat which is developed. Not shown in the drawing are the control devices necessary to operate the reactor and the safety and backup systems necessary for safe operation. Finally, of course, the reactor is enclosed in massive shielding to contain the ionizing radiation.

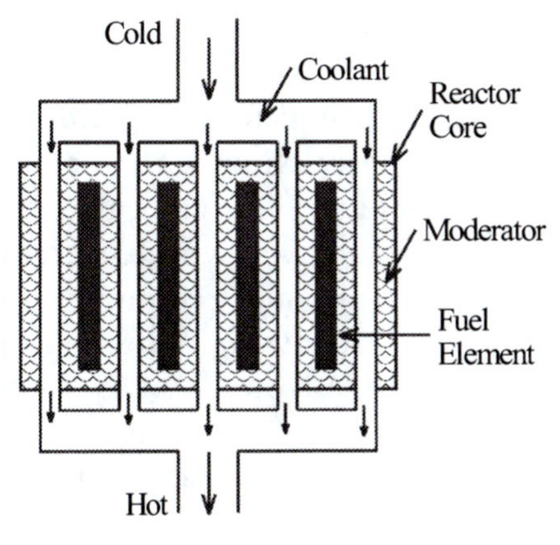

Figure 11-1 Nuclear Reactor

In the following sections all these systems will be discussed separately.

Fuel elements

As we saw in the last chapter, most reactors operate with uranium whose ^{235}U content has been enriched from the natural level of 0.7% to between 3 and 4%. The UF_6 from the enrichment process is converted into solid pellets of UO_2, a refractory material that withstands high temperatures. The pellets, about 1 cm long, are then encased in metal alloy fuel rods, and a number of fuel rods are collected together in a fuel assembly (or bundle) that will be inserted into the reactor core. A Candu reactor fuel bundle is shown in Fig. 11-2.

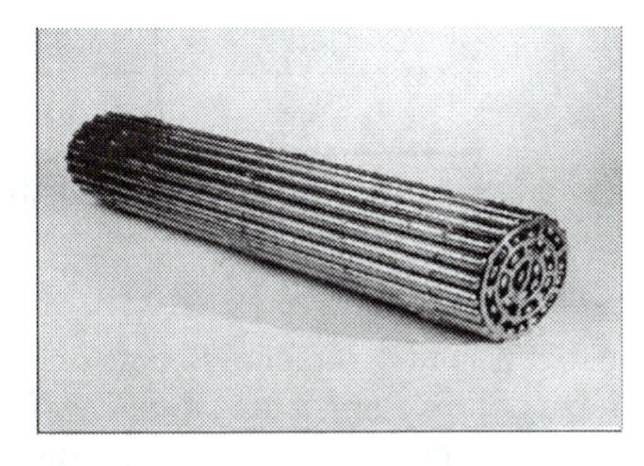

Figure 11-2 Nuclear Fuel Bundle

Moderator

As we already know, fast neutrons are ineffective in producing fission in ^{235}U; only slow, or *thermalized*, neutrons are effective. On the other hand the fission process, which produces the neutrons in the first place, emits fast neutrons with MeV energies. Some method must be found to slow the neutrons down so that most of them are neither absorbed nor lost by escaping. This process is called *moderation.*

The neutrons are slowed down by allowing them to collide with other nuclei and give up some kinetic energy at each collision. Most collisions of neutrons with other nuclei are perfectly elastic, that is, momentum is conserved in the collision and the neutron loses energy only by transferring it to the colliding partner. (In other words, kinetic energy is also conserved by the colliding pair.) If we apply conservation of momentum and kinetic energy to the simple case of a *head-on* collision between a neutron (mass = 1, speed = v) and a nucleus (mass = A, speed = 0) it is easy to show (see Problem 11-5) that the *moderating ratio* of the initial and final kinetic energies of the neutron is given by

$$KE_{\text{final}}/KE_{\text{initial}} = [(A-1)/(A+1)]^2 \qquad \textbf{[11-1]}$$

From this equation it follows that effective moderators (i.e., those which make this ratio small) have small values of A. The most effective moderator is hydrogen with $A = 1$ giving a value of zero for the ratio, that is, the neutron can lose all its energy in only one (head-on) collision. The value of the moderating ratio is given for several nuclei in Table 11-1 along with the most common practical form of the moderator. The last column of the table gives the average number of collisions that is necessary in practice to thermalize the neutron. Why is the number for hydrogen not 1 as mentioned above? This is because very few of the collisions are head-on; this number includes the average over all the angles of impact, a much more complicated analysis.

Table 11-1: Moderating Ratio

Nucleus	Common form	Ratio	No. of collisions to thermalize
Hydrogen	Normal Water	0	19
Deuterium	Heavy Water	0.11	32
Carbon	Graphite	0.72	110

Hydrogen is not a feasible choice because it is a gas; the moderator must be liquid or solid to achieve sufficient density of material to do its job. Water then becomes an obvious choice in view of its hydrogen content, purity and low cost. However, not all reactors use water, because the choice of moderator is in practice intimately linked to the choice of fuel – enriched or natural – and to the choice of coolant. Water has the advantage of being able to do double duty, serving also as the coolant for the reactor. Unfortunately, the hydrogen nucleus, which is a proton, has a rather large cross section for reaction with the neutron to produce a deuteron (see Section 10.9 and Problem 10-16) according to the equation,

$$ {}^{1}_{1}H + {}^{1}_{0}n \rightarrow {}^{2}_{1}H \qquad \textbf{[11-2]} $$

Thus, although water is an efficient moderator, it also removes sufficient neutrons from the reactor core as to make it impossible to achieve a chain reaction with natural uranium, which has only 0.7% of fissionable ^{235}U. If water is to be used, the amount of ^{235}U in the fuel must be increased, i.e., enriched fuel is required.

Deuterated or heavy water (D_2O) is the ideal moderator for a reactor that uses *natural* uranium as its fuel, because the cross section for the reaction

$$ {}^{2}_{1}H + {}^{1}_{0}n \rightarrow {}^{3}_{1}H \qquad \textbf{[11-3]} $$

is very small and so neutron loss is small. The disadvantage is that heavy water is expensive because of the large capital investment required in equipment for its separation from normal water and the large running costs of this energy-intensive process. Another problem lies in the end product of Eq. [11-3]. A contamination of ${}^{3}_{1}H$, or tritium, slowly accumulates in the form of tritiated water DTO. Because tritium is radioactive and water is ubiquitous in living organisms, the tritium must be carefully controlled or removed.

Graphite is not so effective a moderator as water but it has the advantages that it is inexpensive, is structurally useful in the reactor since it is a solid, and does not strongly absorb neutrons (particularly if it is very pure). Enrico Fermi's very first reactor constructed at the University of Chicago in 1942 used a graphite moderator. Its disadvantages are that it will burn in air and cannot also serve as the reactor coolant.

Coolant
Just as in a fossil fuel electric generator, the coolant's role is to transfer the heat from the reactor core to the turbines. Obviously the coolant has to be a liquid or a gas. We have already seen that both normal and heavy water are effective coolants that can serve simultaneously as moderators. Some reactor designs employ gases (helium or CO_2) as coolants; these obviously have insufficient density to also serve as moderators.

11.2 LIGHT WATER REACTORS.

Most of the power reactors in the world are of this type. These include many of the reactors of the USA, France, Great Britain, Germany, Russia and Japan. These reactors are mostly based on original USA concepts. Any country choosing this option is forced to develop isotope enrichment technology or to buy fuel from some country which has. This technology is closely related to isotope separation technology essential for nuclear weapons, so it is not surprising to find that all the nuclear-weapons powers have light water reactors. Light water reactors are of two basic types: the simpler Boiling Water Reactor (BWR) and the more sophisticated Pressurized Water Reactor (PWR).

Boiling Water Reactors.

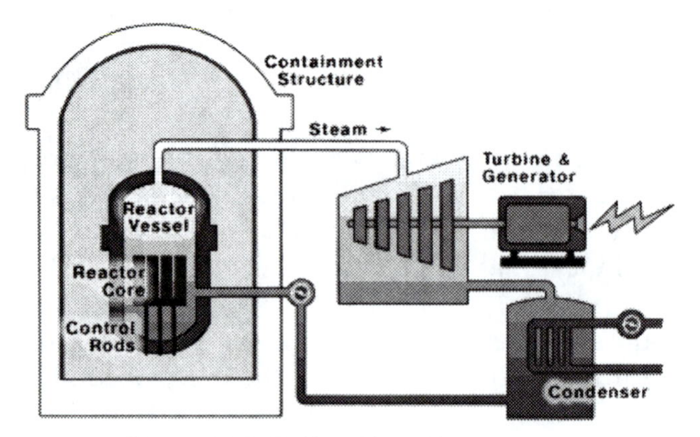

Figure 11-3 Boiling Water Reactor
Courtesy, Nuclear Energy Institute

A schematic drawing of a BWR is shown in Fig. 11-3. Fuel elements in the form of long rods containing enriched uranium are held in an open array in a stainless steel tank filled with water. The water, which acts as both moderator and coolant, is heated by the fission process and boils. The steam at a pressure of about 7 MPa (70 atmospheres) and temperature 285 °C is passed, after drying, to the turbine to drive the generators. It is then condensed, usually with cooling water from a river or lake, and pumped back into the reactor vessel.

Pressurized Water Reactors.

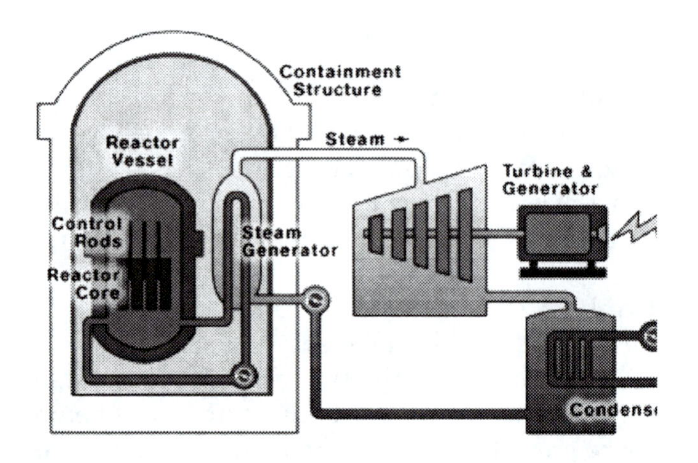

Figure 11-4 Pressurized Water Reactor
Courtesy, Nuclear Energy Institute

In order to increase the efficiency of a power plant the temperature of the steam must be raised. The only way to do this is to raise the pressure. The schematic of a PWR is shown in Fig. 11-4. The reactor core is contained in a vessel which operates at a pressure of 14 MPa (140 atmospheres). At this pressure the primary coolant water is prevented from boiling and remains liquid at 320 °C. This hot pressurized water is pumped through a heat exchanger where it gives up its heat to the secondary coolant loop making steam for the turbines. In the PWR the primary coolant water does not leave the reactor vessel and so keeps radioactive contaminants inside.

In both the BWR and the PWR, the vessel can be opened only from above; this is done about once per year to renew the fuel bundles. Both types of pressure vessel are housed within a steel containment vessel, which in turn is inside a reinforced concrete shield. Early reactors were in the 400-800 MWe range but newer ones are usually in excess of 1000 MWe.

Because the steam temperatures are relatively low, light water systems generally have quite low efficiency, typically 30-33%, compared with about 40% for a fossil fuel plant. This results in greater shedding of heat to the water body used to condense the steam. For a 1000 MWe plant operating at 30% efficiency about 50 m^3 per second of water is drawn and raised through 10 °C before being discharged. On the other hand there are no emissions of sulphur or nitrogen oxide, or particulates, and to supply the fuel only a few truckloads of uranium are needed per year as opposed to trainloads of coal per day.

Light water reactors continuously release radioactivity to the environment. Most of the fission products remain within the fuel elements and only small quantities of radioactive material diffuse out into the water; these include various inert gas isotopes and tritium. A further contribution arises from traces of corrosion products and other impurities in the water; these include iron, cobalt, and manganese. Solids are continuously removed by filtration, dissolved materials by passage through ion exchange resins, and dissolved gases by gas strippers. In a BWR several short-lived rare gas isotopes are carried by the steam to the turbines; these are held back for a period of time to allow the shortest lived ones to decay before being released up a stack.

11.3 PRESSURIZED HEAVY WATER REACTORS (CANDU)

The Canadian heavy water reactor is very different in concept from the enriched-uranium, light-water reactors. Instead of one vertical pressure vessel containing all the moderator-coolant, it has several hundred small horizontal pressure tubes in a honeycomb structure called a *calandria* which, in turn, is contained in a concrete vault as shown in Figs. 11-5 and 11-6. The heavy water coolant and moderator are separate systems; the pressure tubes are surrounded by moderator but coolant heavy water flows at a pressure of 10 MPa through the pressure tubes which contain the natural uranium fuel elements. The coolant leaves the pressure tubes at a temperature of 310 °C and its heat is transferred in a heat exchanger to normal water to make steam. A small annular space on the outside of the pressure tubes carries a flow of cooled nitrogen gas to prevent heat transfer from the cooling system to the moderating system.

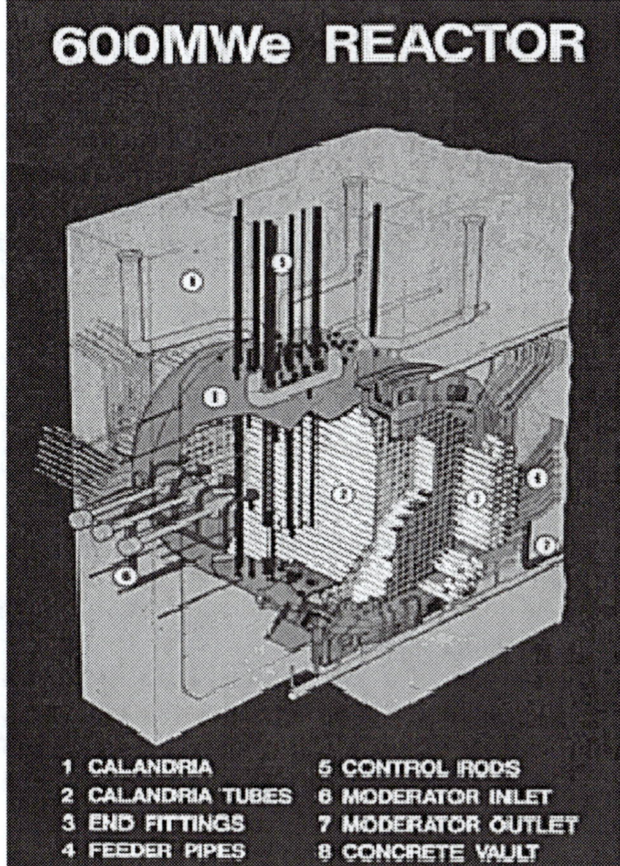

Figure 11-5 Candu Reactor

Courtesy, Canadian Nuclear Association

Earlier versions of the CANDU could stop the fission reaction in 30 seconds in an emergency by dumping the moderator into a tank below the reactor. Unlike the light water reactors, depriving a CANDU of its moderator does not deprive it of its cooling through the pressure tubes, but the moderating power of the cooling water is not enough to sustain the chain reaction. Newer versions such as those at Darlington Ontario retain the moderator fluid for additional cooling in an emergency but inject a solution of gadolinium nitrate into the moderator to "poison" it. Since gadolinium is strong absorber of neutrons (see Table 10-3) this method effectively, and permanently, stops the nuclear reactions.

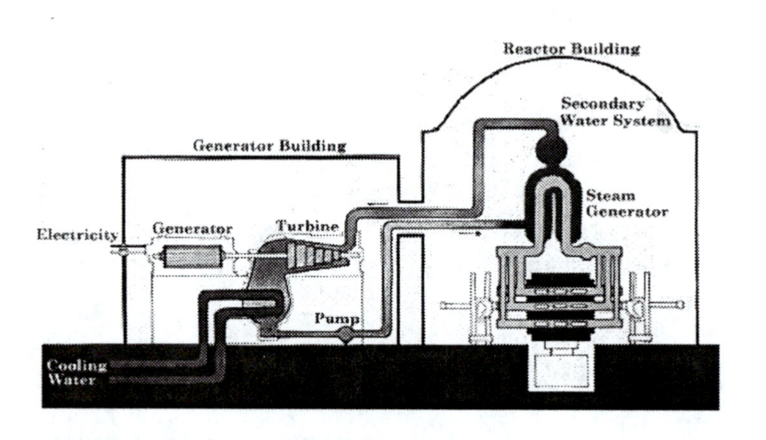

Figure 11-6 Candu Reactor. *Courtesy, Canadian Nuclear Association*

Each of the fuel bundles in the pressure tubes can be replaced independently without shutting down the reactor. The fuel elements are even moved from tube to tube in a planned sequence to maximize the utilization of the fuel. Partly as a result the CANDU has the lowest fuelling cost of any reactor system.

Other countries that also operate heavy water reactors include India, South Korea, Argentina and Romania; almost all are based on the Canadian technology.

11.4 GRAPHITE REACTORS

Power reactors using ultra-pure graphite (carbon) as the moderator have been used in several countries, particularly Britain and Russia. Of course some fluid must also be used as the coolant and the British reactors use carbon dioxide (CO_2) gas at high pressure. Hot CO_2 is corrosive to many materials and so the fuel elements must be clad in a-non corrosive material capable of withstanding high temperatures. The first generation of British reactors used a beryllium/magnesium alloy called "Magnox" which lent its name to the reactor. The CO_2 coolant enters the reactor at about 250 °C and leaves at about 400 °C. at a pressure of 2.8 MPa.

A second generation of British reactors called "AGR or Advanced Gas Cooled Reactor" was developed, which had an upper temperature of 675 °C at a pressure of 3 MPa. This was achieved with the use of slightly enriched uranium fuel elements. The efficiency of these reactors was as high as 42%, but the UK ended construction of AGRs in favour of PWRs.

In the United States work was done on high temperature gas-cooled reactors (HTGR) but this model was not taken to the level where it could be installed in the power grid. This design uses elaborately clad fuel elements in an ultra-pure graphite moderator. The cooling is by means of helium gas which is raised to a temperature of 770 °C. The efficiency approaches 40%. The safety of the HTGR is enhanced by its materials and construction. There is no metal in the core; all structural elements are graphite which increase in strength as the temperature rises. Nothing in the core will melt or vaporize at temperatures up to 3600 °C.

Russian reactors of the type (RBMK) formerly used at Chernobyl in the Ukraine use graphite as the moderator. Water is passed directly over the fuel elements to turn it into steam for use in the turbines, eliminating the need for heat exchangers to extract the heat from the reactor. These reactors were designed to produce plutonium as a by-product of electricity.

11.5 BREEDER REACTORS

In a conventional reactor some of the ^{238}U is converted to ^{239}Pu which is fissionable. After a year or so of operation of a CANDU reactor more of the heat produced actually comes from the ^{239}Pu than the ^{235}U, since the latter is depleted whereas the neutron flux continuously builds up the former. This suggests the idea of building reactors that produce more fuel than they consume. There are two so-called "fertile" materials that can be used: ^{238}U and ^{232}Th. The reactions are

$$^{238}_{92}U + n \rightarrow {}^{239}_{92}U \rightarrow {}^{239}_{93}Np + \beta^- \rightarrow {}^{239}_{94}Pu + \beta^- \quad [11\text{-}4]$$
$$T_{1/2} = 24 \text{ min} \qquad T_{1/2} = 23 \text{ d}$$

$$^{232}_{90}Th + n \rightarrow {}^{233}_{90}Th \rightarrow {}^{233}_{91}Pa + \beta^- \rightarrow {}^{233}_{92}U + \beta^- \quad [11\text{-}5]$$
$$T_{1/2} = 22 \text{ min} \qquad T_{1/2} = 27 \text{ d}$$

Breeding can be achieved if 1 fission event releases, on average, 1 neutron to maintain the chain reaction, about 0.2 neutrons to account for escape and absorption losses, and 1 more for breeding, for a total of 2.2. Fig. 11-7 shows a graph of the number (n) of neutrons per fission as a function of the neutron energy.

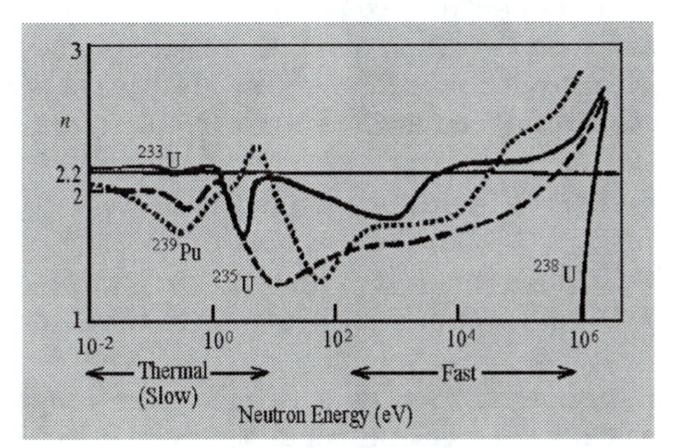

Figure 11-7 Number of Neutrons per Fission

In the low energy region (thermal neutrons) the curves

are mostly at, or below, the critical value of 2.2. This means that breeding with thermal neutrons is inefficient. However at high energies the curves, for all reasonably fertile materials, rise rapidly above 2.2 with increasing energy and breeding becomes efficient. The obvious choice for a fertile material is ^{238}U, the large portion of natural uranium that does not contribute substantially to the operation of thermal reactors.

Unfortunately, breeding cannot be done in a moderated reactor since the moderator quickly thermalizes the neutrons. A special "fast" reactor which operates with fast neutrons must be constructed to do this. Of course, water cannot be used as a coolant in these reactors since water is effective in thermalizing neutrons and that would defeat its breeding with fast neutrons. The coolant to date has been liquid sodium metal which has given rise to considerable controversy as regards safety; sodium burns spontaneously on contact with water.

The operation of fast breeders is much more controversial than thermal power reactors because of their reliance on plutonium, a highly toxic substance and a nuclear weapons material. The two experimental fast breeder reactors at Dounreay in Great Britain have been shut down, and the US program was terminated in 1983. In France, the Super-Phenix at Creys-Malville was shut down in 1998 after a minor incident which caused public concerns about the fire hazard. The original Phenix reactor continues to generate electricity and is also used for studies of the transformation of the waste material into shorter-lived isotopes. The only other fast breeder currently generating electricity is in Russia.

Breeding of thorium represented by Eq. [11-5] has been investigated in a more limited way, theoretically and experimentally, mostly in Canada and India. With the shut-down of fast breeder programs, interest in slow thermal reactors with thorium breeding is re-emerging.

11.6 REACTOR CONTROL AND STABILITY

As discussed in Section 11.5, each fission event produces, on average, more than one neutron. These fast fission neutrons are thermalized in only 10^{-4} seconds, denoted by τ. This increase of neutrons could apparently cause the power of the reactor to rise very rapidly as each generation of neutrons produces even more in the next generation. A reactor is controlled by adjusting the absorption of neutrons through control-rods of neutron-absorbing materials which are inserted into, or withdrawn from, the core.

We can define the multiplication factor k as

$$k = \frac{\text{Number of prompt neutrons in a given generation}}{\text{Number of prompt neutrons in the prior generation}} \qquad \textbf{[11-6]}$$

For constant power, k would need to be maintained at a value of 1.0. For k values just in excess of 1.0, the neutron population will grow as $N = N_o\, k^{\,n}$ where n is the number of generations. Let us define the Reactor Period as the time T in which the neutron population (and the power) grow by a factor e. Then $n = T/\tau$. Using the exponential growth equation, we can write

$$\ln (N/ No) = n \ln k$$

and so, using Taylor's expansion of the logarithm for values just above 1.0,

$$1 = (T/\tau)\,(k - 1)$$

or

$$T = \tau \,/\, (k - 1) \qquad \textbf{[11-7]}$$

It appears then that with a k value, for example, of 1.001 (and $\tau = 10^{-4}$), the reactor period T would be 0.1 s, and then the power would double every 0.07 s. If changes in the neutron population can take place this quickly then how is it possible – using slow-moving mechanical devices – to control them? Fortunately only 99.3% of the neutrons are the so called "prompt" neutrons; 0.7% are delayed, on average, for 14 seconds, which gives an average neutron lifetime τ of 0.1 s [$(99.3 \times 10^{-4} + 0.7 \times 14)/100 = 0.1$]. It is thanks to these "delayed" neutrons that mechanical control can be exercised. With this effective value of τ, we now find that the period becomes $0.1/(1.001 - 1) = 100$ s.

A very important aspect of the different nuclear reactors is their stability in power output when fluctuations occur in certain parameters. For example what happens in a reactor in the event that there is a local upward fluctuation in temperature? Will the reactor naturally damp the fluctuation out, or will it require outside intervention to control it? If it is the latter, then it is important that the fluctuations be slow enough that the mechanical devices (control-rods etc.) can operate in time. Compare for a moment the light water reactors and the water-cooled graphite-moderated reactors, as this is important for an understanding of the Three Mile Island and Chernobyl accidents discussed in Appendix VIII.

PWR and BWR reactors are inherently stable against temperature fluctuations. If the temperature should rise suddenly, the water decreases in density or even boils so that there are bubbles or voids. This reduces the moderating effect and with fewer thermalized neutrons the fission rate decreases. The reactor is said to have a *negative void coefficient.* This stability is so absolute that research reactors have been built in which the control rod is driven out very rapidly by compressed air so that the reaction rate will rise rapidly, however the reactor just as quickly shuts itself down. The objective here is to produce, in a perfectly safe manner, short but intense pulses of neutrons for research purposes.

A Chernobyl-type reactor is moderated by graphite but cooled by water flowing directly over the fuel rods. There is not enough water in the reactor at any one time to contribute significantly to the moderation but it does act as a significant *neutron poison*, removing neutrons by the reaction of Eq. [11-2]. If the temperature should rise and the water boil, this poison is reduced and the neutron flux will increase thus increasing the fission rate. Such a reactor is inherently unstable to temperature fluctuations and has a *positive void coefficient.* So long as the time constant of the reactor is maintained sufficiently long by keeping it critical on the delayed neutrons then the reactor can be operated in perfect safety. If however something is done to make the reactor *prompt critical* then a fluctuation will grow and outrun any mechanical control (see the discussion of the Chernobyl accident in Appendix VIII).

11.7 SAFETY SYSTEMS

The safety of power reactors is discussed specifically in Chapter 12. Safety systems in reactors are not there to prevent nuclear bomb-type explosions; that type of accident is impossible at the low ^{235}U enrichment levels used. The various types of failure that can occur can have only one final result of public importance and that is the release of radioactive material. Safety systems are designed to eliminate or minimize this.

All reactors are designed with safety systems to cope with every conceivable type of failure. The most serious accident that can happen to any power reactor is to have a *loss of coolant accident* (LOCA). For light water reactors, with their negative void coefficient, there is no danger of the reactor running out of control; the chain reaction can be shut down immediately. The necessity is to provide continued cooling of the core to remove the heat of radioactivity of the fission fragments, otherwise the build-up of heat in the core can melt or warp some of the fuel elements and structural members. The initial radioactive thermal power is about 7% of the operating thermal power. So for a typical 1000 MWe reactor the initial heat level is about 200 MW. After one day this falls to about 16 MW and after five days to 9 MW; these figures demonstrate the importance of *emergency core cooling systems* (ECCS) after any emergency shut-down. These are discussed for both present and future reactors in Chapter 13.

11.8 URANIUM RESOURCES

Uranium is a rather common element in Earth's crust. It is present in all rocks and soils as well as rivers and oceans. Granite, which accounts for about 60% of Earth's crust contains about 4 parts per million (ppm) and some rocks have as much as 400 ppm; this is particularly true of some phosphates which are used for fertilizers. Coal contains about 3 ppm, a not insignificant amount considering the enormous volume of coal which we mine and burn, putting many contaminants (including uranium) into the atmosphere in the process.

Natural uranium exists as the oxide U_3O_8, and any ore with a concentration greater than 1000 ppm is considered economic to mine for its uranium content, although some of the most spectacular deposits, in Canada, range up to 20%. Uranium is mined by both shaft and open-pit techniques. It has all the difficulties, hazards and environmental costs of any hard-rock mining. One added hazard is a higher concentration of radon gas, particularly in underground shaft mining. As a result, uranium mines are copiously ventilated and workers use respirators in some cases. The total amount of radon ventilated from uranium mines is negligible however compared to the natural out-gassing of the continental rocks into the atmosphere.

A uranium mine produces two kinds of waste: rock and water. The rock is deposited in controlled areas and contains no uranium and is no more radioactive than other types of rock. The water used to wash the ore does have traces of uranium and radium. The radium is valuable so the water is usually treated with barium chloride to remove it and the treated water is mixed with the discarded part of the ore called *mill tailings*.

Extensive data on uranium production, requirements and reserves are provided and regularly updated by the Nuclear Energy Agency (NEA) of the Organization for Economic Cooperation and Development (OECD). From its 2005 report, the yearly requirement and production of uranium by various countries or areas in the world is assembled in Table 11-2. The requirement is roughly proportional to the amount of nuclear-generated power. As expected, the US is the largest generator of nuclear power although it only accounts for 20% of domestic electricity. France (75%) and Japan (36%) are the next largest producers with all the other countries producing significantly less.

Table 11-2: "Reasonably Assured Resources" (RAR), Production and Use of Uranium [1]

	RAR (units of 1000 tonnes)	Production in 2004 (t)	Tonnes used in 2004
Africa	530	6284	280
Americas			
Canada	345	11957	1700
USA	102	878	24145
Rest	164	300	750
Asia			
China	38	730	1260
Japan	0	0	7140
Khazakstan& Uzbekhistan	438	5806	0
Rest	46	270	3505
Australia	714	8982	0
Mid-East	30		0
Europe			
France	0	6	7185
Russian Fed.	132	3280	4740
Ukraine	58	800	2220
Rest	29	583	13565
TOTAL	2626	40263	66490

The production of uranium follows a completely different pattern as seen in the table. By far the largest producer is Canada, which supplies close to one-third of the world's uranium. Few nations, in fact, are self-sufficient in uranium and the largest users (USA, Japan, Europe) must import all or most of their supply. Australia does not generate nuclear electricity but is a major exporter of natural uranium. Except for the uranium used in Canada's CANDU reactors almost all of the remainder must be enriched in the ^{235}U content. If the user country does not have enrichment facilities (and only a few do) then the uranium must be exported to a country which does

[1] "*Uranium 2005: Resources, Production and Demand*, OECD Nuclear Energy Agency and IAEA, Vienna (2005).

enrichment and then re-exported to the final user; Canada and Australia do not have enrichment facilities.

At the present time, newly mined uranium is providing about 60% of the world's needs. The remaining 40% is coming from reprocessed fuel and from the dis-assembly of nuclear weapons.

Turning to uranium reserves and resources, the classifications used by the NEA are very complex. The numbers shown above as RAR are "Reasonably Assured Resources" with a recovery cost of (US) $80 per kg. The NEA also provides estimates of "Inferred Resources" (IR), and they sum RARs with IRs to provide a quantity called the "Identified Resource;" all these data are then presented for different upper price limits.

Table 11-3: Various NEA Estimates of World Uranium Resource [a]
Units are 1000 tonnes

Price US$/kg	RAR	IR	Identified =RAR+IR
< 130	3297	1446	4743
< 80	2643	1161	3804
< 40	> 1947	> 799	> 2746

[a]: Ref. 1

It can be seen that the range in estimates is considerable, but this approach does have the advantage of demonstrating the sensitivity to price, i.e., the influence of demand upon supply.

The NEA also presents estimates of "Undiscovered Resources", including "Prognosticated Resources" and "Speculative Resources"; we are not summarized here, but together they comprise about 10,000 units, i.e., 10^7 tonnes. The most simplistic summary of the situation is to say that at the $80 price, the identified resource in the table would meet the current demand for about 60 years; the total identified and undiscovered resource would meet a demand 3-4 times greater for the same period.

11.9 THE NUCLEAR FUEL CYCLE

The complete fuel cycle for a typical power reactor using Lightly Enriched Uranium (LEU) is shown in Fig. 11-8. (A CANDU reactor would omit the conversion and enrichment processes.) When a fuel element has been so poisoned in the reactor that it is absorbing more neutrons than it produces, it is removed. Its useful life, however, is not necessarily over. After a mandatory storage period to permit the short-lived radioactive isotopes to decay, the element may go to longer-term storage (and eventual disposal) or be reprocessed.

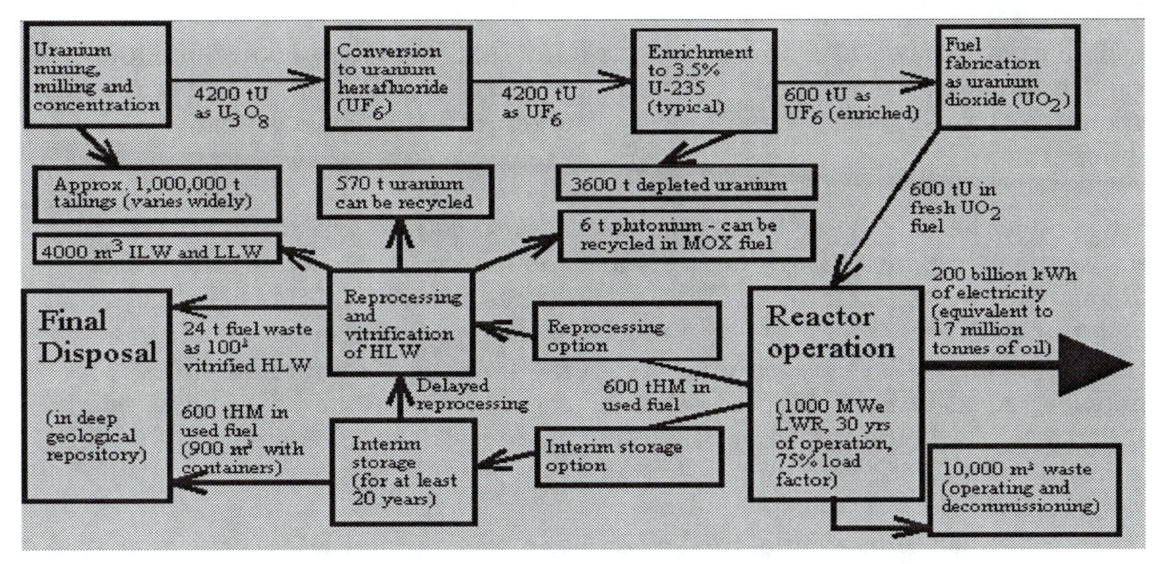

Figure 11-8 The Uranium Fuel Cycle. *Courtesy, Uranium Institute.*

Figure 11- 8 indicates that in practice the complete "burn-up" in an LWR of 600 tonnes of uranium enriched to 3.5% provides 2×10^{11} kWh of electrical energy. If we calculate (see Problem 11-7) the thermal energy released and assume a conversion efficiency of 30%, we find a smaller value, viz., 1.4×10^{11} kWh. Is this a discrepancy? We appear to get more thermal energy than we would expect from fissioning every single uranium nucleus. In fact not all the ^{235}U is "burned" and a considerable amount of ^{238}U is converted to ^{239}Pu which undergoes fission and provides more energy. So there is no discrepancy.

Reactor designers define a basic unit for burn-up as the gigawatt-days (GWd) of thermal energy obtained per tonne of uranium used up. The unit is GWd/t and you can show that 1 GWd is 8.64×10^{13} J. Complete burn-up of a tonne of 3.5% enriched uranium would provide 33 GWd. With recent improvements in reactor design and operation the actual values of burn-up have reached about 36 GWd/t for BWRs and 43 GWd/t for PWRs. [2]

From the beginning of the nuclear-power era it was recognized that spent fuel elements, if reprocessed, were a valuable source of additional energy. Reprocessing of spent fuel first of all removes the non-fuel radioactive fraction (mostly fission fragments) which must be put into a waste disposal stream. The fissionable plutonium which has been manufactured in the fuel but not yet fissioned can be used to make *Mixed Oxide* (MOX) fuel elements, i.e., fissionable plutonium in an enriched uranium oxide matrix, and these can be used directly in the reactor in place of LEU fuel. The remaining uranium portion, stripped of its poisons and then re-enriched can also re-enter the fuel cycle as so-called RepU fuel. These two approaches offer the possibility to displace nuclear fuel fabricated from fresh uranium.

Despite the growing stock of spent fuel, this aspect of the fuel cycle has not developed as rapidly as was at first anticipated due to many factors: environmental, technical, economic, and political. Reprocessing is carried out in only a handful of countries and

[2] D. Bodansky, Nuclear Energy: Principles, Practices and Prospects, Springer, New York (2004) p. 207

has a small impact on world uranium production in that it displaces only 2000 tonnes of natural uranium each year, mainly in France and Germany. Reprocessing plants are in operation in France, Japan, UK and Russia. The reprocessing issue is further discussed in Chapter 13.

After the end of the Cold War, the US and Russia entered into agreements to reduce nuclear weapon stockpiles. The plutonium and the highly enriched uranium (HEU) from dismantled warheads form a valuable energy resource. The Pu is being used to manufacture MOX fuel as in the case of fuel reprocessing described above. The two countries agreed in 2000 that each would convert 34 tonnes of Pu into MOX fuel elements for use in specially adapted reactors. A pact between the USA and Russia in 1993 provides a market in the US for LEU fuel made by "down-blending" military HEU with natural or depleted uranium.

As of early 2006, 262 tonnes out of the 500 tonnes of HEU promised by Russia had been down-blended, eliminating 10467 warheads and providing 7670 tons of LEU fuel. Of the 151 tonnes promised by the USA, 73 had been dealt with by 2005. In that year the USA undertook to remove another 200 tonnes of HEU from weapons, using most of it instead for powering navy ships.

Because of the enrichment in ^{235}U required for most of the world's reactors there is a very large inventory of depleted uranium, i.e., ^{238}U with the concentration of ^{235}U reduced from the natural level of 0.72% to about 0.2 to 0.3%. Depleted uranium has 4/3 the density of lead, is only slightly radioactive, and has about three times the chemical toxicity as lead. There is no problem associated with its storage but its chemical and physical properties have not yet produced a substantial market. About 1000 tonnes per year are used for radiation shielding, counterweights, and conventional weapons; all these uses simply take advantage of its high density.

PROBLEMS

11-1. For a reactor that is critical with delayed neutrons find the value of k that gives a reactor response time of 100 s.

11-2. (a) What mass of natural uranium would be required to operate a pressurized heavy water reactor for one day based solely on the fission of ^{235}U and ignoring any Pu isotopes that are created? The reactor produces 1.00×10^3 MW of electrical power at 30% efficiency. Assume that the energy available from one fission event is 200 MeV. Natural uranium is 99.28% ^{238}U and 0.72% ^{235}U, by mass.
(b) Repeat this problem using a LWR and enriched uranium (3.5% ^{235}U) with the same assumptions.

11-3. In the reactor of problem 11-2(a), how many radioactive fission product nuclei are created in 1 day?

11-4. A reactor produces 3500 MW of heat.
a) How many fissions are occurring each second? Assume 200 MeV (3 significant digits) per fission event.

b) Actually 16 of every 100 neutrons captured by ^{235}U do not produce a fission event but nevertheless use up a fissionable nucleus. How many atoms of ^{235}U are consumed each second?
c) For every 10 atoms of ^{235}U consumed, 3 atoms of ^{239}Pu are bred. How much plutonium is produced in one day?

11-5. A neutron (mass=1) collides head-on, at speed v, with a nucleus of mass A at rest. The neutron rebounds with speed v'. If the collision is elastic (energy is conserved) show that $v/v' = (A-1)/(A+1)$, thus verifying Equation [11-1].

11-6. The nuclear fuel cycle depicted in Fig. 11-8 shows that 4200 tonnes of natural uranium at 0.72% ^{235}U is enriched to 600 tU at 3.5% ^{235}U. The depleted 3600 tU has not had all of its ^{235}U removed. What percent ^{235}U remains in the depleted uranium? This is called the "tailings assay".

11-7 (a) Consider the 600 tons of enriched uranium (3.5% ^{235}U) that is used as fresh fuel in Fig. 11-8. If all the ^{235}U content undergoes fission and if the Pu that is created is ignored, what would be the total energy produced in kWh, assuming an efficiency of 0.3? Why is your result somewhat less than the number of kWh stated in the figure?
(b) Show that the total electrical energy output stated in Fig. 11-8 is compatible with the other data there, viz., 1000 MWe for 30 years with a 75% capacity factor.

11-8. In 2000 there was 350 GWe of installed nuclear power capacity in the world and most of these reactors used LEU fuel enriched to an average of 3.5%. Assume they all used 3.5% enriched fuel, take the average thermal efficiency to be 30%, and take the average capacity factor to be 80%. Assume that all the thermal energy is derived from ^{235}U and none from Pu isotopes (these two assumptions balance off approximately).
(a) How many tonnes of uranium metal were required in that year?
(b) Use Fig. 11-8 to ascertain what is the equivalent mass of U_3O_8 ore that was consumed.
(c) Why do you think your answer for (a) is somewhat greater than the value in Table 11-2.

ANSWERS

11-1. 1.001
11-2. (a) 486 kg; (b) 100 kg
11-3. 1.8×10^{25}
11-4. **a.** 1.09×10^{20}, **b.** 1.3×10^{20}, **c.** 1.33 kg
11-6. 0.26%
11-7 1.43×10^{11} kWh
11-8 (a) 10300 t; (b) 72000 t

CHAPTER 12

HEALTH AND ENVIRONMENTAL IMPACTS OF NUCLEAR ENERGY

When alpha, beta or gamma radiation enters matter it interacts with the atoms and molecules of the matter via numerous processes. In these interactions the energy of the radiation is transferred to the atoms and molecules, which may become altered in important ways, such as the ionization of atoms or the production of free radicals from molecules. We will look first at the practical matter of shielding against radiation, and then at the mechanisms of energy loss and radiation damage at the atomic and molecular level, and resulting health impacts. The later part of the chapter will examine the environmental effects of nuclear energy, from the initial mining of uranium, through a reactor's operation, to the final decommissioning of a reactor.

12.1 PROTECTION AGAINST RADIATION

Radioactive materials are always stored behind sufficient shielding material (or at least they should be) to reduce the level of radiation outside the container to a tolerable level. Shielding is important in the manufacture and use of radioactivity and, of course, proper shielding of accumulated fission product wastes from nuclear reactors is crucial. In Chapter 10 we saw that α- and and β- particles travel only short distances – they can be stopped by a centimetre or so of plastic. However, when alpha or beta activities get to the order of 10^{19} Bq or greater, two other problems arise. First, the material is weakened by the constant damage to its structure, and second the kinetic energy deposited by the particles as they slow down in the shielding ends up as heat and the shielding may warp or even melt. Some fission wastes have to be stored in metal tanks that are constantly cooled.

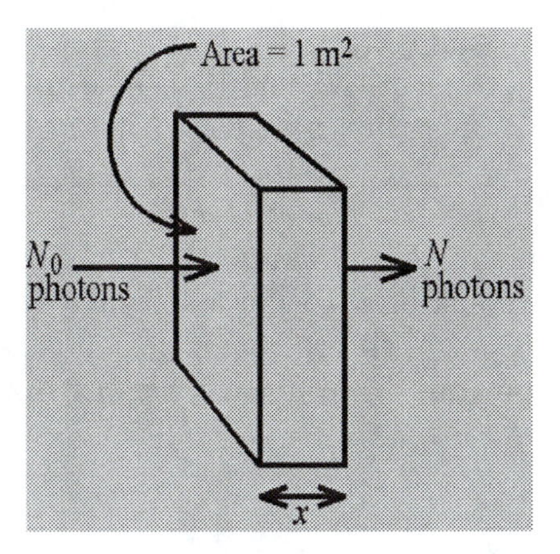

Figure 12-1. Attenuation of a Gamma Ray Beam

It is gamma rays that present the greater shielding problem. Being electromagnetic radiation these are much more penetrating than α- or-β particles. Consider a beam of high-energy photons striking a steel wall as shown in Fig. 12-1. There is no question of them progressively slowing down; rather, there are several processes – see below – which remove photons from the beam, so that as the beam proceeds through the material the number of photons in it steadily decreases; the technical term is that the beam is attenuated. Actually the number decreases exponentially with distance so that if N_0 photons strike a unit area of the wall per second, then the number surviving a distance x downstream is

$$N = N_0 \, e^{-\mu x} \qquad \textbf{[12.1]}$$

where μ is a constant called the *linear attenuation coefficient* and is a characteristic of the material and the photon energy. This is analogous to the radioactive decay law. Here a number of photons are attenuated by thickness; in the former case unstable nuclei were attenuated by time. If a thickness $L_{1/2}$ is defined which removes one half of the photon beam then, analogous to radioactive decay, this *half thickness* is given by

$$L_{1/2} = 0.693/\mu \qquad \textbf{[12-2]}$$

Material of high atomic number Z has a larger μ and smaller $L_{1/2}$ than material of low Z. It is for this reason that lead is an effective and convenient shield against γ-rays.

In the case of stored radioactive liquids, lead is too soft to be structurally sound and stainless steel tanks are preferred. Such tanks can be surrounded by lead to reduce the intensity of γ-rays that escape the steel.

From this discussion it is important to realize that heavy lead shielding in radiation structures and equipment is there not for the α- and β-particles but for the γ rays. If the gammas are adequately shielded the alphas and betas will automatically be taken care of.

EXAMPLE 12-1.

What thickness of lead is needed to reduce a flux of 1.5 MeV γ-rays by a factor of 100? (For energy $\mu = 55 \text{ m}^{-1}$ in lead).

SOLUTION

Since $N/N_0 = 0.01$, we have $0.01 = e^{-\mu x}$, or $-\mu x = \ln(0.01) = -4.6$
Therefore, $x = 4.6/\mu = 8.4 \times 10^{-2}$ m = 8.4 cm.

In the energy range of 1 to 2 MeV the linear attenuation coefficient μ is roughly proportional to the material's density ρ. Radiation workers prefer to use tabulated values of the ratio μ/ρ, which is called the *mass attenuation coefficient* μ_m, and which does not differ greatly among materials. Equation [12-1] then becomes

$$N = N_o \exp(-\mu_m \rho x) \qquad \textbf{[12-3]}$$

In Table 12-1 the mass attenuation coefficient at 1.5 MeV for a number of materials is given. It is evident that it is almost constant and $\mu_m \approx 0.0050 \text{ m}^2/\text{kg}$.

Table 12-1: Mass Attenuation Coefficients

Material	ρ (kg/m)	μ_m (m^2/kg)
Air	1.3	0.0058
Al	2700	0.0053
Cu	8900	0.0047
NaI	3200	0.0047
Pb	11000	0.0052

EXAMPLE 12-2

Repeat Example 12-1 for steel which has a density of 7800 kg/m^3.

SOLUTION

Using Eq. [12-3], $0.01 = e^{-0.0050 \cdot 7800x}$

$\ln 0.01 = -0.0050 \cdot 7800x$

$x = 0.12\text{m} = 12$ cm

Neutrons require a rather different analysis; they have no electric charge so they do not ionize atoms as do alphas and betas. They are not electromagnetic radiation so our discussion of photons is not relevant. In fact, fast moving neutrons are very hard to stop; slow moving neutrons, on the other hand, ($E \sim 1$ eV) are copiously absorbed by the nuclei of certain elements (e.g., boron and cadmium) as a result of peculiarities of the nuclear structure in these cases. The task then reduces itself to slowing the neutrons down, i.e., moderating them, as described in Chapter 11, and then absorbing them with boron or cadmium.

12.2 RADIATION ABSORBED DOSE

When fast-moving alpha or beta particles travel through material they undergo a very large number of collisions with the bound orbital electrons of the atoms but they rarely collide with the nuclei. As a result, nuclear reactions are rare in irradiated matter; the flux of the radiation would have to be very large, such as is found in the core of a nuclear reactor. The reason for this is, of course, purely geometric; the electrons are spread out over a volume of the dimension of an atomic diameter (~10^{-9} cm), whereas the nucleus is tiny by comparison, having a diameter of ~ 10^{-13} cm.

The particle is slowed down by these collisions and is eventually brought to rest. In each collision with an orbital electron, the electron receives some of the kinetic energy of the incoming particle and may even receive enough to be torn from its parent atom; in other words the atom is ionized. This type of damage can be classified in two ways in living matter.

- The ionization process may break a valence bond in a macromolecule such as DNA. The resulting rearrangement of bonds in the molecule, or subsequent chemical reactions with the disturbed site on the molecule may then disrupt the proper functioning of the molecule and the cell.
- Much of the material in a cell is water. Incoming particles may disrupt the water molecule leaving molecular fragments called *free radicals* such as H and OH; these are chemically very reactive and attack biological molecules doing great damage. In fact, most radiation damage to living material is of this type.

The *Linear Energy transfer* (LET) is the energy deposited by the particle per unit length traveled in the biological material. An α-particle is stopped in a few microns (10^{-6} m) of tissue, a β-particle in a few mm. The LET of alphas is large; that of betas less and for gammas smaller still. For charged particles it is measured in units of keV/μm. Both alphas and betas have fairly well defined ranges which increase with the initial particle energy. Both deposit all their energy within that range.

Absorbed Dose (*D*) is based upon the amount of energy deposited per unit mass. When the latter quantity is 1 J/kg, then the absorbed dose is 1 gray. The rate at which dose is incurred is important and so the Absorbed Dose Rate is expressed in grays per second. (An older unit of dose is the *rad* which is defined as 1 rad = 0.01 J/kg = 10^{-2} Gy).

The situation is more complex with gamma rays. Since these are radiations whose speed is always that of light, they are not slowed down by a multitude of collisions as charged particles are. Also, since they are so very penetrating, many can pass through the body without interaction. When they do interact, there are several possible mechanisms. In the photo-electric effect, the gamma ray is absorbed by an orbital electron which then moves off at high speed in the tissue, carrying a kinetic energy equal to the gamma energy less the binding energy of the electron concerned. In the Compton effect, the gamma ray scatters off an atomic electron (which again is ejected) and continues in a different direction with reduced energy; this scattered gamma can then either escape the material, undergo a photo interaction elsewhere or under go a second Compton interaction elsewhere.

In either case the end-product is a fast-moving electron which will cause ionization events along its short track in precisely the same manner as a beta particle. But in this case the electron tracks will occur widely throughout the target material, and the radiation damage will be widespread.

12.3 EQUIVALENT DOSE AND EFFECTIVE DOSE

When radiation enters living tissue the damage it produces has various consequences for the organism. Some damage is repairable; it would be remarkable if this were not so since living organisms have had to evolve in an environment that includes natural sources of ionizing radiation. But some is not.

A simple calculation of the absorbed dose (Gy) does not tell the whole story as regards biological effects, because these effects also depend on the spatial distribution of the energy deposited. For every electron knocked out of its parent atom the incident α- or β-particle uses up an amount *w* of its kinetic energy, where *w* is about 60 eV.[1] If the initial kinetic energy is *E* then the number of ionizing events caused by a single alpha or beta is $n = E/w$. Thus one particle with energy 1 MeV produces about 17,000 ionizing events. This is true for both the α- and the β-particle. The difference is that the easily stopped α does all this damage in a region a few microns deep, whereas the damage caused by the beta is more sparsely distributed over a depth of a few mm. In the alpha case, a cell can be damaged at several places so that it is unable to recover, but repair of the less concentrated damage caused by the beta may occur.

The quantity that accounts for this difference in the relative effect of radiation is the *Radiation Weighting Factor* (w_R). This is a dimensionless multiplication factor that

Table 12-2: Values of the Radiation Weighting Factor w_R *

Radiation	w_R	Radiation	w_R
Photons, all energies	1	Neutrons, energy < 10 keV	5
Electrons, all energies	1	Neutrons, 10 –100 keV	10
Protons, energy > 2 MeV	5	Neutrons, 100 keV – 2 MeV	20
α-particles, fission fragments	20	Neutrons, 2 – 20 Mev	10

* Adapted from reference [2]

indicates the relative effectiveness of a given absorbed dose in producing biological damage. The quantity w_R has been determined by comparing the dose needed to produce a specific effect in comparison with some arbitrary standard. This standard is taken to be the dose of X-rays and γ-rays which are among the least damaging of radiations; the w_R of this standard is taken to be 1.[3] We expect then that values of w_R will be numbers equal to or greater than 1. With this standard it is found that alphas require only one twentieth the dose to produce the same damage as X-rays, therefore for alphas $w_R = 20$. The values of w_R are summarized in Table 12-2 for all the radiations of biological interest and concern.

The *Equivalent Dose* (H_T) is defined using w_R, and is a more accurate measure of the biological damage than is the absorbed dose. The equivalent dose must be summed over all the radiations incident on the organ or tissue (T) of interest. Thus

$$H_T = \Sigma_R[\, w_R \cdot D_R] \qquad \textbf{[12-4]}$$

[1]The most probable value is 22eV and the average is 60 eV. See *Radiation Biophysics* by Edward L. Alpen, Prentice Hall (1990).

[2] Recommendations of the International Commission on Radiological Protection, ICRP Publication 60, Pergamon Press (1991).

[3] The exact definition is that w_R=1 for all radiation that has an LET of 3.5 keV/μm or less.

The unit of Equivalent Dose, for a dose in Grays, is the "Sievert" (Sv).

A further complication arises because different radiations affect the different tissues of the body in different ways. For example, high energy γ-rays pass completely through every tissue in the body and so expose them all. On the other hand α-particles, when incident on the body externally, have a small effect since many of them are absorbed in the dead layers of the skin, whereas the same particles from radon gas inhaled into the lungs are much more damaging. It is not realistic to simply add up all the radiation to which the body is exposed and use that number as a measure of risk. The irradiation of the body's most sensitive tissues must be taken into account; this is done by multiplying each equivalent dose by a *tissue weighting factor*, w_T, and summing over all the tissues of the body that are irradiated to produce the *Effective Dose* still measured in Sieverts. The Effective Dose E is given by

$$E = \Sigma_T [w_T \cdot H_T] \qquad \textbf{[12-5]}$$

where the summation is over all the tissues that make up the body. In Table 12-3 the values of w_T are given.

Table 12-3: Values of the Tissue Weighting Factors w_T *

Tissue or Organ	w_T
Gonads	0.20
Bone Marrow, Colon, Lung, Stomach (each)	0.12
Bladder, Breast, Liver Oesophagus, Thyroid (each)	0.05
Skin, Bone Surface (each)	0.01
Remainder	0.05

* Adapted from reference [2]

12.4 SOURCES OF ENVIRONMENTAL RADIATION

Living things on the planet are subject to a continual irradiation by high energy particles and γ-rays. There are a number of natural sources, such as cosmic rays, and in the modern technological era we have added a number of ubiquitous artificial sources. Under all normal circumstances, the dose delivered by natural sources is greater than that from artificial sources. (It is worth emphasizing that "natural" radiation is just as damaging as "artificial" radiation, but more plentiful. This should be obvious, but the point is often missed by the opponents of nuclear power.)

Table 12-4 lists a number of sources of radiation, both natural and artificial with their annual average dose for a resident of the United States. It shows us that the annual

Table 12-4: Sources of Radiation and Annual Doses (for average person)

Source	Dose (mSv)	Rank	Source	Dose (mSv)	Rank
Smoking	13	1	Domestic water	0.05	8
Indoor radon	2.0	2	Bomb tests	0.02	9
Medical	0.5	3	Fertilizer	0.02	10
Internal nuclides	0.4	4	2nd hand smoke	0.01	11
Cosmic rays	0.3	5	Airline travel	0.01	12
Rock, soil	0.3	6	Occupational	0.008	
Buildings	0.07	7	Nuclear power	0.004	

Ranked items [4]: last 2 items [5]

exposure to radiation from the natural internal sources (mostly ^{40}K) is larger than any artificial exposure except for medical X-rays. The exposure from the nuclear fuel cycle is un-measurable and all artificial sources account for less than one fifth of the total. One natural source of radiation - recognized only in recent years - is radon which contributes over half of our natural exposure. The exposure due to radon infiltrating houses from the underlying soil is now seen as a significant contributor to lung cancer in non-smokers; radon can also be present in domestic drinking water since radon diffuses from the rocks into the ground-water which is taken up in the water supply; from there its daughter nuclides can find their way into consumer products. Smokers have an additional exposure from natural ^{210}Po in tobacco, which contributes a relatively high dose (~0.2 mSv/yr) to the bronchial epithelium; the American Cancer Society has estimated that this is the cause of a large fraction of lung cancers. [6]

A source of radiation not included in the table is that encountered by passengers and aircrew in high-flying aircraft. While flying at an altitude of 10,000 m a passenger in northern latitudes can receive a dose of 0.05 mSv in an eight hour trip. This is about equal to a diagnostic chest X-ray. Air crew can receive between 3 and 17 mSv annually; half of this is contributed by neutrons. This dose is above that recommended for radiation workers; epidemiological studies are now underway for aircrew.

Well-defined rules have been set up by national and international bodies such as the ICRP (International Commission on Radiological Protection) regulating the maximum radiation dose that an individual should receive. Related tables have also been compiled defining the maximum amounts of various radionuclides which can be ingested "safely"; of course no one is supposed to ingest any. The objective of the ICRP system of dose limitation is to promote the use of safety precautions whenever radioactive materials are handled and to insure that external and internal doses are

[4] Health Physics Society Newsletter, June (1998).
[5] *Health Effects from Exposure to Low Levels of Ionizing Radiation*, BEIR VII, National Academies Press, Washington, D.C. (2005).
[6] R.S. Yalow, Interdisciplinary Science Reviews 16, No. 4, (1991) p. 351.

controlled within the ICRP recommended limits for occupational exposure. These limits are in Table 12-5.

Comparison of Tables 12-4 and 12-5 show that the recommended maximum level of 1 mSv/yr is lower than the environmental sources now encountered on average in, say, the USA. This average value is dominated by the large figure for radon which is very variable from location to location and indeed even from house to house. In certain areas modern well-insulated and well-sealed houses are likely to have high, and even unacceptably high, concentrations of radon [7]. This is a problem that has only recently been recognized.

Table 12-5: Recommended Annual Dose Limits [8]

	Dose Limit for Radiation Worker	Dose Limit For Public
Effective Dose	20 mSv[a]	1 mSv
Equivalent Dose in:		
Lens of the Eye	150 mSv	15 mSv
Skin	500 mSv	50 mSv[b]
Hands and Feet	500 mSv	-

[a]: Averaged over 5 years; not to exceed 50 mSv in any one year. Additional restrictions apply to pregnant women.

[b]: Averaged over any 1 cm^2 of skin regardless of the area exposed.

12.5 HEALTH EFFECTS OF RADIATION

We know much more about the health effects of very large one-time doses of ionizing radiation than about those of very small ongoing doses. Among the effects that are visible shortly after irradiation are blood changes, vomiting, and skin damage. Doses up to 0.25 Sv (250 mSv) produce no detectable effect. Between 0.25 and 1 Sv there is blood and bone-marrow damage, but the person feels little immediate effect. Above 1 Sv, treatment with antibiotics may be necessary to cope with malaise, vomiting and blood changes (i.e., radiation sickness). Above 3 Sv, internal hemorrhaging and diarrhea may occur and blood transfusions may be necessary. At the level of 5 Sv there is a 50% chance of death.

Survivors of large doses (> 1 Sv) or people exposed to doses at a somewhat lower level (0.1 – 1 Sv) can develop cancer and leukemia; leukemia does not usually appear until five years or more after the exposure and cancer 20 years or more. The vast majority of

[7] A. Nero, Physics Today, April (1989) p. 32.

[8] See ref. 2.

data that we possess on such delayed health effects from large doses are for the 120,000 Japanese A-bomb survivors; data are also available from 30,000 Canadian women treated with X-rays for tuberculosis between 1930 and 1952, 14,000 British spondylitis patients treated by spinal irradiation between 1935 and 1954, and other smaller groups. The relative risk of cancer incidence due to a specific dose is defined as

$$RR = \frac{Incidence\ rate\ \lambda\ in\ exposed\ population}{Incidence\ rate\ \lambda\ in\ unexposed\ population} \quad \textbf{[12-6]}$$

And the excess relative risk is defined as $ERR = RR - 1$ **[12-7]**

So $\lambda_E = \lambda_U\ (1 + ERR)$ **[12-8]**

From the studies of the A-bomb survivors, an almost linear relationship has been established between *ERR* and absorbed dose as illustrated in Fig. 12-2 for the specific case of radiation-induced cancer. (The data can be fitted equally well to a quadratic relationship). Can we use this relationship to predict the effect of a continuing low dose of 1 mSv per year on the North American population? On the figure, the *ERR* at 1 Sv is about 0.6. Let us very simplistically extrapolate downwards to a dose of 1 mSv, getting an *ERR* of 6 x 10^{-4}. Hence λ_E/λ_U is about 1.0006. We know that the rate of cancer incidence λ_U in the USA is about 82,000 per 10^5 population, so there are 50 cancers from the 1 mSv dose, and about 25 deaths. If this dose is repeated every year, the number of cancer deaths is 70 x 25 or 1750.

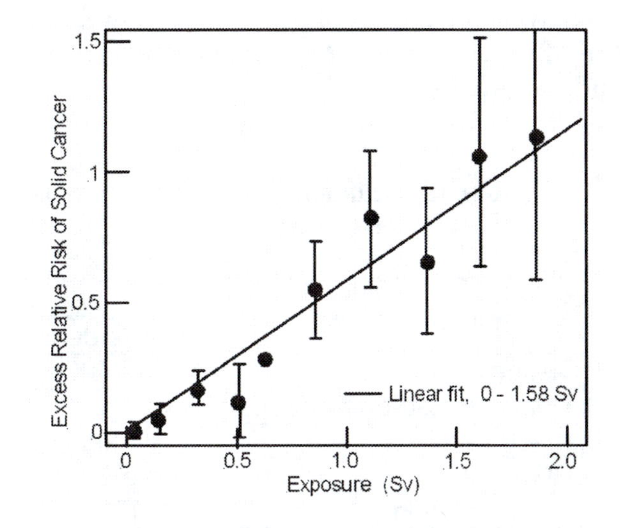

Figure 12-2 ERR for Radiation-Induced Cancer in the Japanese Atomic Bomb Survivors. *Data from ref. 5.*

The reality is much more complex than our simple extrapolation. The national and international bodies who have studied dose-effect relationships do indeed recommend a linear hypothesis, and they assume that there is no threshold dose towards the lower end of this linear relationship, below which any damage is automatically repaired and no permanent outcome transpires. This LNT (linear – no threshold) approach is the basis of the regulations laid down by radiation protection organizations. However there is much uncertainty as to the real effects of very low ongoing doses and much research remains to be done. It is rather obvious that, ethical reasons aside, there is no way we can do statistically viable experiments on humans that might provide the answer.

Some studies have suggested that there is indeed a threshold dose below which some damage is repaired by the body, leaving no lasting effect. This is probably true for cases where whole cells are killed in an organ which continuously eliminates dead cells and generates new ones. An example of this type would be the depression of the blood count after irradiation and subsequent recovery from the radiation-induced anemia as new cells are produced. There is a growing body of evidence from mouse studies that ongoing radiation exposure at very low levels actually stimulates defence mechanisms, so that the effects of a subsequent single large dose are much reduced. But from the overall body of research the experts cannot conclude definitively that there is evidence for either beneficial or deleterious effects from doses below 0.1 Sv.

Various very sophisticated analyses of the A-bomb survivor and other data have been made over time by both government and private organizations with the aim of predicting the lifetime attributable risk (*LAR*), i.e., the lifetime probability of fatal cancer per Sv of effective dose. Account must be taken of gender-dependence, age-dependence, and the different susceptibility of different populations to cancer. "Transport" of Japanese data to the North American population is rendered complex by the very different natural incidence of cancers, especially stomach cancer. For high doses, published values of *LAR* are in the range 0.08 – 0.12 per Sv. For low doses and dose rates, most organizations, but not all, introduce a correction which is typically a factor of 2, giving *LAR* values in the region of 0.05 per Sv. Table 12-6 presents a few of the extensive conclusions of the BEIR VII Report of the US National Academies. We see that the lifetime number of cancer deaths per 10^5 people (viz., 710) is indeed about half our crude prediction.

Table 12-6: Lifetime Attributable Risk of Cancer Incidence and Death per 10^5 Population: *assembled from data in BEIR VII* [9]

Scenario	Cancer		Leukemia	
	Incidence	Deaths	Incidence	Deaths
Baseline	82400	39600	1420	1240
0.1 Gy to mixed-age population	2110	1020	172	121
0.1 Gy at age 10	3860	1690	206	123
0.1 Gy at age 30	1600	810	146	115
1 mG/y for lifetime	1520	710	126	85
10 mG/y from age 18 to 65	6630	3580	630	510

Normal cancer incidence varies greatly, depending on cancer type, age and gender. For any one type of cancer there is a great variation among countries and even districts within countries; the λ_U value for stomach cancer incidence varies by a factor more than ten among different countries. Many factors contribute, probably in a synergistic manner, to the incidence of types such as bronchial and lung cancer; these are presently increasing due to environmental factors such as chemicals, smoking, air pollution, etc. It is interesting to look at populations who live in regions of higher than normal background. Americans living in the Rocky Mountain states experience nearly twice the average background of other U.S. states but they have the lowest cancer rates in the nation. Indians living in Kerala experience a background that is some seven times higher than that in the U.S., but display cancer rates that are 25% lower.

[9] See ref. 5

It follows from the table that the natural background radiation can play only a very small role in cancer incidence, as do levels similar to background. The current effective dose due to nuclear power and processing plants is less than 0.01 mSv per year; in fact, around most plants the dose is un-measurable and has to be estimated on the basis of known releases of radioactivity. Using the tabulated risk factor, we can estimate the resulting cancer plus leukemia fatalities in a population of 10^5 as about 7, which is not discernible against the natural rate of about 40,000. This risk level is dwarfed by many human activities such as driving, smoking, etc.

There have been reports in the media that increased incidence of certain cancers have been detected in populations living around nuclear installations. The most controversial of these issues may have been the incidence of leukemia in children living in the village of Seascale near the Sellafield nuclear reprocessing installation in England where six leukemia deaths in children aged 0 to 24 years occurred between 1968 and 1974; 0.6 was the expected value. Four of the fathers were Sellafield employees, and one hypothesis advanced was that pre-conception radiation doses to the fathers, caused deleterious effects that were passed on to the children via the germ cells. However other data [10] contradict this idea. More than 92% of the births to Sellafield employees occurred outside Seascale and there no excess leukemia cases occurred (the paternal hypothesis would have suggested 53). A subsequent study of childhood leukemia around Canadian nuclear facilities found no increase in the rate of the disease comparable to that found near Sellafield. The incidence near the plant at Pickering, Ontario was slightly above the Ontario provincial average both before and after it started operation. [11] And no excess of leukemias was seen in the children of the Japanese atomic bomb survivors.

There are several difficulties with studies of this kind. One of the major ones for the epidemiologist and the statistician is that the population to be studied is defined after the effect is discovered and not before. What should be the boundaries of the population under study? Should it be expanded or contracted? If this is done will the effect vanish? In proper epidemiological studies the population is defined before it is studied. Another difficulty is that the site - in this case a nuclear station - has been singled out for attention because of what it is. Other installations, let's say automobile manufacturing plants, are not under the same scrutiny. If they were, would we find unusual statistical fluctuations in certain diseases around them as well? Further it is not clear that all the relevant facts have been identified. Is the fact that it is a nuclear site the important parameter, or is it because it is a newly collected group of people of a certain narrow age and socio-economic status? The latter is just as likely since similar excesses in childhood leukemia have been found around *prospective* sites for nuclear plants as well as other places where there is a large population influx. It should be noted that Sellafield is a nuclear reprocessing plant, something far different from a nuclear power plant; is it clear that the nuclear aspect is the cause? (Reprocessing plants formerly used large amounts of benzene, a known carcinogen.) Finally it should be noted that there are far more nuclear installations which show no such effects, and these

[10] R. Doll, H.J. Evans, S.C. Darby, Nature 367, (1994) p. 678.
[11] *Childhood Leukaemia around Canadian nuclear facilities*, Atomic Energy Control Board, (1991).

observations must not be excluded when reaching overall conclusions. Epidemiological problems like this are very difficult to resolve and require very careful and lengthy study and analysis. This problem is similar to the one concerning health effects of electric and magnetic fields (Section 9.3).

12.6 URANIUM MINING AND PROCESSING

Uranium mines present all the hazards of any hard-rock mining, with the additional danger of radioactive radon gas which deposits radionuclides on the lung surface. However, since the energy content of uranium ore is greater than that of coal by about a factor of 30, the total conventional hazard per unit of energy produced is considerably lower in uranium-mining than in coal-mining. [12] The radon hazard gave uranium miners in the past a higher than normal chance of contracting lung cancer; nowadays extensive precautions are taken to avert the danger of breathing radon.

The mill tailings (the ore discarded after removal of the uranium) are slightly radioactive because of long-lived daughter isotopes from uranium that have already decayed in the ore. This material is usually a liquid sludge, and after drying it is susceptible to dispersal to the environment through erosion (by wind and water) and by leaching. Therefore, the preferred method of disposal is to place it in empty mines or in lined pits dug for the purpose. Another problem associated with tailings is the emission of radon to the atmosphere (Problem 12-11), which can be inhibited by covering the tailings with compacted earth, clay, concrete, or asphalt. Since the half-life of radon is 3.8 days, the radon has to be retained for only 25 days to reduce its activity by a factor of 100.

The concentrated ore from the mills is sent to refineries (then to ^{235}U-enrichment facilities if the uranium is to be used in LWRs), next to fuel-fabrication plants, and finally to reactors. At all stages in this transportation, there are strict requirements on the design of shipping containers to prevent escape of radioactivity to the environment in the case of an accident. In the various plants, there are rigorous accounting procedures to ensure that no uranium is lost, stolen, or makes its way into the general environment. These procedures are in place for a number of reasons: the uranium is expensive, it could be important as a source of material for weapons if enriched sufficiently in ^{235}U, and it is radioactive, although not strongly. (Because of uranium's long half-life, its activity is relatively low. The activity of 1 kg of natural uranium is 1×10^{7} Bq , whereas the activity of 1 kg of ^{60}Co, an isotope of half-life 5.3 yr that is used in cancer therapy, is 4×10^{16} Bq)

---

[12] W.F. Vogelsang and H.H. Barschall, "Nuclear Power," in *The Energy Sourcebook*, Ed. by R. Howes and A. Fainberg, American Institute of Physics, New York, (1991), p. 142

12.7 POLLUTION AND RADIOACTIVE WASTE FROM REACTORS

Although a nuclear reactor does generate radioactive waste, it produces no greenhouse gases, no SO_2, no NO_x, no ozone, and no particulate matter. Hence, in terms of these pollutants, a nuclear plant is much "cleaner" than a fossil-fuel electrical plant. However, it creates thermal pollution of the cooling waters (more than does a fossil-fuel plant), as discussed in Section 5.2.

All energy-generating technologies have waste products to cope with, and the waste from nuclear reactors has attracted much attention from the public and media. What is not often appreciated, however, is the physical scale of the problem for the various technologies. Let us first, then, compare the volume of waste which must be handled in a coal-burning and a nuclear plant of 1000 MWe capacity. The nuclear fuel is in the form of UO_2 which has a density of 11 g/cm^3, whereas coal has density 1.5 g/cm^3. Since only nuclear energy is extracted from the UO_2, the mass and volume of the waste is almost the same as the fuel. For coal, the ash comprises about 10% of the volume but of course the other 90% is exhausted up the stack in the form of CO_2; we must never forget that large amount of invisible but very important greenhouse gas, not to mention the nitrogen and sulphur oxides. Table 12-7 summarizes the situation for the two cases, for solid wastes alone.

Table 12-7: Fuel and Waste from a 1000 MWe Power Plant

Type of Plant	Annual Fuel		Annual Waste
	(Tonnes)	(m^3)	(m^3)
Coal	3000000	2000000	200000
Nuclear	200	20	20

Every year the coal-fired plant must dispose of 200000 m^3 of toxic ash — this is equivalent to the volume of a large office block (~60 m on a side). The nuclear plant must deal with 20 m^3 of very toxic used fuel occupying a volume about the size of a small office (~2.5 m on a side). The nuclear waste problem can be dealt with in two stages: short-term and long-term storage.

Short-term Storage

After the fuel elements are first removed from the reactor it is necessary to store them in such a manner that they are secure and adequately cooled. Heat is still being generated in the fuel by radioactive fission fragments. The elements are usually stored in ponds under re-circulated water which provides both shielding and cooling. There are a large number of very short-lived isotopes in the spent fuel and the activity decreases by about 60% in the first hour. After that, the activity is dominated by the longer-lived isotopes. After three months, one tonne of spent fuel has an activity of

6.3×10^{17} Bq and generates heat at a rate of 27 kW. Of this activity, 98% is due to fission products and 2% to actinides (plutonium generated by breeding, for example). After ten years, the activity is 8.9×10^{16} Bq and the rate of heat production has decreased to 1 kW. At this stage the necessity for close-monitored cooling is over and long-term storage can be considered.

Long-term Storage

No nation except the U.S.A. has yet adopted an agreed method of long-term storage of spent reactor fuel. There are many reasons for this delay as there are several conflicting interests; some people favour reprocessing to strip out of the spent fuel all the remaining fissionable fuel (^{235}U, ^{239}Pu, etc.) and hence maximize the efficiency of the power generation; this is an elaboration of the fuel enrichment technology that has grown out of the nuclear weapons programs. With concerns about nuclear terrorism, plutonium toxicity, and weapons proliferation, reprocessing has met with strong opposition in some quarters. Reprocessing plants have been built in England and France, but the MIT Report of 2003 [13] strongly recommended that the USA opt for "once-through" use of uranium followed by long-term storage of the waste.

It seems likely that most nuclear nations will opt for some form of long-term storage of the fuel with little, or no, preparation and processing. One option is to do nothing other than provide supervised above-ground storage. Supporters of this option argue that "out-of-sight is out-of-mind" and it is better to keep the material in view and look after it properly. Detractors of this view argue that it is unlikely that institutions can be put in place to tend it for 1000 or even 10000 years.

The most probable option will be to develop some sort of permanent storage in below-ground vaults, either supervised or more likely sealed, deep in continental rock. There seems to be no technological reason that such vaults and containers cannot be designed and constructed to contain the wastes for a sufficiently long time to reduce the radiation to levels consistent with that of the rock itself. Several countries are engaged in this type of research.

Canada has been involved since 1978 in an active research program to study long-term storage of nuclear waste, and has created an Underground Research Laboratory (URL) located 420 m below the earth's surface in previously undisturbed granitic rock in Manitoba. Experiments are being performed to determine how rock and groundwater behave at depth, and how they would be affected by heat given off by used fuel elements. Other experiments study the mobility of weakly-radioactive tracers released underground. Switzerland, Sweden and Finland are also planning disposal in granite, The U.S.A. has approved use of the Yucca Mountain tuff formation in Nevada. Tuff is a compressed volcanic ash over ten million years old, and at this site the water table is so far below the surface that the possibility of water entering the storage caverns in the remote feature is vanishingly small. Nevertheless, Yucca Mountain continues to be the focus of intense argument.

---

[13] *The Future of Nuclear Power: an Interdisciplinary MIT Study,* Pergamon Press (2003).

Low-Level Radioactive Discharges from Reactors

The radioactive isotopes created in reactors are of three types:

- fission fragments;
- isotopes produced from neutron bombardment of impurities and corrosion products in the coolant-moderator water as it passes through the reactor core;
- isotopes produced in the entire reactor structure by neutron bombardment.

Most of the fission products remain in place within the fuel bundles, but some can diffuse or leak through the fuel cladding into the coolant-moderator. The most common elements that diffuse are isotopes of cesium, iodine, xenon, krypton, rubidium, and bromine, [14] as well as tritium. The radionuclides in the coolant-moderator from fission products and from impurities and corrosion can exist in gaseous form, as dissolved solids, or as suspended solids. The suspended solids are removed by filtration, and the dissolved solids by ion-exchange beds; these materials can then be disposed of, typically by burial. Gaseous waste, e.g., isotopes of krypton and xenon, can be stripped from the coolant and/or moderator, and stored for up to three months to allow short-lived isotopes to decay, after which it is much less radioactive and is vented to the atmosphere (Problem 12-12). The coolant-moderator water, after having the solids and gases removed, can be re-used in the reactor or discharged (with only slight radioactivity) to the environment. The isotopes that are bound within the structure are not released to the external environment

The tritium that is generated in reactors usually becomes bound in water molecules, replacing one of the hydrogens to produce HTO, which would be relatively difficult to remove. However, the concentrations of tritium in normal emissions of tritiated water from a reactor present a negligible health hazard. In part, this is because tritium emits only low-energy beta radiation, and does not become concentrated in any way in biological systems. (However, a large leak of heavily tritiated water would be considered a serious problem, because the tritiated water could become incorporated into plants and animals in the area.)

Occasional leaks (particularly of radioactive gases) from valves, pipes, etc. also result in discharges of weakly radioactive materials to the environment. The radiation dose to the general public as a result of all these emissions is extremely small — recall from Table 12-4 that all aspects of the nuclear power industry contribute at most 0.4% of a person's average annual effective dose.

12.8 NUCLEAR ACCIDENTS AND SAFETY

Probably the most serious concerns about nuclear reactors are related to the possibility of a nuclear accident. The total inventory of radioactive materials in a reactor is so large that it would represent a serious threat to life if released into the environment, and could render a large area of land uninhabitable. Recall that a reactor cannot

[14] M. Eisenbud, *Environmental Radioactivity from Natural, Industrial, and Military Sources (Third Ed.)*, Academic Press, Orlando, 1987, p. 210.

undergo a nuclear explosion — the concentration of ^{235}U is are required to make bombs. However, the heat produced by a reactor could, under certain thirty times less than the levels that circumstances, produce melting of the fuel, thus allowing fission products into the coolant (and also making the reactor inoperable). In an extreme situation, a steam explosion could result.

Safety Mechanisms

A reactor is a producer of heat, and if the heat produced cannot be removed from the reactor core, then overheating can lead to a serious accident. Reactors have various safety devices in place to prevent this occurrence. In case of a problem with a reactor, there are two issues that need to be addressed: shutting down the fission process (which produces most of the heat), and then removing the continuous residual heat produced by the fission products. The various types of reactors have safety mechanisms that differ slightly in their details, but all reactors share some common features in their approaches to safety.

In order to stop the nuclear fissions, neutron-absorbing rods are inserted into the reactor core. These rods can be special shutoff rods, or just the usual control rods that are normally used to adjust the reactor's output. By absorbing neutrons, these rods cause the number of fissions to decay, shutting down the reactor. In the event that the rods fail to operate properly, a neutron-absorbing material (a “neutron poison”) can be rapidly injected into the coolant-moderator.

Once a reactor is shut down, normal circulation of the coolant removes heat produced by decay of fission products. If this circulation is somehow interrupted in a loss-of-coolant accident, an emergency core-cooling system is turned on that provides a completely separate circulation of cooling water. A final safeguard is provided by the structure of the containment building surrounding the reactor; this building is typically constructed of thick concrete with a steel-plate lining, and is designed to withstand a steam explosion. (The Chernobyl reactor had a much flimsier containment building, and this feature contributed to the severity of the accident there.

Accidents and consequences

The first accident at a nuclear power reactor occurred in March 1979 at a PWR plant on Three Mile Island (TMI) in the Susquehanna River in Pennsylvania. (A small number of accidents had occurred previously [15] at experimental reactors, or at reactors producing plutonium for weapons.) This was a loss-of-coolant accident. It occurred as a result of a combination of mechanical problems, design features that gave inadequate monitoring and display of reactor conditions to the operators, and human errors. The reactor core was destroyed due to insufficient provision of emergency cooling. From estimates of the total radioactivity released, the investigating commission's best estimates [16] of fatal cancers were as follows: 50% chance of none; 35% chance of one fatal cancer; 15% chance of more than one.

---

[15] D. Bodansky, Nuclear Energy: Principles, Practices, Prospects, Springer 2004.

[16] Report of the Kemeny Commission on the Accident at Three Mile Island ,Pergamon Press, NY, 1979.

The disaster in April 1986 at the Chernobyl reactor in the Ukraine region of the former USSR was much more serious. This was a criticality accident, in which the neutron population and the power level increased uncontrollably. It was caused by inadequate training, bad design, total lack of containment, and repeated violation of safety rules. An explosion and fire released a great deal of radiation to the environment, contaminating the region around Chernobyl which had to be permanently evacuated. The accident produced 134 cases of acute radiation doses in reactor workers and fire-fighters. Within four months after the accident, 31 people had died — one from the explosion, 28 from radiation, one from steam burns, one from a heart attack. Ten years later, there had been 14 additional deaths in this group of 134, but not all of these deaths were directly attributable to radiation exposure. Extensive research is continuing on health and environmental outcomes, the most serious of which was widespread non-fatal thyroid cancer in children; this could have been prevented had the government issued iodine pills to suppress uptake of radioactive iodine from milk. However the USSR government chose to suppress information about the accident until it was too late.

Long-term fatalities due to cancer are very difficult to predict, but attempts have been made to do so [17], based on the approaches described in this chapter. The largest radiation doses (average 100 mSv) were incurred in 1986/7 by the 240,000 clean-up workers and military personnel officially classified as "liquidators"; the expected background cancer deaths in this group are 41,500 and the estimate of the excess that might be caused by radiation dose is 2000, i.e., a 5% increase. For the 116000 evacuees and the 5.27 million people living in contaminated areas, the background and excess rates are given as 865,000 and 6250 respectively, i.e., an increase of 0.7%. Additional leukemia deaths would be about 700 overall.

The Chernobyl disaster showed what can happen when a reactor and its containment are not designed with enough attention to safety, when operators are given insufficient training for accidents, and when operating procedures can be violated easily. Extensive steps have been taken to improve safety in Chernobyl-type reactors. Better training is being given to operators, more effective displays of information are being installed in reactor control rooms, and much more stringent controls are being placed on the disabling of safety systems. The structure and operation of shutoff rods have also been improved, and the composition of the fuel has been changed (enriched further in ^{235}U) to decrease the magnitude of the positive void coefficient.

Appendix VIII provides more detailed accounts of what is believed to have happened in these two accidents.

---

[17] Contribution by E. Cardis et al to the International Conference: Chernobyl - *Looking Back to Go Forwards to a United Nations Consensus on the Effects of the Accident and the Future*, IAEA, Vienna, 2005.

12.9 DECOMMISSIONING OF NUCLEAR PLANTS

Nuclear reactors have useful lifetimes of 30 - 60 years, and then need to be shut down permanently. The remaining fuel, containing the radioactive fission products, will be removed and disposed of, but the entire building structure is also somewhat radioactive from neutron bombardment, and pipes, valves, etc. are contaminated with corrosion and wear products that have been neutron-activated, and also with fission products that have leaked from the fuel assemblies. What can be done with all this radioactive material? There are essentially four options, ranging from immediate dismantling of the plant to using the normal reactor containment structure as a long-term storage facility.

Immediate dismantling would require precautions to ensure that the personnel engaging in the work do not receive large doses of radiation, since the radioactivity will have had little time to decay. Some of the work would probably have to be done by robots. The material would have to be stored elsewhere in a secure site, likely underground. Proponents of quick dismantling argue that this approach minimizes radiation exposure to the general public.

Another option is to wait perhaps 30 to 50 years before dismantling, to allow the radioactivity to decrease. This would reduce both the danger to workers and the amount of radioactive material that would have to be stored permanently. An obvious disadvantage is that the site would have to be made secure during the waiting period.

A third alternative is to entomb the entire reactor building in a strong encasement, and simply leave it for a very long time until the radioactivity has decayed to a level where the site poses no danger to anyone. Again, site security and maintenance of the structure would be required. A fourth choice, which is a variant of this entombment, would simply be to seal the existing reactor containment structure and use it as the storage site. Supporters of this option point out that it would be the least expensive. Both of the entombment approaches result in long-lasting eyesores on the landscape.

In the USA, the Nuclear Regulatory Commission has endorsed the first two options as being preferred, and has required that plant operators set aside sufficient funds for eventual dismantling. The amount required varies from about $350 million for a PWR to $450 million for a BWR.

12.10 NUCLEAR POWER AND NUCLEAR WEAPONS

Although a reactor, due to the low levels of ^{235}U enrichment used, cannot undergo a nuclear explosion, there is an indirect connection between nuclear power and nuclear weapons. Any fissionable material such as ^{235}U or ^{239}Pu can be used to create a nuclear bomb; however, the concentration of the fissionable material must be enriched to 90% or more. In the case of ^{235}U, this enrichment is neither easy nor inexpensive. However, the ^{239}Pu created by neutron bombardment on ^{238}U in a nuclear reactor is easily separable by chemical means from the uranium fuel and fission fragments, and could be used to construct a nuclear weapon. Thus, any country with nuclear power plants could produce nuclear weapons. (In addition to plutonium's danger in nuclear weapons, it is an extremely toxic chemical.)

In 1957-58 a Nuclear Non-Proliferation Treaty was signed by more than one hundred countries. One of the main objectives of the treaty was to require the processing of used fuel elements to come under the surveillance of the International Atomic Energy Agency in order to help prevent the spread of nuclear weapons. However, a number of countries did not sign the treaty; examples include India, which has an extensive nuclear power program.

There is also the danger of construction of a crude nuclear weapon from plutonium in used fuel assemblies stolen by a terrorist group, for example. Since the September 2001 attacks in the USA, concern about such a venture has heightened, driven also by concern as to the fate of plutonium stocks at the time of the disintegration of the U.S.S.R.

EXERCISES

12-1. Table 12-8 shows the principal fission-product radionuclides that are contained in a LWR core after the reactor has been shut down for one day after two years of steady operation[18]. If these radionuclides are allowed to decay for two years, which ones will then be making the most significant contributions to the overall activity? To answer this question, do not perform any calculations; just look at the original activity for each isotope in Table 12-8, and think about whether this activity will decrease a little or a lot in two years, given the half-life.

Table 12-8: Principal Fission-Product Radionuclides after a 1-Day Shutdown Following 2 Years of Operation of a LWR (1000 MW thermal)

Isotope	Activity (Bq)	Half-life	Isotope	Activity (Bq)	Half-life
^{3}H	4.8×10^{14}	12.3 y	^{134}Cs	6.7×10^{16}	2.06 y
^{85}Kr	2.8×10^{16}	10.7 y	^{132}Te	3.8×10^{18}	78 h
^{89}Sr	2.7×10^{18}	51 d	^{133}I	2.4×10^{18}	20.8 h
^{90}Sr	2.0×10^{17}	28.9 y	^{136}Cs	8.1×10^{16}	13 d
^{90}Y	2.0×10^{17}	64 h	^{137}Cs	2.7×10^{17}	30.2 y
^{91}Y	3.6×10^{18}	58.8 d	^{140}Ba	5.1×10^{18}	13 d
^{99}Mo	4.4×10^{18}	66.6 h	^{140}La	5.4×10^{18}	40.2 h
^{131}I	3.1×10^{18}	8.06 d	^{144}Ce	3.9×10^{18}	284 d
^{133}Xe	6.0×10^{18}	5.3 d			

[18] Ref. 14, p. 207

PROBLEMS

12-2. Concrete (density= 2900 kg/m^3) is a relatively inexpensive absorber of high-energy gamma rays from radioactive sources of large physical size. What thickness of concrete would reduce the intensity of radiation from a source of 1.7-MeV gamma rays by a factor of 100? The linear absorption coefficient of concrete for 1.7-MeV gammas is 0.11 cm^{-1}.

12-3. When a piece of wood 5.0 cm thick is placed between a low-energy gamma ray source and a detector, it reduces the count rate by a factor of 12. What additional thickness is required to further reduce the intensity by a factor of 4.0?

12-4. For 40-keV X-rays, the mass attenuation coefficient for bone is 0.31 cm^2/g, and for muscle it is 0.068 cm^2/g. An upper arm is being X-rayed so that the muscle and the bone show up beside each other on the photographic film. If the bone is 2.0 cm thick and the muscle is 4.0 cm thick, and if the densities of bone and muscle are 3.0 g/cm^3 and 1.3 g/cm^3 respectively, what is the ratio of the intensity of the beam transmitted through the bone to that through the muscle? Ignore absorption by other tissues.

12-5. A quantity of a fall-out radioactive beta emitter is taken up by a growing plant and distributed uniformly throughout the tissue. A 200 gram plant is found to contain 3700 Bq (0.1 μ Ci). What dose does the plant receive in one day if the average beta energy is 0.10 MeV?

12-6. A tiny radioactive source is embedded in flesh a few millimetres under the skin. The source is a 6.0 MeV α-emitter of 3.7×10^{-2} Bq (1 pCi) activity. What is the dose rate and the equivalent dose rate? The range of 6 MeV alphas in water is 3.9×10^{-5} m. Why is this dose rate so large?

12-7. The same α-source considered in Problem 12-6 is uniformly distributed throughout the lungs (mass ~ 1 kg). Repeat the calculations of Problem 12-6 and comment on the difference.

12-8. A research reactor has an accident in which a beam port is incorrectly opened for 5 s with a worker standing in the beam. The beam consists of three components: γ-rays, fast (1 MeV) neutrons, and thermal (100 eV) neutrons. The dose rates are respectively 0.9×10^{-4} Gy/hr, 1.1×10^{-4} Gy/hr, and 0.8×10^{-4} Gy/hr. What is the equivalent dose and dose rate? What fraction is this of the worker's allowed dose?

12-9. The US regulations for the maximum permissible concentrations of radionuclides in water give the values:

Species	Max. Conc.	$T_{1/2}$
^{131}I	0.011 Bq/cm^3	8.1 days
^{239}Pu	0.19 Bq/cm^3	2.4×10^4 year
^{90}Sr	0.011 Bq/cm^3	28 year

Calculate the corresponding concentrations of these elements by mass in parts per million (ppm).

12-10. International couriers make many more long distance air flights than even air-crew. If a courier makes 200 transatlantic flights in a year (8 hr flying time) what would be the radiation dose the courier would receive in a year? How does this compare to the recommended annual dose for the general public and radiation workers?

12.11. The US Environmental Protection Agency has set a limit[19] on the emission of radon-222 from mill tailings: 0.75 $Bq{\cdot}m^{-2}{\cdot}s^{-1}$. For an uncovered pile of tailings of dimensions 35 m ×25 m × 5.0 m, sitting on the ground and emitting radon-222 at the maximum rate allowed, what mass of radon-222 is emitted to the atmosphere per day? The half-life of radon-222 is 3.8 d.

12.12 . BWRs continually release several Kr and Xe radionuclides to the atmosphere. Table 12-9 shows the Kr isotopes that are normally released. (An "m" if front of the Kr indicates an excited nucleus that is normally released. (An "m" if front of the Kr indicates an excited nucleus that is metastable, i.e., one that has a sufficiently long half-life to be studied independently from the ground-state nucleus). The isotopic half-lives are listed, along with the typical activity[20] that would be released from a 3400-MW (thermal) BWR in one year if the gases are held up for only 30 min. Charcoal beds are often used to hold the gases for an additional 45 h. Calculate the activity of each isotope that would be released each year after this additional hold-up, and comment on the benefit (or lack of it) of this hold-up to radioactive contamination of the atmosphere.

Table 12-9: Typical Annual Releases of Krypton from a BWR after 30-min Hold-up

Radionuclide	Half-life	Ci/yr emitted
^{83m}Kr	1.9 h	$1.6{\times}10^{15}$
^{85m}Kr	4.4 h	$3.1{\times}10^{15}$
^{85}Kr	10.8 y	$1.1{\times}10^{13}$
^{87}Kr	76 min	$8.9{\times}10^{15}$
^{88}Kr	2.8 h	$1.0{\times}10^{16}$
^{89}Kr	3.2 min	$1.0{\times}10^{14}$

ANSWERS

12-1. ^{144}Ce, ^{137}Cs, ^{90}Sr
12-2. 42 cm
12-3. 2.8 cm
12-4. 0.22
12-5. $2.6{\times}10^{-5}$ Gy
12-6. $1.4{\times}10^{-4}$ Gy/s, $2.9{\times}10^{-3}$ Sv/s
12-7. $3.5{\times}10^{-14}$ Gy/s, $7{\times}10^{-13}$ Gy/s
12-8. $3.5{\times}10^{-3}$ mSv, $25{\times}10^{-4}$ Sv/hr, 0.002%
12-9. $2.4{\times}10^{-12}$ ppm, $8.2{\times}10^{-5}$ ppm, $2.1{\times}10^{-9}$ ppm
12-10. 10 mSv
12-11. 1.6×10^{-11} kg
12-12. $1.2{\times}10^{8}$ Bq/yr for ^{83m}Kr, etc

[19] Ref. 13, p. 179
[20] Ref. 13, p. 214

CHAPTER 13

NUCLEAR ENERGY: 21st CENTURY

13.1 END OF THE FIRST NUCLEAR ERA

In the 1960s and 1970s, the new technology of nuclear power was seen as a source of abundant cheap electricity, and the growth in number of installed nuclear reactors was rapid. But the accidents at Three Mile Island and Chernobyl, coupled with the very low cost of fossil fuels and uncertainty as to nuclear waste disposal, effectively ended the promise of this first generation of nuclear power reactors. In some countries, politicians and the public lost faith. In Germany and Sweden it was decided to phase out nuclear reactors. Yet in France, the nuclear commitment strengthened to the point where the nuclear industry now provides close to 80% of that country's electricity.

By 2006, there were 443 reactors (about three-quarters being LWRs) around the planet, an installed capacity of about 370 GWe. These are often referred to as "Generation II" reactors (Generation I being the early prototypes of the 1950s and 1960s). The growth rate in Europe and North America is very small; the average age of reactors is therefore increasing and in the absence of a renewed nuclear strategy the number in these regions will steadily diminish as they reach the end of their useful lives. On the other hand, in the emerging economies of Asia nuclear growth continues and plans to accelerate it are in place; of the 24 reactors under construction in 2006 most were in Asia.

However, the global energy and environmental context has changed significantly. Oil and natural gas prices have been escalating dramatically as reserves run down; the inevitability of Hubbert's model is at last being accepted in many quarters (see Chapter 3). The value of oil and gas for home-heating, and the manufacture of materials such as drugs, fertilizers and plastics is such that the massive use of fossil fuels to generate electricity is being queried. The air pollution issue and its impact on human health are now front and centre in many jurisdictions. An example is the Canadian province of Ontario, where as Fig. 13-1 shows, the number of officially declared "smog days" in the capital city of Toronto has soared due to increasing air pollution and shutdown of nuclear generating capacity. Above all, the climate change issue has become crucial; the evidence for

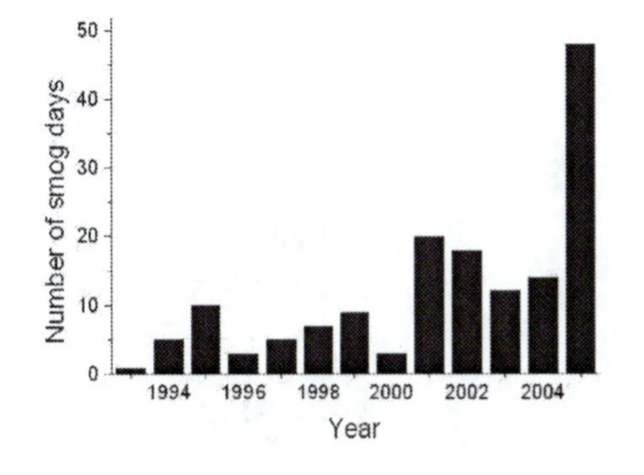

Figure 13-1 Annual Smog Days in Toronto

global warming induced by CO_2 emissions is now accepted by the scientific community with very few exceptions. Yet, despite the Kyoto Accord, very little progress is being made in reducing CO_2 emissions.

In this situation, it is important to carefully examine the pros and cons of nuclear energy, and to try to understand if a second nuclear era with enhanced safety offers the possibility of a major contribution to the Kyoto and air pollution problems.

13.2 FROM ONE GENERATION TOWARDS THE NEXT

Obviously, public perception is strongly influenced by the two major nuclear accidents that have occurred at Generation II reactors. These two accidents were very different in kind. In the Three Mile Island accident, the core was partially melted, rendering the reactor unusable, yet the escape of radioactivity was so small that the predicted number of additional cancers in the entire surrounding population ranges from zero to two. In this case containment worked. At Chernobyl, there were no containment structures and hence a huge escape of radioactivity from the RBMK reactor ensued when the core melted; to make things much worse, the event was initially concealed from the public.

The two main conclusions we would draw from these two events are: (i) nuclear power is not compatible with a political culture that does not place human safety first; (ii) the containment strategy of western reactors succeeded in its one major test, but the need for improvements in worker training and for re-examination of maintenance processes and safety systems became manifest.

There is no doubt that these events changed the entire nuclear industry culture. One example of this is the capacity factor data. Averaged over the lifetime of the reactors, unplanned shutdowns have often resulted in disappointingly low capacity factors. Table 13-1 shows data for 402 reactors (about 90% of the world total). The CANDU subset of PHWRs is an interesting case; in their early years of operation these reactors

Table 13-1: Reactor Average Capacity Factors [1,2]

Reactor type	Lifetime CF (%)	2005 CF (%)
PWR	73.1	81.8
BWR	70.3	77.2
PHWR	68.7	67.7
Candu-6	85.0	87.4
Magnox	58.6	56.5
RBMK	59.5	73.5

[1] Nuclear Engineering International, May 2006
[2] CANDU Owner's Group Station Performance Reports, December 2005

led the world with capacity factors exceeding 90%, but the performance fell off and in fact several of them had to be shut down in the 1990s; in contrast, the newer CANDU-6 model has re-established a good performance. Overall in the last two decades there have been remarkable improvements. Figure 13-2 shows the average performance for all the U.S. reactors; the mean capacity factor has increased steadily from about 50% in the 1970s to over 92% at the present time.

Again taking US data, the mean occupational radiation dose at nuclear power plants fell by a factor 5 between 1985 and 2001. These numbers reflect better training, better equipment and better procedures. They augur well for the closely-linked goals of reliability and safety.

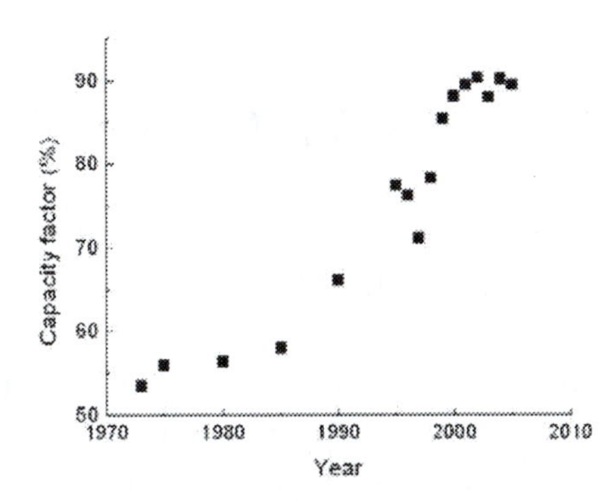

Figure 13-2 Capacity Factors for U.S. Nuclear Reactor Fleet. *Data from Department of Energy website.*

These data are however for Generation II. We have to ask now what is on the drawing boards, and whether new designs show major advances in safety. The crux here is to augment active safety systems with passive ones, and to move towards systems that provide operators with sufficient time to diagnose problems without undue pressure. Examples of active systems are valves that must be opened or closed by application of power, or emergency diesel generators that will switch on automatically if electrical power for the cooling systems is lost. Valves that revert without intervention to their safe position in the case of loss of electrical power, and emergency cooling supplied by gravity or natural circulation instead of by pumps are examples of passive systems.

Capital and operating cost is the next major issue we need to consider. The costs of the nuclear fuel and of its transportation constitute only a tiny component of the overall cost of a nuclear generating station. It is the cost of building and operation, aside from fuel costs, that have rendered nuclear power uncompetitive with natural gas in recent years; a necessary caution regarding this argument is that natural gas costs are increasing rapidly. Reactor construction times in North America have been excessively long, ten years being not unusual. In the first "learning" phase of nuclear power successive reactors often differed, making it difficult for their operators to share information and experience, which led to higher operating costs. These lessons appear to have been learned. For example, Atomic Energy of Canada standardized in the 1990s on the CANDU-6 design, and all 10 of these reactors (in five different countries) are operating at high capacity factors (average 85%).

Economies of size are another issue. Today's reactors are typically about 1000 MWe. Evolving designs for the next generation (see below) are somewhat larger. For example, Westinghouse designed its innovative, new AP600 reactor (600 MWe) and deduced that its final electricity costs would not be competitive with natural gas. By scaling the design up to 1000 MWe (the AP1000) it reduced that cost by one quarter. On the other hand, more is being heard now from those who argue for much smaller reactors -at the level of 100 MWe. These could be built in factories and transported to the site, instead of being built on-site. A large power plant would simply have enough of them to do the job, and could expand rather easily. In addition, an accident would be confined to one much smaller unit.

These are some of the considerations that are driving the third generation of reactor design. Some of these are ready for deployment; others are very close. In the next few sections we mention a subset of the rather large group of designs and programs that are in progress.

13.3 INNOVATIVE DESIGNS FOR LARGE REACTORS

The first Generation III reactors were designed in the 1990s, with the design focusing on enhancements in both safety and economics. As mentioned above, they are standardized, modular, and easier to operate, and have projected lifetimes of 60 years. Passive safety systems that can handle the first phase of a loss-of-coolant situation without immediate need for operator intervention are a significant new feature. Several Generation III reactors have been installed. As an example we can take the four U.S.–Japanese Advanced BWR reactors (AWBR) that have been operating reliably for several years in Japan. Each of these was built in 48 months; more are under construction. This model, whose design evolved from a successful line of conventional BWRs, is certified for use in the U.S. and in Europe, and is likely to be deployed soon in the USA. A key change is that the circulating water pumps are located inside the reactor vessel, which enabled a reduction in the size of the emergency core cooling system and therefore of the containment building.

Some sources use the term Generation III+ to classify designs that are now evolving and that we can expect to see in operation within a few years time

We cannot examine every new design that is at some stage of evolution. Every one of these shows improved safety and reduced accident risk. The common themes are standardization, simpler and more rugged design, greatly reduced construction time, high capacity factor and much longer lifetime, higher fuel burn-up, reduced environmental impact and reduced possibility of core meltdown. A few examples will be chosen to convey an overall picture of major progress.

Light water reactors

The AP600 reactor design, which, as was said above, has been scaled up to the AP-1000, provides us with some examples of passive safety systems in a PWR. It is shown in Fig. 13-3. The construction time for this highly modular unit is 3 years and the operating lifetime is estimated at 60 years. Both the reactor vessel and the steam generator are housed within the heavy steel pressure vessel, which in turn is within the concrete containment building. The first achievement here is to simplify all the systems: compared to a conventional 1000 MWe PWR there are 36% less pumps, 50% less valves, 83% less piping and 87% less electric cabling – much, much less equipment to go wrong. Newly designed passive safety systems employ gravity, natural circulation and compressed gas, instead of pumps, fans, diesel generators, chillers or other machinery.

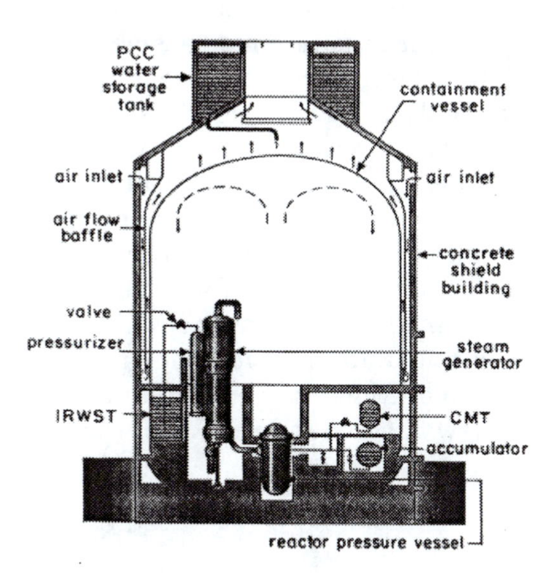

Figure 13-3 Schematic of AP600 Reactor. *From D. Bodansky, Nuclear Energy: Principles, Practice and Prospects, Springer, New York, 2004, with permission of Springer Science and Business Media.*

There are three sources of coolant in the event of a breakage or leak in the primary cooling circuit of the reactor. The first is the *core makeup tanks* (CMTs), containing water in which a compound of boron, a neutron absorber, is dissolved. These tanks have valves whose natural position is open, but which are held closed by the reactor pressure and water-level; so if the latter quantities fall below prescribed levels, the valves immediately revert to open, and the emergency coolant is applied to the reactor. The CMTs are supplemented by *accumulator tanks* from which the water is driven out by stored high-pressure nitrogen gas. For longer-term cooling in the case of an accident, there is a much larger water source – the *in-containment refueling water storage tank* (IRWST), which comes into action when the water level in the CMT has dropped sufficiently. As the reactor heat boils the emergency cooling water, the steam condenses on the inside surface of the containment vessel; its outside surface can be cooled for three days by water falling under gravity from the passive containment cooling (PCC) tank above the containment vessel; after that period, this cooling can be provided by a forced air flow or by externally provided water.

Another innovative design, in the BWR category, is the 1500 MWe ESBWR (ES means "Economic and Simplified"). The next stage of BWR design work by General Electric was a much smaller simplified SWBR with 600 MWe output, incorporating major emphasis on passive safety systems. That design was then incorporated into a new ESBWR design (Economic and Simplified BWR) with 1400 MWe output. This re-direction in size mirrors the path taken by Westinghouse with its PWRs. The ESBWR has reduced the number of valves, pumps and motors by 25%, and its passive safety system removes decay heat directly to the atmosphere.

The new 1600 MWe European Power Reactor (EPR) developed by France and Germany builds on a line of successful European PWRs, introducing better materials, improved control systems, and increased redundancy in safety systems; in this sense it can be described as evolutionary. It incorporates some passive safety measures, but to a lesser degree than the AP series. It is designed for a 60-year service life, a 92% capacity factor and a very high efficiency of 36-37%. This model, less modular than the AP1000 or the ACR designs, is claimed to provide electricity more cheaply than gas plants even in the absence of a carbon tax. The design includes a separate compartment in which the molten core would be trapped and protected in case of an accident. In the event of core melting, the heat would melt a steel plug, and the core material would progress into this "core-catcher", where it would spread itself over a large surface area of concrete containing channels through which cooling water circulates. The material would be cooled, passively, by water falling under gravity from a tank located within the containment vessel and by evaporation. It is expected that the cooled core would solidify completely within a few days. Construction of the first EPR model commenced in 2005 at Okiluoto, Finland.

Heavy water reactors

The design for the next generation of CANDUs sees major changes from the CANDU-600. The heavy water coolant is replaced by light water, which necessitates Canada importing slightly enriched uranium (1.5 - 2%). The moderator remains as heavy water. This design, with a smaller core, results in a smaller calandria, a 75% reduction in heavy water costs, improved burn-up of fuel, and a 25% reduction in fuel costs. The capital cost is reduced to a level (40% less than the established CANDU-6) at which this Advanced CANDU Reactor ACR-700 is expected to compete in electricity price with combined cycle gas turbine plants. The safety problem with tritium in the coolant is eliminated. A negative void coefficient and passive safety features provide further safety enhancement.

The other major user of HWRs is India, whose advanced heavy water reactor (AHWR) design is similar to CANDU as regards moderator and coolant. However its fuel assemblies will contain Th - ^{233}U elements and Th-Pu elements, and the pressure tubes of the calandria are oriented vertically rather than horizontally. The size is much smaller than CANDU, with 300 MWe output. The role of the AWHR in India's overall program is described in Section 13-5 below.

MOX fuel

With the possibility of reprocessing in mind, all of the above designs can use mixed-oxide fuel. In the ACR-700 and EPR cases, the MOX fraction could be anything up to 50%.

13.4 INNOVATIVE DESIGNS FOR SMALL REACTORS

An interesting recent concept is the modular "Pebble Bed" reactor (MPBR) that is being designed in South Africa, USA (MIT) and China. The original concept is German, and a demonstration unit operated there for 22 years. Figure 13-4 provides a schematic of the MPBR. The LEU fuel and the graphite moderator are intimately combined in tennis-ball sized spheres with an outer graphite coating and an inner fuel zone, and a multiple barrier coating to lock in the radioactivity. About 4% of each sphere is actual uranium. The reactor would contain about half a million spheres, three quarters being fuel spheres and the rest graphite moderator; these would be cycled repeatedly through the reactor, with some being removed at the base every day and new ones fed in at the top; a single sphere would spend three years in the reactor. The heat would be removed by high-pressure helium gas, which cannot be rendered radioactive in the reactor, and the helium would drive the turbines directly with no intermediate steam circuit. The multiple-barrier fuel concept is also intended to provide very long-term containment when the waste is eventually stored. The system uses cadmium control rods to govern the level of the chain reaction.

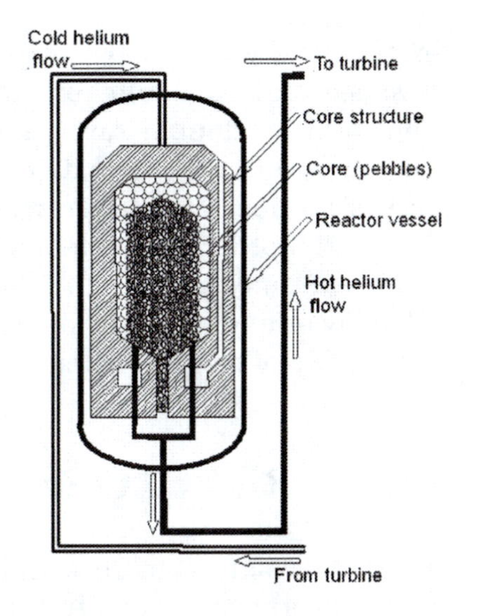

Figure 13-4 Schematic of Pebble Bed Reactor Concept

Figure 13-5 gives more details about the fuel+moderator spheres. Each sphere has a central fuel zone and an outer 5 mm graphite layer protecting it. The central zone has 15000 fuel kernels embedded in a graphite matrix. Each of these kernels has a tiny 0.5 mm sphere of UO_2 at its centre protected by successive shells of porous carbon, silicon carbide, and pyrolitic graphite – this is tougher than diamond and the SiC barrier retains its integrity up to 1700 °C, so that gaseous and metallic radioactive species cannot escape. This idea offers many advantages. The higher temperature combined with use of plus magnetic turbine bearings results in 45% efficiency. The continuous re-fuelling means that there need be no shutdowns. If there is an upward temp excursion or a loss of helium, the chain reaction slows down due to increased neutron absorption in ^{238}U. The fuel burn-up is higher than in a PWR.

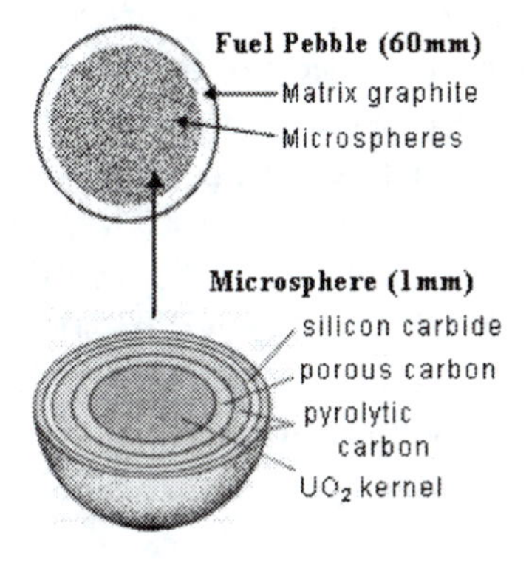

Figure 13-5 MPBR Fuel

Although China's main nuclear thrust is in PWRs, it also has a MPBR development program. It sees this small unit as ideally suited to power generation in remote areas with small demand, process heat for heavy oil recovery and the petrochemical industry, and process heat for conversion of coal to gas. (This raises the interesting question of why Canada is using its valuable natural gas resource for tar sands exploration instead of shifting to nuclear power for process heat).

On first examination, the MPBR looks very promising in terms of safety. Nevertheless safety issues have been raised by one expert [3] who points to the lack of a containment building in the event of accident or terrorist attack. He believes that further work is needed to fully demonstrate the integrity of the fuel spheres. And while the designers claim that ingress of air and water is prevented, it has to be borne in mind that the combination of air with hot graphite means fire. A final issue is that the volume of waste is more than ten times greater than for an equivalent PWR; but the heat density in the spent fuel is lower and so closer packing of waste is likely to be possible. In this case there is no question of re-processing.

13.5 BREEDER REACTORS

The French, British, German, Japanese and U.S. fast breeder reactor (FBR) programs were terminated or severely curtailed in the 1980s and 1990s. The issues were the complexity of the liquid sodium heat transfer system (which had leakage problems), the very high capital cost compared to ordinary reactors, and above all, concerns as to diversion of plutonium into weapon manufacture. Despite these problems, efforts in this area are re-commencing, mainly in Japan, China, Korea and India. The driving force is long-term sustainability of energy supply: FBR's offer the possibility of sixty times more energy per kg of uranium. If greenhouse gas concerns do result in a second era of conventional reactors, then it is likely that the pressure on uranium resources will eventually re-stimulate interest in FBRs in western countries.
An interesting direction is found in India, which is rich in thorium reserves, and has a three-stage long-term plan. Its existing fleet of twelve CANDU-like PHWRs supply electricity to the grid and produce plutonium as a by-product. In the second phase, it is intended to burn this Pu in FBRs, with a blanket of thorium plus uranium around the core; in this blanket both ^{233}U and more Pu will be created. To this end, construction of a 500 MWe FBR was started in 2004, based upon experience with a ten times smaller test FBR that has run successfully for 20 years. The third phase will involve the AHWR mentioned above, fuelled with the ^{233}U and Pu.

13.6 FUEL REPROCESSING AND SECURITY ISSUES

Despite large stocks of spent fuel, very little has been done in. N. America towards extracting fissionable plutonium to make mixed oxide fuel elements MOX. There is however a major project underway, mentioned in Chapter 11, to convert weapons-grade

[3] E.S. Lyman, Physics and Society 30, 16 (2001)

HEU and Pu into reactor fuel. In contrast, France and Germany are using MOX fuel in their reactors, averting the need for about 1400 tons of natural uranium per year. Different countries hold different views as to the merits of reprocessing, and underlying the debate is the concern about weapons material or even just highly radioactive material falling into the hands of terrorists or rogue states.

One school of thought is, roughly, as follows. Spent fuel is waste material; efforts need to be focused towards safe long-term storage of that waste. For the moment, uranium resources are plentiful and so there is not an urgency to re-process waste. Leaving Pu within the highly radioactive spent fuel makes it extremely difficult if not impossible for terrorists to extract Pu, whereas building up stocks of separated metallic plutonium would create a target for theft. Alternatively, with the integrity of the longest-term portion of the waste in mind, the Pu could be separated and converted into ceramic form. This would then be embedded within a highly radioactive matrix to deter theft or diversion.

The other side of the debate is very different. Here the theft or diversion of Pu is also of great concern but incorporation of the separated PU into MOX fuel provides a material that can be burned in the coming generation of reactors, which can handle 50-100% MOX. This approach not only gets much more energy out of the uranium resource but destroys the Pu that has been bred.

The Canadian Association of Physicists [4] studied the Pu separation issue and concluded that both the options - immobilization and MOX burning - are feasible and safe, whereas the greatest risk would be doing nothing, i.e., leaving the separated Pu in metallic form.

The authoritative MIT study [5] concluded that current international safeguards are not capable of dealing with future global growth of nuclear power, and that reprocessing as practiced in Europe, Japan and Russia present risks of proliferation. It believed that uranium resources were adequate to proceed without reprocessing for 50 years, and recommended that the USA adopt the policy of once-through use of fuel only, with no re-processing. It recommended a stronger focus on waste storage.

In contrast with this view, the industrializing Asian countries are proceeding with plans that involve extensive re-processing in the long term.

13.7 HIGH-TEMPERATURE REACTORS FOR PROCESS HEAT

A group of experts from ten countries and the European Union is now examining the long-term possibilities for the fourth generation of nuclear power. Of course, improved economics, enhanced safety, minimal waste and proliferation resistance are front and centre in their considerations. But in addition a major emphasis is the use of nuclear

[4] *Disposition of Weapons-Grade Pu as MOX fuel in Canadian Reactors*, Can. Assoc. Physicists, 2000.
[5] *The Future of Nuclear Power*, Massachusetts Institute of Technology, 2003.

reactors for hydrogen production. We have already talked earlier in this book about the use of hydrogen in fuel cells both for vehicle propulsion and for distributed electricity generation, and we will have more to say in Chapter 18 about the "hydrogen economy" concept. But for the moment, we need only remark that a major transition to hydrogen as a fuel will not benefit the greenhouse effect if the hydrogen is derived in the first place by using fossil fuels.

A reactor could produce hydrogen by using off-peak electricity to electrolyze water. Alternatively the heat energy could be used directly to drive chemical reactions that release hydrogen. One example is such a thermochemical process is the decomposition of sulfuric acid at 800 °C:

$$H_2SO_4 \rightarrow H_2O + SO_2 + \tfrac{1}{2}O_2 \qquad [13\text{-}1]$$

Then the following two reactions are employed:

$$I_2 + SO_2 + 2\,H_2O \rightarrow 2\,HI + H_2SO_4 \text{ (at 120°C)} \qquad [13\text{-}2]$$

$$2\,HI \rightarrow H_2 + I_2 \text{ (at 350 °C)} \qquad [13\text{-}3]$$

The net reaction is simply the conversion of water to hydrogen and oxygen, with all the reagents being recycled so that there are no effluents.

High-temperature gas-cooled reactors are seen as the appropriate technology for these hydrogen-generating applications. The pebble-bed reactor is an example, but other HTGR designs are being brought forward.

13.8 NUCLEAR FUSION

In Section 10.8 it was mentioned that there was a second method by which energy can be extracted from some nuclei. If very light nuclei are combined the binding energy of the resulting nucleus is increased and energy is released; this is called *nuclear fusion.* The process has not yet been achieved in a controlled way, but much research effort and money has been expended on it in the hope of developing controlled fusion reactors. These may provide a clean, almost boundless source of energy for the future, but that future is probably still quite distant.

The Physics of Fusion

In Fig. 13-1 it can be seen that if two very light nuclei, such as two protons, are combined into a heavier nucleus, the binding energy per nucleon is increased and energy is released. For reasons which we will see later, the reaction which has been most extensively studied is that of deuterium (${}^2_1H$) with tritium (${}^3_1H$).

$${}^2_1H + {}^3_1H \rightarrow {}^4_2He + {}^1_0n \qquad [13\text{-}4]$$

Using Table 10-4 the energy released is 4 × 7.074 – (3 × 2.827 + 2 × 1.112) = 17.6 MeV.

How do we know that this is even possible? This and many other fusion reactions have been observed in nuclear research laboratories using particle accelerators, and the reaction of Eq. [13-4] has been made to work on Earth albeit in a destructive way, that is, in the hydrogen bomb. In addition there is very strong evidence that the energy produced in the Sun and most stars results from the fusion of hydrogen into helium. The technical difficulty arises from the fact that the nuclear forces which bind the two hydrogen nuclei together into a helium nucleus – releasing energy in the process – have a very short range. Opposing the attractive nuclear force is the electrical repulsion caused by the positive charge on each of the hydrogen nuclei. Very large forces must be exerted on the two nuclei in order to push them close enough to each other so that the very much stronger but short–range nuclear force can take over and bind them together.

One of the methods chosen to accomplish this is to raise the temperature of a 50%–50% mixture of tritium and deuterium to very high levels. As the temperature (T) is raised the mean speed (v) of the particles rises according to the equation

$$v = (3k_B T/m)^{1/2} \qquad \textbf{[13-5]}$$

where m is the mass of the particle and k_B is Boltzmann's constant. At sufficiently high temperatures enough of the particles will make sufficiently strong head–on collisions to overcome the electric repulsion and fuse. The temperatures required to accomplish this are very high; it requires tens and even hundreds of millions of degrees to produce a significant number of fusing collisions.

This requirement of extremely high temperatures causes other severe difficulties. A relatively high density of material is required in order to get a significant number of fusions, but gases at high temperature exert high pressure if the densities are high. This can be seen from the Ideal Gas Law

$$pV = nk_B T \qquad \textbf{[13-6]}$$

where n is the number of particles in volume V at pressure P. Very high pressures would require a very strong container for the plasma which would contribute so substantially to the heat loss that the temperature could not be maintained. In fact, it is not possible to use a material container at all!

Since it is not possible to work at very high densities then some arrangement must be made to keep the particles confined in a restricted volume for a long time so that there is ample opportunity for the particles to make many collisions and thus increase the chance of having one of the rare head–on fusing ones. But the high temperature has made this difficult as well; any increase in the particle's speed will also increase its tendency to escape the reaction area. The problem reduces itself to one of finding a way to have a combination of a sufficiently high temperature, density (n, in particles per m^3) and confinement–time (τ in s). A condition known as the *triple product criterion* states:

$$T\tau\, n > 6\times10^{28}\ \mathrm{s{\cdot}m^{-3}{\cdot}K} \qquad \textbf{[13-7]}$$

If this condition is satisfied, self–sustaining nuclear fusion can occur. The technical solution will be to achieve high temperatures and long confinement times at as high a density as possible. So far this has been beyond the capabilities of modern technology.

Fusion Technology

At temperatures where fusion will take place, all materials are vaporized and completely ionized, that is, all of the orbiting atomic electrons have been stripped away from the nuclei. In ordinary matter the material is electrically neutral at the atomic level. Now, although the material is still electrically neutral in bulk, the individual particles are not. Matter in this form is called a *plasma*. The technology that has been most pursued is that of *magnetic confinement*. It takes advantage of the fact that charged particles in the presence of a magnetic field are trapped on the magnetic field lines and so can be contained without actually using a physical container. The most advanced device of this type is called a "*Tokamak*". [6] The elements of a tokamak are shown in Fig. 13-6. The plasma is free to move in a highly evacuated 'doughnut'–shaped toroidal chamber. The toroidal and vertical field coils carry electric currents that produce a magnetic field keeping the plasma confined along the axis of the toroid and preventing it from touching the walls and losing its heat. The plasma, being conducting, is itself a single loop of electrical conductor and, inserted through the centre of the toroid, is another magnet acting as a transformer (see Fig. 13-6). As current rises in the transformer primary coil, it induces a rising magnetic field in the magnet which in turn induces a current in the plasma heating it up by 'ohmic heating', i.e., its resistance. In addition to this method other technologies are used to heat the plasma such as injecting electro–magnetic waves. Many tokamaks have been built in ever–increasing

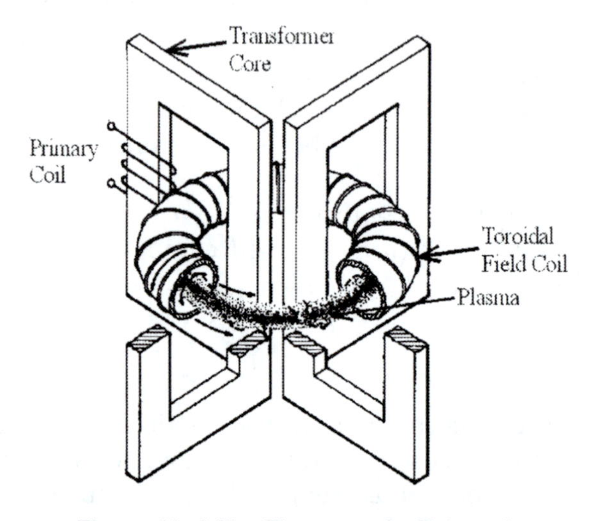

Figure 13–6 The Elements of a Tokamak.

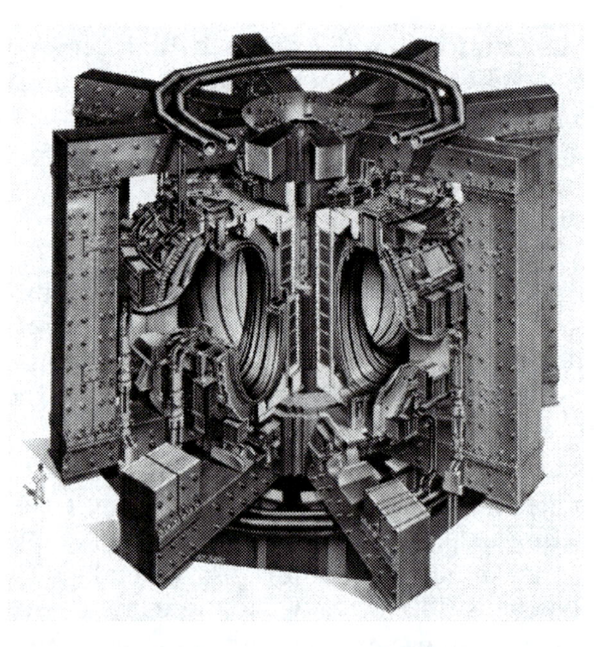

Figure 13–7 Drawing of the JET Tokamak. Note the figure of the man (lower left) for scale.

[6] "Tokamak" is a Russian acronym for "a toroidal chamber in a magnetic field". Such devices were first built in Russia.

size and complexity. None of these devices has been built to actually produce useful power. They have only been built to achieve certain objectives. The first was the objective of *breakeven*, that is, to produce as much heat energy from fusion as was actually put into the device to run it. The most recent devices are very large as can be seen from Fig. 13-7 of the Joint European Torus, (JET), at Culham, England. The best performance of some of these devices is shown in Fig. 13-8 where the plasma temperature is plotted against the triple product; the region of breakeven is indicated.

In 1997 JET was very near breakeven, and it and all the other large machines have since passed that barrier. Breakeven alone does not make a fusion power reactor. For that it is necessary first to achieve *ignition* where the reaction is self–sustaining, that is, all external heating power can be shut off and the plasma will maintain its own temperature. In terms of the triple product of $T\tau n$, calculations predict that ignition requires $T\tau n \geq 6\times10^{28}$ with T in the region of 100–300 million K. The region of ignition is also shown on Fig. 13-8. In spite of much effort and expenditure this has not been achieved. It is clear that we are a long way from a working fusion reactor. The 1997 performance of JET is typical of the largest tokamaks throughout the 1990s and further progress will require an even larger and much more expensive machine. This has given rise to the ITER (International Thermonuclear Energy Reactor) collaboration, a consortium of nations,[7] to design and build the next generation of machine. The consortium has been stalled in cost and location arguments for years and the prospects for its implementation seem poor in the first years of the 21st century.

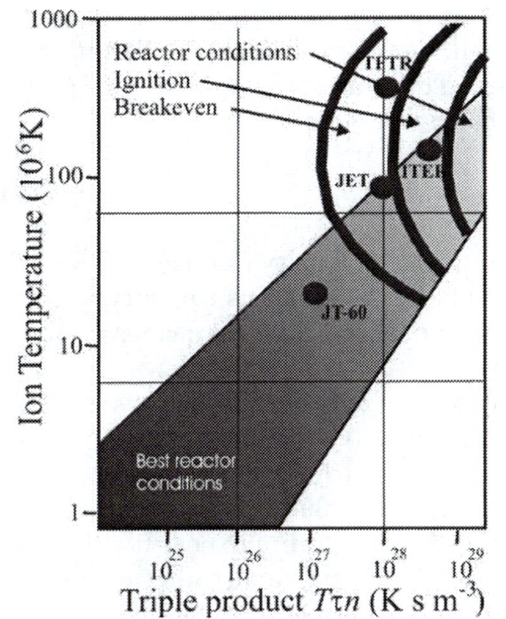

Figure 13–8 The Performance of various Tokamaks and the Design Target for ITER

Even if all of the foregoing were possible by whatever method of confinement, there is still a formidable problem in extracting the energy from the device. When the reaction given in Eq. [13-4] takes place, most of the energy is taken by the neutron. When the D and T nuclei collide head on they do so with almost zero total momentum. Since momentum is conserved, the fusion products ^{3}He and $^1_0$n also have zero total momentum. This means they recoil from each other but with unequal speeds because of their unequal masses. A simple calculation (see Problem 13-10) shows that the velocity of the neutron is four times that of the helium nucleus and it carries 80% of the reaction energy. It is hoped that the 20% of the energy remaining with the ^{3}He will sustain the heat of the plasma and the neutron's energy can be extracted. Since the neutron has no charge it is not confined by the magnetic field and passes out of the plasma. As we have seen earlier, neutrons are difficult to capture but if this is accomplished then the neutron's energy will be transferred to the capturing nucleus and appear as heat in the

[7] The European Union, Japan, China, Korea, India, Russia and the USA.

bulk material. How this is to be effected bears on the subject of *fusion fuel* and is discussed below.

Inertial Confinement

A completely different method for achieving fusion has been proposed and received considerable funding and attention in the 1970s and '80s: *inertial confinement.* The idea here is to irradiate a small spherical pellet of solid D–T mixture from all sides with a very powerful pulse of light from a laser. The explosion of material from the surface of the pellet creates a shock wave that compresses and heats the remaining core of the pellet. The compression must take place in less than 10^{-9} s and the relevant criterion can be expressed as

$$\rho R > 30 \text{ kg}\cdot\text{m}^{-2} \qquad \textbf{[13-8]}$$

where ρ is the density (kg/m^3) and R is the initial radius of the pellet. For pellets of the order of 10^{-3} m in radius this means an increase in density of between 10^3 and 10^4. Enormous lasers have been constructed to investigate this possibility but such compressions are still a long way off. Research in this area has not been well funded in recent years and there is, at present, an air of disenchantment with inertial confinement. A very large array of 192 lasers called the "National Ignition Facility" (NIF) is under construction at the Lawrence Livermore Laboratories in the USA for completion in 2009. It is intended to direct an energy of 1.8 MJ onto a 2 mm sphere of solid D–T. The main purpose of NIF, however, is for weapons research and only secondarily to continue studies of controlled fusion.

Fusion Fuel

It is clear that if the present research on fusion ever bears fruit the fuel will be a mixture of deuterium (D) and tritium (T). One of the popular misconceptions about fusion on Earth is that when its proponents talk about "hydrogen fusion" they mean the fusion of ordinary hydrogen as takes place on the Sun. This is definitely not the case and some of the proponents have been disingenuous in not making this clear in their publicity. Of course ordinary hydrogen is almost limitless in supply in ordinary water but the reaction that takes place on the Sun is, in fact, quite different in that its reaction depends on a different aspect of the nuclear force. The solar reaction will probably never be within the grasp of technology for very fundamental reasons; only the Sun (and other stars) can do what the Sun does.

The proposed fusion fuels D and T are isotopes of hydrogen but are much rarer; in fact only one of them, D, occurs naturally in any reasonable quantity.

- **Deuterium** is the stable isotope of hydrogen and occurs in about one part in 7000 in normal hydrogen, and so about one in 3500 water molecules on Earth is HDO (see Problem 13–2). Only one in 50 million water molecules actually is D_2O. Deuterium is extracted from ordinary water by a series of hydrogen exchange reactions to build up the concentration of D_2O to about 6% and then to further enrich it to 99% or higher by fractional distillation. The latter is possible since the boiling points of ordinary and heavy water are slightly different. All of these processes are very energy intensive and often use the waste heat from a power plant to carry them out. The price of heavy water is rather high because of the large amount of energy required and the complexity of the separation plants. Among the nations of the world Canada is the most advanced in the technology of producing heavy water because D_2O is required as the moderator and coolant in the CANDU power reactors.

- **Tritium** does not occur in nature except for a very small amount produced by the action of cosmic rays on water in the atmosphere. It is, in any case, radioactive with a half life of 12.3 yr. so it must be manufactured continuously for use as a fusion fuel. For experimental and start–up purposes, Canada could again be a major source, as tritium is produced in the CANDU nuclear reactors (see Section 11.1) and a tritium separation facility has been built in conjunction with the Darlington Nuclear Power Station in Ontario.

In a fusion power reactor it is proposed to solve the problem of energy capture and tritium fuel production in one step. It is proposed to blanket the reactor with thick layers of some lithium compound. About 7% of lithium is ^{6}Li and this nucleus will absorb a neutron according to the equation

$$^{6}_{3}Li + ^{1}_{0}n \rightarrow ^{3}_{1}H + ^{4}_{2}He \qquad \textbf{[13-9]}$$

This reaction also produces an extra 4.8 MeV of energy. The net result is to heat up the lithium and to produce one of the necessary fuels ($^{3}_{1}H$) for further operation. Both the heat and the $^{3}_{1}H$ must be extracted from the lithium blanket by continuous processing. The 93% of ^{7}Li can also assist via the reaction

$$^{7}_{3}Li + ^{1}_{0}n \rightarrow ^{3}_{1}H + ^{4}_{2}He + ^{1}_{0}n \qquad \textbf{[13-10]}$$

In this case, however the reaction extracts energy from the neutron but with 80% of the fusion reaction energy the neutron has more than enough. Needless to say, none of this technology has been carried out on a large scale.

It is often stated that fusion energy production will be the clean, safe way to produce limitless supplies of energy. "Clean" is interpreted to mean "without the radioactive waste" of fission reactors. While it is true that the only waste of the fusion reaction is the harmless inert gas helium, the process is not without its radioactive hazard. The tritium budget of a fusion reactor will be very large and an accidental release could be quite serious. Most fission reactor wastes are not gases but tritium is, and so it is easily dispersed. In the form of radioactive water it would be very damaging in the biosphere. In addition, the intense neutron fluxes of fusion reactors will render many structural members of the reactor radioactive, and it is a challenge to technologists to find ways to minimize or control this. In fact some estimates indicate that per unit of electricity produced, the quantity of radioactive waste will be comparable to that from a fission reactor. The major difference is that it will not have very–long–lived waste products that make fission wastes so controversial and difficult. Nevertheless some estimates indicate that the decommissioning costs of a fusion reactor will be comparable with those of a fission reactor.

The Future For Fusion

At the beginning of the twenty–first century the future utilization of fusion looks farther away than ever. The proponents of the technology have had a long history of presenting optimistic predictions of the imminence of a working system – this has always met with disappointment. Every step in research or development has been progressively more costly. Many governments have ended financial support owing to the discouraging results so far. Those nations still interested have come to realize that only international co–operation can supply the money needed to continue and that success is probably further away than has been realized. The scale of research has been reduced everywhere particularly as the energy crises of the 1970s and '80s receded. Useful, controlled fusion energy may yet become available but late in the twenty-first century at the earliest.

Cold Fusion

The sheer weight, size and complexity of fusion technology are very daunting; it is not now clear whether it will ever be economically feasible. Imagine the surprise, shock and even incredulity on the part of physicists and engineers when, on March 23 1989, two chemists announced that they had carried out nuclear fusion in a test tube at room temperature with a large release of energy! The form of the announcement was a press conference, a form that angered and alienated almost all of the scientific community right at the beginning. The accepted way of announcing scientific results is in the scientific literature; results are submitted for scrutiny by scientific peers before public announcements are made, especially truly earth-shaking ones.

The claim was made that by passing an electric current through heavy water (which contains deuterium) between electrodes made of palladium, more heat was evolved than could be accounted for from the electric energy input. The claim was made that only the fusion of deuterium nuclei within the palladium lattice could account for this energy. There were also additional claims that some of the signatures of fusion were also present such as the emission of nuclear particles and radiation. Many scientists were sceptical but at the same time there was a rush to the laboratory on the part of electrochemists and nuclear physicists everywhere; the promise of cheap boundless energy from apparatus that cost almost nothing was too good to pass up.

In the first few weeks there were many experiments that failed to confirm the claim but a few that provided partial confirmation. As time went on, however, the sceptics were proved correct as one after another of the positive results was explained away. There may indeed be something peculiar about electrochemistry with palladium and heavy water but there seems to be no substantial evidence that nuclear fusion is involved. The exact nature of the initial experiments has never been fully disclosed, and the motives of the proponents of cold fusion in choosing the method of press releases over peer review, has never been explained. Although a few die-hard believers still claim that the effect is real (although much weaker than first proposed) the vast majority of scientists have dismissed the claims and relegated the phenomenon to one of the more bizarre episodes in science.

EXERCISE

13-1. What is the energy released in each of the following reactions? (Use Table 12-1)
a) ${}^{2}_{1}H + {}^{2}_{1}H \rightarrow {}^{3}_{1}H + {}^{1}_{1}H$
b) ${}^{2}_{1}H + {}^{2}_{1}H \rightarrow {}^{3}_{2}He + {}^{1}_{0}n$
c) ${}^{2}_{1}H + {}^{3}_{2}He \rightarrow {}^{4}_{2}He + {}^{1}_{1}H$

PROBLEMS

13-2. Of the two reactions involved in making tritium by neutron capture on lithium, one produces energy and the other takes energy from the neutron. Calculate the energy for each case (use Table 10-4).

13-3. Deuterium (D) occurs as 1 atom in 7000 in normal hydrogen (H). Show that for every molecule of D_2O in normal water there are 14×10^3 molecules of HDO and 49×10^6 molecules of H_2O.

13-4. How many molecules of D_2O and HDO are in 1 m^3 of ordinary water?

13-5. What mass of D_2O, HDO and D is in 1 m^3 of normal water?

13-6. If all of the D in 1 m^3 of normal water was fused with tritium how much energy would be released?

13-7. If all the D in 1 m^3 of normal water was fused with tritium how long would it run a 1000 MWe power plant operating at 33% efficiency?

13-8. Considering the power plant in problem 13-7, what would be the activity of the tritium required to run it for one year? The half life of tritium is 12.3 yr.

13-9. What temperature is required so that two nuclear particles, each bearing one positive electric charge, and of masses m_1 and m_2, will be brought within 10 fermis (1 fm = 10^{-15} m) of each other in a head-on collision if they are moving at their mean speed? (Remember the electric potential energy of two charges separated by a distance r is given by $U = kq_1q_2/r$ and $k = 9 \times 10^9$ N m^2 C^{-2}).

13-10. The performance of JET, illustrated in Fig. 13–7, was a result of operation with the following parameters:

$$T = 310 \times 10^6 \text{ K}$$
$$\tau = 1.4 \text{ s}$$
$$n = 2.1 \times 10^{19} \text{ m}^{-3}$$

By how much would the confinement time have to be increased at the same temperature and density to achieve ignition? (You will have to scale data from Fig 13–8.)

13-11. A tritium and a deuterium nucleus, with a total momentum of zero, fuse to produce a helium nucleus (mass = 4u) and a neutron (mass = 1u). They recoil from each other with speeds v_{He} and v_n. Show that

$$KE_n/KE_{Total} = 0.8$$

13-12. a) What mass of deuterium would be required daily to power a reactor based on D-D fusion (see Exercise 13–1(a))? The reactor produces 1.00×10^3 MW of electrical power at 50% efficiency. Assume that the energy released in one fusion event is 4.0 MeV.
b) In this reactor, what mass of tritium is created daily? What is the activity of this freshly- produced tritium? The half life of tritium is 12.3 years.

13-13. The oceans contain about 1.3×10^{24} cm^3 of water. Deuterium constitutes 0.028% by mass of natural hydrogen. What is the total energy (in joules) available from this deuterium via D-D fusion? Assume 4.0 MeV per fusion event. For how many years could fusion reactors of 50% efficiency supply a power requirement of 2.0 million MW? (This requirement is expected to be reached early in the 21st century.)

13-14. The NIF laser array will use 192 lasers to deliver 1.8 MJ onto a target of solid D-T which is 2 mm in diameter. If all of the D-T is fused (an unreasonable assumption) what percentage of the expended energy will be recouped as fusion energy? The densities of solid deuterium and tritium are 196 and 309 kg/m^3.

ANSWERS

13-1. **a.** 4.03 MeV, **b.** 3.27 MeV, **c.** 18.4 MeV
13-2. 4.79 MeV; -2.5 MeV
13-4. 6.8×10^{20}; 9.6×10^{24}
13-5. 2.3×10^{-5} kg; 0.30 kg; 0.032 kg
13-6. 2.7×10^{13} J
13-7. 2.5 hr.
13-8. 6.1×10^{19} Bq
13-9. 550 million K
13-10. 6.2 s
13-12. **a.** 1.8 kg, **b.** 1.3 kg; 4.8×10^{17} Bq
13-13. 3.9×10^{30} J; 3×10^{10} yr!!!
13-14. 0.01%

CHAPTER 14

RENEWABLE ENERGY I

A general recognition that the carbon dioxide build-up in the atmosphere due to fossil fuel burning must be brought under control, combined with concerns (some justified, others not) about the nuclear alternative has focussed public attention towards other sources of energy, especially renewable ones that minimize pollution. In this chapter and the following one, these are explored – solar radiation, winds, waves, tides, geothermal and biomass. Without question the renewable source that can provide the greatest power is the sun. The ground-level solar power incident on the USA is estimated [1] to be about 500 times greater than the total present US power consumption. The next strongest source available is the wind – the estimated power in surface winds in the USA is estimated to be about 30 times US power requirements. All other sources, such as photosynthesis (to produce biomass), geothermal energy, tides and waves would each make a much smaller contribution. Because wind energy is the renewable alternative that is making the greatest inroads we start with it.

14.1 WIND ENERGY

The kinetic energy of the wind has been harnessed for centuries by sailing ships, windmills, etc. Now wind energy is emerging rapidly as a non-polluting source of electrical energy. One important consideration when contemplating such an alternative energy source is whether the power it can provide is large enough (on a global or national basis) to replace a significant amount of fossil fuel. A second is whether it has a high enough capacity factor to provide baseload power.

Figure 14-1 5 MW REpower Wind Turbine installed in Germany

In the typical wind turbine (see Fig. 14-1), the blades rotate a small turbine which in turn generates electricity. Most models that are now in operation have outputs of 1 -2 MWe. However, the wind industry is now ramping up the size, and the largest model known to us at this time is the 5 MWe device shown in the figure.

[1] J.M. Fowler, Energy and the Environment (2nd Ed.), McGraw-Hill, New York (1984) p.364.

14.2 CALCULATION OF POWER FROM A WIND TURBINE

For simplicity we will look at the case where the wind blows at right angles to the blades of a horizontal axis turbine like the one shown in Fig. 14-1. If the blades sweep out a circle whose radius is r, and the velocity of the wind is v, then in one second a cylinder of air having length v and radius r passes through the blade area. The volume of air is then $\pi r^2 v_1$ and and if the density of the air is ϱ, then the mass is $M = \pi r^2 v_1 \varrho$. The kinetic energy of this piece of air is $\frac{1}{2} M v_1{}^2 = \frac{1}{2} \pi r^2 v_1{}^3 \varrho$. Because this is the energy arriving per second, this expression is the input power to the wind turbine, i.e.,

$$P_{in} = \tfrac{1}{2} \pi r^2 v_1{}^3 \varrho \qquad \textbf{[14-1]}$$

Because the blade radius is squared and the wind velocity is cubed in this equation, the merit in using large blades and locations that have high wind speed is very clear. The equation tells us that increasing the wind speed from 8 to 10 ms^{-1} (i.e., by 25%) will nearly double the electrical output from the turbine that is spun by the blades. Not only is the wind speed 60 m above the ground 20-60% greater than at 20 m height, depending on the local terrain, but the air is colder and has about 10% greater density.

A few more steps are necessary before we reach the output power. All the power given by Eq. (14-1) cannot be converted to electricity. If we assumed this to be the case, it would imply bringing the approaching air to a complete standstill, in which no further air could pass through the blades. It is clearly impossible to convert all the incoming kinetic energy into rotational energy of the blades. In practice, only a fraction of the incoming energy can be converted, and the wind speed is thus decreased from v_1 to v_2. Let us make the assumption now that the actual air speed at the plane of the blades is the average value $v_m = \frac{1}{2} (v_1 + v_2)$. Then the mass of air passing through per second needs to be re-written as $M = \pi r^2 v_m \varrho$. The kinetic energy extracted per second from the wind and converted to rotational energy of the blades is

$$P = \tfrac{1}{2} M v_1{}^2 - \tfrac{1}{2} M v_2{}^2 = \tfrac{1}{2} \pi r^2 v_m \varrho (v_1{}^2 - v_2{}^2)$$

and so

$$P = \tfrac{1}{4} \pi r^2 \varrho (v_1 + v_2) (v_1{}^2 - v_2{}^2) \qquad \textbf{[14–2]}$$

This can be compared to the incoming power given in Eq. [14-1] and their ratio R written in terms of the velocity ratio $x = v_2/v_1$ as

$$R = 0.5 (1 - x^2) (1 + x) \qquad \textbf{[14-3]}$$

This expression will take its maximum value when $dP/dx = 0$; if we differentiate R with respect to x, and set the result equal to zero, we find that $x = 0.33$. This leads to an R value of 0.59. This outcome was first calculated by Betz and is known as Betz's Law. The theoretical maximum efficiency of a wind turbine is 59%. The maximum rotational power is

$$P_{max} = 8 \pi r^2 \varrho v_1{}^3/27 \qquad \textbf{[14-4]}$$

In practice, of course, there remain inefficiencies such as friction in conversion of the incoming kinetic energy to rotational kinetic energy. The practical efficiency factor η for converting the input power of Eq. [14-1] to output rotational power will always be less than 0.59. And in the conversion of the rotation energy to electrical energy there will be a further efficiency factor η_e which is only slightly less than 1.0. So we can finally just write the output electrical power as

$$P_e = \tfrac{1}{2}\, \pi\, r^2\, v_1^3\, \varrho\, \eta\, \eta_e \qquad \textbf{[14-5]}$$

The most efficient designs at present have overall efficiency values around 45%.

EXAMPLE 14-1

In winter, a typical southern Ontario house uses about 40 kWh of electrical energy daily. (This includes water heating but not space heating). The annual average wind speed is about 5.5 ms^{-1} (a moderate breeze) at a height of 10 m above the ground. In order to provide all the electrical power for this home using a wind generator with 35% overall efficiency, what rotor blade diameter is needed? Assume that the wind blows continuously, which is obviously not so.

SOLUTION

Power = Energy/Time = 40 kWh/ 24 h = 1670 W

Now use Eq. [14-5], with P = 1670 Js^{-1}, v = 5.5 ms^{-1}, ρ = 1.3 kgm^{-3}, and $\eta\ \eta_e = 0.35$; this gives r = 3.74 m. So the required diameter is 7.5m, which is not a practical value for a typical backyard. We have not included any shadowing effects due to buildings, trees or other wind turbines.

14.3 PROFILE OF THE AIR FLOW THROUGH A WIND TURBINE

The fact that the incoming cylinder of air slows down as it passes through the blades has the consequence that the diameter of the cylinder is increased downstream, as shown in Fig. 14-2. The mass of air in the incoming cylinder at the left is $M_1 = \pi\, r_1^2\, v_1\, \varrho$, and the mass in the exiting cylinder at the right is $M_2 = \pi\, r_2^2\, v_2\, \varrho$. Because the mass is unchanged, i.e., $M_1 = M_2$, these two expressions are equal. Thus

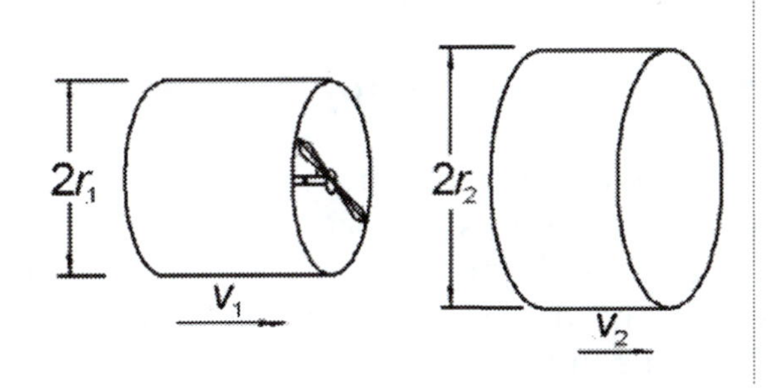

Figure 14-2 Behaviour of Wind Profile as it passes through Turbine

$$r_1^2\, v_1 = r_2^2\, v_2$$

But $v_1 < v_2$, and so $r_2 > r_1$. This effect is one of the factors that determine the spacing of turbines in a large-scale wind farm.

14.4 WIND SPEEDS AND LOCATION CRITERIA

Weibull distribution

At any place the wind speed varies from day to day throughout the year. Fig 14-3 shows a plot of the probability distribution of wind speeds at a hypothetical site; this function is called the Weibull distribution. The area under the entire curve is 1.0. Because very high winds are less likely than weak winds, the curve is asymmetric, and so the average wind speed is not equal to the median wind speed (defined as the speed at which half the area lies to the right and half to the left). In the example shown, the mean speed is 7 m s^{-1} and the median value is 6.6 m s^{-1}.

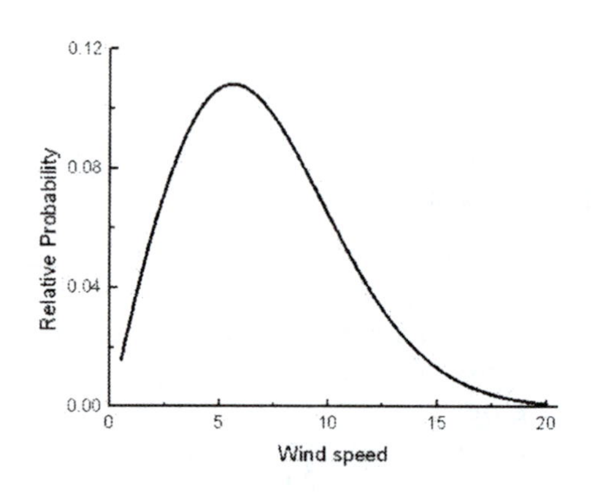

Figure 14-3 Weibull Distribution with Mean Wind Speed 7 m/s

If we substituted the average wind speed at our site into our equation for the power produced, we would make a big mistake, seriously under-estimating the actual power that a turbine would produce. This is because the wind speed appears in our equations to the third power, i.e., as v^3. If we double our wind speed from, say, 5 to 10 ms^{-1}, we get *eight* times more power. The power produced depends very much on the shape in the upper portion of the Weibull distribution that describes the highest wind speeds at the site. What we have to do in practice is write a computer program that breaks the speed axis into small intervals; then for each interval $v + dv$, we have to find the corresponding contribution to the power, which is $P(v) = Prob(v)\ \frac{1}{2}\ \pi\ r^2\ v^3\ \varrho\ \eta\ \eta_e$. Finally all these contributions are added to account for all velocity intervals.

Wind speed classification

Obviously, wind turbines must be sited at locations where the wind speeds are high. Table 14-1 shows the formal classification of land areas in terms of their average wind speed at a height of 50 m above ground.

Table 14-1: Wind speed classes

Wind power Class	Average wind speed ms^{-1}	Wind power density W/m^2
1	< 5.6	0 – 200
2	5.6 – 6.4	200 – 300
3	6.4 – 7.0	300 – 400
4	7.0 – 7.5	400 – 500
5	7.5 – 8.0	500 - 600
6	8.0 – 8.8	600 – 800
7	> 8.8	> 800

Wind maps for different countries can be drawn using different colours or shadings to depict areas that fall into the different classifications in the table. Ontario's wind speeds are comparable to and in many places higher than those in Denmark and Germany, the two countries which are leading the world in terms of wind-based electrical generating capacity. For example, the average wind speed at Danish turbines is 6.3 ms^{-1}, while the prevailing south-west winds along a good portion of the shores of the Great Lakes are 6 to 8 ms^{-1}. In the USA, there are 232000 km^2 of Class 4 land within 10 km of existing transmission facilities, which is eight times more land than all the Class 5 and 6 land put together.

The last issue here is the variability of the direction the wind blows from. How much power is available from the various points of the compass? This information is often presented in a diagram called a "wind rose", where a compass is divided into 36 sectors and the radius of each is equal to the mean wind speed from that direction. A site is preferable when the wind tends to come from a fairly small range of directions, as is often the case on a sea-coast. Wind turbines are equipped with direction sensing devices, which generate signals that instruct machinery to keep the turbine area oriented at right angles to the wind direction. The wind rose diagram becomes more useful when the quantity plotted is the product of the fraction of time that the wind blows from a particular direction multiplied by the cube of the corresponding mean wind speed. That quantity represents the distribution of actual power from all points of the compass.

14.5 DESIGN ISSUES

Three decades ago efforts were focussed on vertical-axis designs, the most common being the egg-beater style Darrieus rotor (named for the French inventor who patented it in 1920); it is shown in Fig. 14-4. The major advantage is insensitivity to the wind direction, unlike the horizontal-axis design of Fig. 14-1, which needs a mechanism to keep the blades perpendicular to the wind. Another advantage is that the turbine and generator are located at the base, thus providing easy access for maintenance. But there are major disadvantages in the Darrieus design. The rotor is near the ground where the air density is less than at a few tens of metres. Each blade must move against the wind during part of the rotation. Guy ropes are needed to anchor the structure, thus decreasing the land area that is available for simultaneous use.

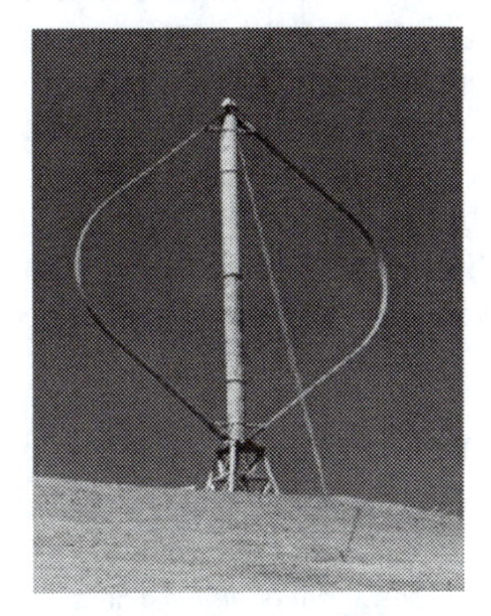

Figure 14-4 Darrieus Rotor

The present focus is therefore entirely on the horizontal-axis design. Sizes have increased dramatically as designers aim for greater height (higher air speed and lower density) and greater blade radius. The swept areas of the large turbines have diameters up to and beyond 100 m, with corresponding heights. Operating plants have typically 1 – 2 MWe power output and new installations reach 5 MWe. They require yaw control systems to maintain the plane of the blades at right angles to the wind direction. Most (not all) operate at constant or nearly constant rotation speed. A variety of engineering approaches, including adjustment of the blade angle, is used to achieve this. The blades need to be light but they also need to be very strong.

14.6 ENVIRONMENTAL IMPACT

The obvious advantages are that no fuel needs to be transported or used and no air pollution or waste material is created. Wind energy is renewable in the sense that the resource is not depleted. In the early stage of development noise was a concern but modern turbines make little noise above the rush of the wind; at 200 m distance they can scarcely be heard. Metallic blades caused TV interference but most blades are now made from composite materials such as fibre-glass which alleviate that problem. Some concerns have been raised about the effects upon weather of extracting large amounts of energy from the lower portion of the atmosphere, and this is now under examination through modelling studies.

Bird-kills are a concern that is often voiced, but for inland wind farms this problem has decreased with the shift from open steel towers to tubular concrete ones. The European experience is that bird kills are fewer than with standard large office buildings. However this does not remove the need when making siting decisions to give careful consideration to both migratory patterns (a concern with wind farms in high mountain passes in the USA) and feeding/nesting patterns of birds. A three-year study [2] of a large eider duck population near a Danish off-shore wind park concluded that the birds kept a safe distance from the turbines but were not intimidated as regards foraging for food in their preferred areas. More recently [3], there has concern regarding the sea eagles near the 68-turbine Smola wind farm off the north-west coast of Norway; these are Europe's largest eagle species and they had reached extinction in Scotland, where they have subsequently been re-introduced. Four eagles were found dead due to multiple heavy impacts and thirty failed to return to nesting sites. These findings are now an issue as plans are made for major wind development on the north-west coast of Scotland.

The aesthetic issue is also increasingly controversial. Many people view large wind farms as an eyesore and in some countries there is concern about impact on the tourism industry. In fact, agriculture, forestry and recreation can easily co-exist with wind generation. Land can be cultivated within a few metres of the base of the tower. In a typical wind farm only 1 - 2% of the land is occupied by the actual structures, with 12 - 24 turbines per km^2. This co-existence is one of the factors behind the remarkable growth of wind power in Europe. However the move towards off-shore farms does reflect in part the aesthetic concern.

---

[2] *Impact Assessment of an Off-Shore Wind Park on Sea Ducks*, NERI Technical Report 227 (1998). Danish Ministry of the Environment and Energy

[3] Scotsman, 28 Jan 2006

Finally the accident potential lies, prosaically, in the risk of workers falling during servicing activities. Careful precautions, similar to those used for work on any high structure, must be in place.

14.7 PROGRESS, INSTALLED CAPACITY AND COST

Wind turbines can be prefabricated quickly in a variety of sizes and easily installed using large cranes. Construction time is 1 - 2 y, which can be compared to 3 - 4 y for a nuclear reactor. The annual growth rate of installed capacity was an impressive 32% over the period 2000-2004, the global total in 2004 being about 47000 MWe. The great majority of this was in Europe, where the Kyoto accord is taken very seriously. Among the most ambitious countries is Denmark, which plans to generate 30% (5500 MWe) of its power from the wind by 2030, with more than 4000 MWe being generated offshore to capture the steady winds from the North Sea. Offshore generation is also important in Germany, whose wind industry employed 45000 people in 2005. An example is the E112 power plant near Magdeburg, whose 120 m high tower has three rotor blades made of fiberglass- reinforced plastic. Each blade is 52 m long, 6 m wide and weighs 20 tons. The nacelle which houses the axle, generator and windmill mounting weighs 440 tons.

Table 14-2: Some National Installed Wind Power Capacities (MWe) in 2002 and 2004: *Assembled from Website of Global Wind Energy Council*

Region	2002	2004
Germany	12001	16629
Spain	4830	8263
Denmark	2880	3117
Italy		1125
Europe	23056	34205
USA	4685	6740
Canada	236	444
Australia	104	380
China	473	700
Japan	414	740
India		3000

As with any new technology, installation costs are falling as the technology matures and the installed capacity grows. In the USA the basic generating cost per kWh fell from 80c in 1980 to 10 cents in 1991. A similar trend has occurred in Europe, assisted by government subsidy of early installations. It is difficult to pin down the present costs per kWh with a high level of accuracy, partly because these depend strongly on the typical wind speed and hence on the land classification where the turbines are sited; and large wind farms give very considerable economies of scale. The American Wind Energy Association claims a cost of 3.68 c/kWh from a large (50 MW) state-of-the-art farm with a very high wind speed of 8.1 m/s; but a 3 MW farm at this site would have a 5.9 c/kWh cost. Both these costs are after application of the present 1.5 c/kWh federal incentive subsidy. The U.S. Department of Energy has set a goal of reducing the cost for large installations in Class 4 regions (7.0 - 7.5 m/s) to 3.6 c/kWh by the year 2012. After the basic generating cost, one has to consider the cost of linking this variable source into the grid, and the long-term operating costs of repair and maintenance. The

DOE's 2006 estimate [4] of base installation costs for a typical 50MWe farm is $1091 per installed kW. However, for offshore installations, the cost could range up to 100% greater. To encourage wind development , the government of Ontario offered in 2006 to pay suppliers 11c/kWh for wind-generated electricity, a figure which can be compared with a dual householder cost (excluding transmission) of 5.8 c/kWh for the first 600 kWh in a month and 6.7 c/kWh thereafter.

New designs can be expected to reduce the costs further. Examples include advanced airfoil shapes for rotor blades, computerized electronic controls that allow the turbine to turn at optimum speed for high efficiency under a variety of wind conditions, and lighter, stronger materials.

14.8 VARIABILITY OF THE RESOURCE

The European Wind Energy Association [5] claims that the 40.5 GW capacity installed in Europe by the end of 2005 will produce 83 TWh of electrical power; these figures correspond to a 24% capacity factor. This is very low compared to the fossil or nuclear alternatives. On the positive side, in the winter season when electricity demand is very high, driven by demand for heat in buildings, the winds speeds are high and the air is colder and denser. For example 68% of the wind energy production in Ontario would be in the October - March period. But in much of North America, on hot, humid summer days, air-conditioning demand is maximum; again taking Ontario as example, the maximum daily electricity demands occurs in July and August. And on these days there tends to be very little wind.

These observations illustrate the major issue with wind-generated electricity. When wind supplies only a small fraction of the overall power in an electricity grid, this is a useful addition and the intermittent nature can be coped with. But if wind becomes a major contributor, then the loss of a major fraction of power on still days is not tolerable in an industrial economy. In a worst-case scenario, with a heavy reliance on wind power, combined cycle natural gas plants would have to be run up quickly to cope with the demand for power. This suggests that wind installations would have to be duplicated by conventional installations, and the cost of such duplication of plant is clearly a very significant economic issue. The need for research and development of energy storage technologies now becomes obvious. For example, at appropriate sites, one might use wind-power in "on-periods" to pump water up in a hydro-electric storage plant, and then generate power when the wind is still. In the longer term one can envisage using wind-generated electricity to electrolyze water and thus create hydrogen for use in distributed fuel-cell generators.

---

[4] *Assumptions to the Annual Energy Outlook*, Report DOE/EIA-0554(2006), U.S. Department of Energy (2006).
[5] European Wind Energy Association website.

14.9 TIDAL ENERGY

You probably know that in the past there have been grain-grinding mills powered by wind or by falling water in rivers. However, there have also been mills powered by the tides, dating from as early as the eleventh century in England, France, and Spain. One such mill [6] at Woodbridge, England, was mentioned in the parish records in 1170, and operated until 1957. In North America, tide mills were used as early as 1617. Whether the tides power a mill or produce electricity, the basic principles are the same. A dam (often called a *barrage*) is built across a natural tidal basin, with sluice gates to allow the passage of tidal water into the basin. The incoming tide is allowed to enter, and the gates are closed at high tide. After the tide has fallen somewhat, the collected water is allowed to leave the basin, turning a wheel or turbine.

Tides are caused by the gravitational forces of the moon and sun on the waters of the earth. As the ocean waters move relative to the earth in twice-daily tides, energy is dissipated through friction with the coast and also through internal viscous friction. The rate of energy dissipation by the tides has been estimated [7] to be 3×10^{12} W, or 3000 GWW, and since this energy is drawn largely from the earth's rotational kinetic energy, the earth's rotation is gradually slowing down. The length of the day is increasing by about 0.002 s per century. [8]

Only a small fraction of this tidal energy has the potential to be used. A suitably large basin and a drop in water level of several metres are needed to make a tidal project worthwhile. Estimates of what electrical generating capacity might eventually be developed vary greatly, the lower end of the range being about 100 GWe. This is small compared to, say, the estimated worldwide hydroelectric resources [9] of 2000 GW.

Present Tidal Barrages

There are just four tidal barrage projects in operation in the world, and only two of these exceed 1 MWe in output. A 240- MWe plant has operated since 1966 in the la Rance estuary in the northwest of France. At Annapolis Royal in Nova Scotia (Fig. 14-5), a 20-MWe pilot came into operation in 1984, using the 20-m tides of the Bay of Fundy. With the low price of oil, there was little activity throughout the 1990s in developing new tidal plants.

Figure 14-5 Tidal Power Plant at Annapolis Royal. Nova Scotia. *Courtesy, Nova Scotia Power*

[6] G. Duff, *Tidal Energy: or, Time and Tide Wait for No Man*, American Assoc. of Physics Teachers, (1986) p. v.
[7] Encyclopedia of Energy, ed. C. Cleveland, Elsevier Academic Press, (2004).
[8] G. Abell, *Exploration of the Universe (4th Ed.)*, Saunders Publ., Philadelphia, (1982) p. 142.
[9] D. Deudney and C. Flavin, *Renewable Energy: The Power to Choose*, Worldwatch Institute, W.W. Norton, New York, (1983) p. 168.

Advantages and Disadvantages of Tidal Barrages

Tidal energy is renewable, and produces no air pollution or radioactive waste. Although the operating costs are low and the plant lifetime long, these installations are expensive to build. An evident and major drawback is the periodicity of the power output, since high tide occurs only twice per day. This problem could be alleviated somewhat by allowing the incoming water to flow over the turbines, thus giving four power-producing cycles per day, or by a complicated scheme in which the collecting basin is divided into two or more separate chambers, and the flow is arranged so that there is always a tidal head between two chambers or one chamber and the ocean. However, the simple scheme with one basin and one flow per tide is the most economical. Another possibility for tidal plants is the use of electricity from other sources during times of low demand to pump water into the basin, to be utilized by the tidal turbines at peak-demand times.

Environmental problems that can arise from tidal plants are sedimentation, flooding of upstream river areas if a plant is built at a river mouth, and destruction of tidal mudflats with consequent effects on animals and plants. Hydrographic studies have suggested that major development of the Bay of Fundy could significantly alter the resonance of the Gulf of Maine and produce flooding as far away as Boston.

Tidal Current Systems

Some thought is now being given to the idea of placing underwater turbines at sites where the current flow is very fast. This concept is very similar to wind turbines, as is evident from Fig. 14-6; the system would be mounted on a pile secured to the seabed. With a current of 2 m/s, the available power would be about 4 kWh/m^2. This is much larger than in the case of wind, because of the much higher density of water compared to air.

Figure 14-6 The SeaGen Tidal Current Turbine System. *Courtesy, Marine Current Turbines TM Ltd.*

Unlike winds, tides are predictable, which would be advantageous. But the tide varies with a 15-day cycle, and so the power output would change significantly with time, albeit in a predictable manner, which would enable planners to devise appropriate strategies for integration into the electricity grid. The corrosive effects of sea water are an inescapable component of capital costs. Environmental concerns are mitigated by the fact that fast tidal regions tend not to be the habitat of many species. As with tidal barrage systems, attention will need to be paid to the potential impacts upon the transport and deposit of sediments, and this may be the factor that will impose upper limits upon extractable energy. However the underwater turbine approach is clearly much less intrusive than a barrage.

A 300 kW (maximum) test system has operated on the south coast of England since 2003. On the basis of experience with this system, a 1 MW device is to be installed off the Northern Ireland coast in 2006.

14.10 WAVE ENERGY

There is a great deal of energy in ocean waves — for example, it has been estimated [10] that approximately 120 GW of wave power is dissipated on the west coast of the British Isles, about five times the British electrical power requirement.

It can be shown that the power P per unit length l of wave-front in a surface-wave on a liquid of density ρ is given by

$$\frac{P}{l} = \tfrac{1}{4}\rho g A^2 \frac{\lambda}{T}, \text{ with } \frac{\lambda}{T} = \sqrt{\frac{g\lambda}{2\pi}}$$

$$\frac{P}{l} = \tfrac{1}{4}\rho g^{3/2} A^2 \sqrt{\frac{\lambda}{2\pi}} \qquad \textbf{[14-6]}$$

where λ is the wavelength, A is the wave amplitude, T is the wave period, and g is the gravitational acceleration. [11] Inserting the appropriate value of the quantities for water Eq. [14-6] can be written as (See problem 14-6)

$$\frac{P}{l} = 3\, A^2\sqrt{\lambda} \ \text{ kW/m} \qquad \textbf{[14-7]}$$

EXAMPLE 14-2

Find the power delivered by a 1.0 km front of an ocean wave with an amplitude of 1.0 m and a wavelength of 100 m. (Such a wave travels at 40 m/s)

SOLUTION

From Eq. [14-7], $P = 3(1.0)^2(100)^{1/2}(1.0\times10^3) = 30$ MW

Example 14-2 shows that, theoretically, there is a great deal of energy in ocean waves if only it could be extracted without damaging the coastal eco-systems.

At the present time, there are only a few experimental projects investigating the use of wave energy to produce electricity, which is expected to cost about 9-15¢ per kW·h, well above the price of coal-fired or wind-produced electricity.

Most generators designed to use wave energy exploit the vertical motion and most of the research on wave energy is concerned with one of three methods:

[10] D. Ross, *Energy from the Waves, 2nd Ed.*, Pergamon Press, New York, (1981) p 28.
[11] See for Example: A.R. Patterson, *A first course in fluid dynamics*, Cambridge University Press, (1983) pp 289-297.

- *Wave Surge or Focusing Devices* (tapered channel systems);

These have a shore-mounted structure to channel waves, increasing their amplitude. The increased amplitude lifts the water into an elevated reservoir; in other words, kinetic energy is converted into potential energy. Water flows out of this reservoir and generates electricity, using standard water turbine methods and generates electricity.

- *Floats or Pitching Devices*

These devices generate electricity from the undulating motion of a floating object. For example, a number of rafts are hinged together and the flexing action at the hinges is used to pump a fluid through a turbine to generate electricity. Such systems are in the pilot model stage with a 750 kW model called "Pelamis" operating since 2004 in the Orkney Islands north of Scotland.

- *Oscillating Water Columns (OWC)*

These plants generate electricity from the wave-driven rise and fall of water in a vertical shaft. The motion of the water piston drives air into and out of the top of the shaft, turning an air-driven turbine. This is the most advanced technique with the 500 kW "Limpet" generator operating off the Scottish island of Islay since 2001. Also about 1000 floating navigation buoys in oceans around the world are each powered by a 60-W wave-powered air turbine that works on this principle. [12]

Wave energy is renewable, and generates no air or water pollution and no radioactive waste. Large-scale exploitation of wave energy would undoubtedly affect local aquatic and shoreline ecosystems, and perhaps alter ocean currents as well.

14.11 GEOTHERMAL ENERGY

We are all aware of dramatic displays of geothermal energy (for example, volcanoes and geysers — Fig. 14-7), as well as its more passive manifestations, such as hot springs (often used for recreation). Geothermal energy is derived from natural nuclear energy; the radioactive decay of elements such as uranium, thorium, and radium in the earth produce heat, which gradually makes its way to the surface of the earth (and is eventually radiated away into space as infrared radiation). In most areas of the world, this geothermal energy is very diffuse; the rate of heat transfer to the earth's surface is only about 0.06 W/m^2. A related aspect of geothermal energy is the increase of temperature with depth below the earth's surface, with the average thermal gradient being about 30 °C/km.

Figure 14-7 A Geyser in the Napa Valley in California

[12] M. Sanders, *Energy from the Oceans*, in The Energy Sourcebook, American Institute of Physics, New York, (1991) p. 274.

Geothermal energy can be readily exploited in regions of the earth (so-called *thermal areas*) where the rate of heat transfer to the surface is much higher than average, usually in seismic zones at continental-plate boundaries where plates are colliding or drifting apart. Here we have hot rock which is stressed and fractured, and water reservoirs under pressure. For example, the heat flux at the Wairakei thermal field in New Zealand [13] is approximately 30 W/m^2, and there are regions of Iceland [14] where the thermal gradient is greater than 100 °C/km.

Using Sources of Geothermal Energy

There are two general ways that geothermal energy is utilized: direct use for space-heating and indirect use in generating electricity. In 2000, the global estimate [15] for direct geothermal power in 58 countries was about 15,000 MW, having doubled in 15 years; the major users are in Japan, Iceland, and Hungary. The first geothermal electric plant began operation using dry steam at the Larderello thermal field in Italy in 1904, and the plant still operates. The installed global electrical generating capacity [16] is about 8700 MWe, the dominant countries being the U.S.A (2900 MWe) and the Phillipines (1900 MWe). This is equivalent to nine typical nuclear reactors.

Geothermal sources are categorized into various types, and the type determines the way in which the source is exploited:

- *Dry steam reservoirs* — these are the most desirable sources for generating electricity, since the dry steam can be used directly in turbines in a *dry steam power plant*. However they are very rare. The technology is basically the same as for fossil-fuel electric plants, except that the temperature and pressure of the geothermal steam are much lower. For example, in the Geysers geothermal area in California, the world's biggest geothermal source, the temperature is about 200 °C and the pressure about 700 kPa (7 atm.). Hence, from the laws of thermodynamics (Section 4.6), the efficiency of conversion to electricity is less than at a fossil-fuel plant; at the Geysers, it is only 15% to 20%. The generating units are also smaller, ranging in size from 55 MW to 110 MW at the Geysers. This geothermal source is in decline, having reached peak output in 1987 but it still supplies over a million people with electricity.

- *Wet steam reservoirs* — these fields are much more common than the simple dry type. New Zealand's Wairakei field is an example. Such a field is full of very hot water (usually below 200 °C), under such high pressure that it cannot boil. When a lower pressure escape route is provided by drilling, some of the water suddenly evaporates (flashes) to steam, and it is a steam-water mixture that reaches the surface; the steam can be used to drive a turbine in a *flash steam power plant*. These plants use resources that are hotter than 175 °C. The fluid is sprayed into a low-pressure tank where some of it vaporizes (flashes) and then drives a turbine; the residual hot water, perhaps mixed with further water from the ground, can be flashed again to lower pressure, providing

[13] R. Howes, *Geothermal Energy*, in The Energy Sourcebook, American Institute of Physics, New York, (1991) p. 239.
[14] R. Harrison, *Applications of Geothermal Energy*, *Endeavour*, New Series, Vol. 16, No. 1, (1992) p. 31.
[15] J. W. Lund, *Geothermal Direct Use*, in Encyclopedia of Energy, Elsevier 2004.
[16] K.C. Lee, *Geothermal Power Generation,* in Encyclopedia of Energy, Elsevier 2004.

steam for a second turbine. Flashed electric power production is used in many countries, including the USA, the Philippines, Mexico, Italy, Japan, and New Zealand. If the temperature of the original hot water is too low for effective flashing, the water can be pumped to the surface under pressure to prevent evaporation (which would decrease the temperature), and its heat transferred to an organic fluid that evaporates and acts as the working fluid in a turbine. This *binary-cycle power plant* is likely to become the most common approach. It uses resources with temperatures as low as 85 °C.

- *Hot water reservoirs* — these reservoirs contain hot water at a temperature too low for electricity generation. However, the water can be used in district heating systems and for agri-business ventures such as greenhouses, fish hatcheries, etc. This heating can be either direct or through the use of heat pumps. The most extensive use of geothermal heat is in Iceland, which lies in a region of continental-plate activity. Virtually all the homes and buildings in Iceland are heated geothermally the 830 MWt capacity supplies 160,000 people), and there is also a small geothermal electric plant. Other important examples of geothermal heating are in France, both near Paris and in the southwest. During oil-exploration drilling in the 1950s, hot water was discovered in the Paris region, but exploitation did not begin until the 1970s as a result of rapidly increasing oil prices. In France, the equivalent of 200 000 homes are being provided with space-heating and water-heating from geothermal sources. One interesting feature of the French geothermal sources is that they do not occur in regions of elevated thermal gradient.

- *Hot dry rock* — in many regions of the world, hot rocks containing sealed fractures lie near the earth's surface but there is little surrounding water. Exploitation of these geothermal resources requires opening the fractures or creating new ones, and then supplying water which can be pumped through the fractures to extract the thermal energy. This is an *enhanced geothermal system*. In 2006, plans for three EGS plants in the size range 15 - 30 MWe in Nevada (which already has 277 MWe of installed geothermal capacity), were at various stages, with construction approved for the first. A related and imaginative approach is being taken to extend the life of the Geyser dry steam field mentioned above; waste water from the city of Santa Rosa is to be pipe-lined to the Gysers and injected at a rate of 50 million litres per day; this is expected to increase the electrical output by 85 MWe.

- *Normal geothermal gradient* — in principle the normal geothermal gradient produces a useful temperature difference anywhere on the globe. If a hole is drilled to a depth of 6 km (which is feasible), a temperature difference of about 180 °C is available, but no technology has been developed to take advantage of this resource. At this depth, water is unlikely, and the problems of extracting the energy are similar to the difficulties encountered with hot dry rock near the surface.

Advantages, Disadvantages and Costs of Geothermal Energy

The main advantage of geothermal energy is that it can be exploited easily and inexpensively at several locations in and near geologically active zones around the world. However, it is unlikely that geothermal sources will become economical in areas of normal thermal gradient in the foreseeable future. As in the case of wind plants, the site can be developed incrementally and wells brought on-line as they are drilled.

One of the primary disadvantages of geothermal energy relates to dissolved salts and gases in the water and steam. Hot water dissolves salts from surrounding rocks, and this salt produces corrosion and scale deposits in equipment. In order to prevent contamination of surface water, the geothermal brine must be either returned to its source or discarded carefully in another area. A gas that is often found in association with geothermal water and steam is hydrogen sulphide (H_2S), [17] which has an unpleasant odour (like rotten eggs), and is toxic at high concentrations. At the Geysers plant, the past 10 years have seen development of H_2S abatement systems using catalysts, and aerobic and anaerobic bacteria. (In the case of an iron catalyst it is turned into iron pyrite [fools gold]). Geysers is now abated to the level of 99.9% for steady operation. An advantage of the binary cycle approach is that dissolved salts are returned to the reservoir instead of being released to the environment.

Most people tend to think of geothermal energy as being renewable, but in fact one of the major issues in choosing a geothermal energy site is estimating how long the energy can usefully be extracted. If heat is withdrawn from a geothermal source too rapidly for natural replenishment, then the temperature and pressure can drop so low that the source becomes unproductive. Since it is expensive to drill geothermal wells and construct power plants, a source should produce energy for at least 30 years in order to be an economically sound venture, and it is not an easy task to estimate the working lifetime beforehand. Another potential problem, particularly if geothermal water is not returned to its source, is land subsidence. For example, there has been significant subsidence at the Wairakei field in New Zealand.

Just as with wind, the capital cost per installed kWe increases inversely to the quality of the resource, which in this case is the temperature of the water. The figure presently quoted by the U.S. Department of Energy [18] for a 50 MWe plant is $2100 per installed kW, excluding interest charges. Estimates of electricity costs from new plants in the USA span the range 5-8 cents (US) per kWh.

Geothermal Heat Pumps

Another way that heat from the ground can be utilized is through the use of heat pumps (Section 4.8) as shown in Fig. 14-8. While we encounter great variations from season to season in above-ground temperatures, the temperature several metres below ground surface is relatively constant. Pipes containing a fluid are buried in the ground, and heat can be extracted from the earth in

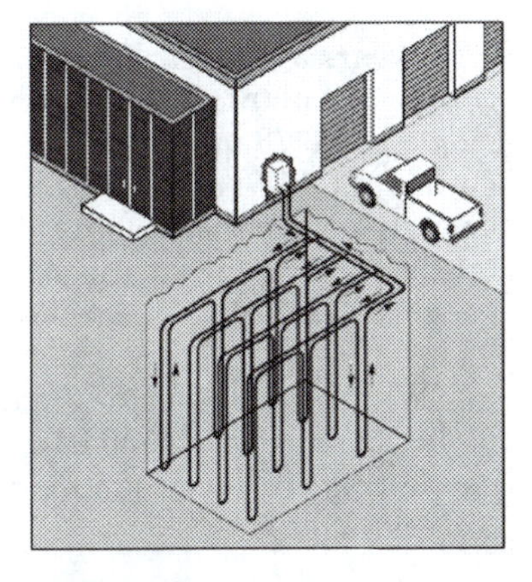

Figure 14-8 Geothermal Heat Pump. *Courtesy, US Dept. of Energy, Geothermal Technologies Program*

[17] This is the same contaminant that makes so much natural gas ("sour gas") unusable without expensive treatment.

[18] Assumptions to the Annual Energy Outlook 2006, report # DOE/EIA-0554 (2006), U.S. Department of Energy

the winter to heat a building, and dissipated in the earth in the summer to provide air conditioning. The main disadvantage is the capital and installation cost of the very large amount of piping that is required.

In the USA there are half a million GHP installations, reducing CO_2 emissions by a million tonnes each year. They are mostly used in residential, commercial and government buildings. A Canadian example dating back to 1985 is a store in Kitchener, Ontario which uses 1000 m of pipe placed in five vertical 20-cm diameter wells drilled for the purpose; the wells extend 40 m below the store's parking lot. The cost to install the system was about $40000; the immediate annual savings were about $8000. A larger-scale example from the U.S.A. is the conversion to GHP of 4000 houses at the Fort Polk army base in Louisiana: electricity consumption was reduced by 32% and natural gas consumption was eliminated; and the peak demand in mid-summer was cut by 43%.

EXERCISES

14-1. The total electrical generating capacity in Canada in 2000 was about 115 GW. If all this power were to be provided by wind-electric turbines each having 500-kW power output, how many such turbines would be required? (If the answer seems like a huge number of turbines, make an estimate of the number of automobiles in Canada; how does the number of turbines compare with the number of automobiles?)

14-2. What is the maximum possible theoretical efficiency of a thermal-electric plant operating at the Geysers thermal field, if the temperature of the steam is 200 °C and the temperature of the condensed water at output is 20 °C? What percentage of the total energy contained in the steam is discharged as waste heat?

14-3 Calculate the power in kilowatts for a reasonably sized windmill ($R = 4.1$ m) in a moderate wind ($v = 5.6$ m/s) for $\eta = 23\%$. The density of air is 1.3 kg/m^3.

PROBLEMS

14-4. Some years ago a European company revealed plans to construct a very large wind electric generating plant. The following data are from their press release. The mill would have a single blade 73 m in radius and will rotate at 17 r.p.m. It would operate at an electrical output of 5 MW, and will generate an average of 17×10^6 kW·h each year. This would be enough to save the burning of 31000 barrels of oil.
(a) If this windmill extracts 50% of the wind energy, what must the wind speed be to generate 5.0 MW if the generator operates at 90% efficiency? The density of air is 1.3 kg/m^3.
(b) What capacity factor did the designers hope to realize? Why was this?
(c) Is the figure of 31000 barrels of oil compatible with 17×10^6 kW·h per year, assuming a generating efficiency of 35% in an oil-burning plant? (Refer to Appendix III for the energy content of oil.)

14-5. The tidal power plant in Nova Scotia has one 20-MWe turbine. If it is operating at its rated power, what volume of water is flowing through it per second? Assume that the water is falling through a vertical distance of 4.0 m (the average tidal range in the Annapolis area is 6.4 m), that the water density is 1.02×10^3 kg/m^3, and that the efficiency of conversion of gravitational potential energy to electrical energy is 95%.

14-6. Show that Eq. [14-7] follows from Eq. [14-6]

14-7. Waves on lakes have considerably less amplitude and wavelength than on the ocean. Typical waves on Lake Ontario on an average summer day might have an amplitude of 20 cm and a wavelength of 2.0 m. What is the fraction of power delivered by these waves compared to the ocean waves in Example 14-2?

ANSWERS

14-1. 230 thousand
14-2. 38%, 62%
14-3. 1.4 kW
14-4. **a.** 10 m/s, **b.** 39%, **c.** yes
14-5. 5.3×10^{2} m^3/s
14-7. 5.7×10^{-3} or 0.6%

CHAPTER 15

RENEWABLE ENERGY II

A general recognition that the poisoning of the atmosphere with carbon dioxide cannot be allowed to continue until the last kilogram of fossil fuel is burned, combined with a reluctance to embrace the nuclear alternative has focussed the public attention on other sources of energy that produce less pollution. In this chapter various other energy sources are explored: the Sun, wind, geothermal sources, tides, waves, and biomass-emphasizing the advantages and disadvantages of each.

15.1 THE ENERGY FROM THE SUN

The sun pours out a prodigious amount of energy, more than we could ever use. Why not capture it and use it directly? Probably no area of alternative energy generation has had so much research and media attention with so little return. The first four sections of this chapter focus on some of the possibilities and the problems with large-scale and small-scale utilization of solar energy. Much of the material follows on from the first three sections of Chapter 6.

As mentioned in Section 6.5, the Sun supplies energy at the position of the Earth at an average rate of 1372 W/m^2 at the top of the atmosphere and about 50% of this, or 700 W/m^2, is delivered to the Earth's surface. For any location on the surface this must be adjusted for the inclination of the surface to the Sun's direction, then averaged over the day and night hours, and finally averaged over the year.

For example, in Ontario, Canada (Latitude 44°N) the annual averaged solar radiation energy (or *insolation* *R*) has a value R=3.5 kW·h/m^2 per day. If this was delivered continuously over the 24-hour day the intensity would be 146 W/m^2; of course, in reality it is a higher value at noon and zero at night. There is a great seasonal variation; in Toronto the daily insolation varies between 1 and 7 kW·h/m^2 as shown in Fig. 15-1. About 90% of this energy is from the direct sunlight and 10% from diffuse light. On normal cloudy days the light is all diffuse and is about 25% of the value on clear days.

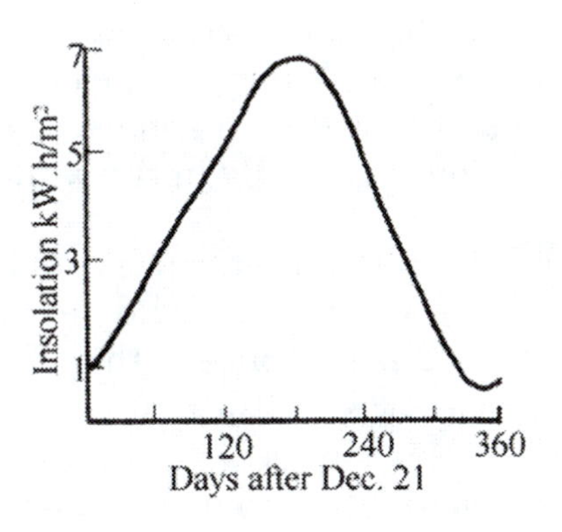

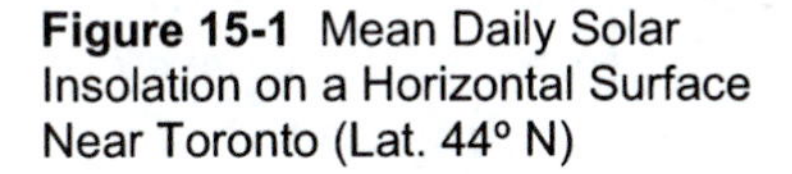
Figure 15-1 Mean Daily Solar Insolation on a Horizontal Surface Near Toronto (Lat. 44° N)

As expected, the insolation is lowest at the winter solstice (Dec. 21) and highest at the summer solstice (June 21).

Figure 15-2 is a recording of the insolation at the University of Guelph's Elora Research Station on June 19, and December 19, 1990, very close to the summer and winter solstice. The details of the shape of the curves have to do with the daily variations of the local weather. For example on Dec. 19 there was precipitation in the morning which reduced the insolation but cleared the atmosphere of particulate matter. After the clouds moved away there was an unusually clear sky with higher than average intensity for that time of year.

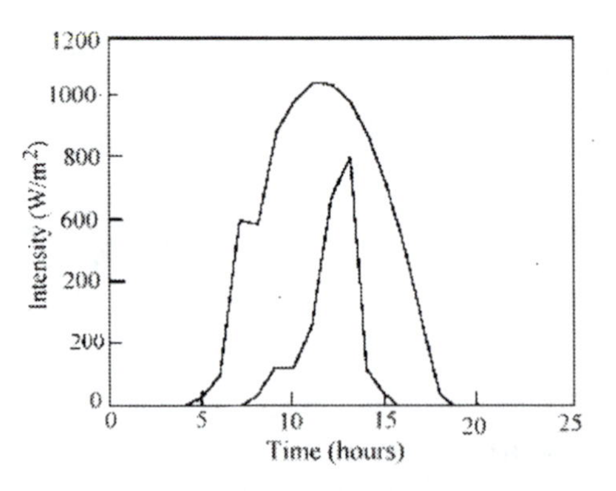

Figure 15-2 Solar Intensity at Guelph Ont. (Lat. 44 N) Near Summer and Winter Solstice. *Courtesy R.H. Stinson.*

Going south from the Canada-USA border R increases from 3.5 to 7 kW·h/m^2 per day at the USA-Mexico border. The USA average is 4.8 kW·h/m^2 per day. At R = 7 k·W·h/m^{2}per day the American southwest has a distinct advantage for the utilization of solar energy.

15.2 SMALL-SCALE DIRECT UTILIZATION OF SOLAR HEAT

Schemes to use solar heat directly generally are of three types: passive heating, active collector panels, and solar ponds.

Passive Heating

Many styles of house have been designed to optimize the direct absorption of solar heat. This, combined with careful insulation, can greatly reduce the amount of additional energy required for comfortable temperatures, at least in temperate climates. The challenge to the architect is to design a house that collects solar radiation maximally in the winter but excludes it in the summer; otherwise undue amounts of air-conditioning energy would be required. Figure 15-3 shows a typical house of this type; it has been designed for a latitude of 38°. At the summer solstice the Sun rises as high as 75° (90° - 38° + 23°) where 23° is the tilt of the earth's axis. Plants in the solarium receive this heat but it is isolated from the house by windows or transparent curtains to admit illumination. At the Winter solstice the Sun rises 90° - 38° - 23° = 29° above the horizon, and the solar windows are designed to maximize entry of this low-angle radiation. Furthermore in summer the radiation falls on reflective floors and is largely returned to the outside. In the winter the radiation falls on massive concrete structures which store it to be given up at night. Houses of this type are more expensive to build but can eliminate 20 to 50 % of annual heating costs. In the early 1980s, when energy costs were

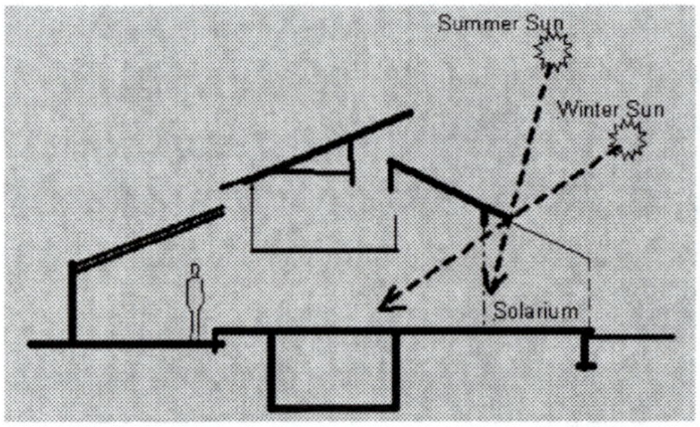

Figure 15-3 A Passively–Heated Solar House.

greater, there was considerable interest in this type of house. As energy costs dropped drastically the payback times became unattractively long and interest has waned. It will surely return as the cost of energy inevitably rises.

Active Collector Panels

Because of high cost there is very little economic incentive to build and install solar collectors that track the Sun, or even those that focus the light for residential or commercial application. For this reason most solar collectors are of the fixed flat-plate type mounted at the best angle and facing south. In Ontario, mounting the plate at 60° to the horizontal reduces the seasonal variation of 1 to 7 kW·h/m² in Fig. 15-1 (which was for a horizontal plate) to the range 1.5 to 4.6 kW·h/m².

Flat plate collectors consist of blackened absorbing surfaces which transfer heat to pipes containing water; an outer glass cover admits visible light but prevents re-radiation of the emitted infrared radiation (c.f., the greenhouse effect in Chapter 6). Efficiency varies with design; for a given collector, the hotter the surface the greater the heat losses and the less the efficiency. Obviously, improving the insulation and even evacuating the space under the glass improves the efficiency but this is costly. The hallmark of small domestic solar collectors is usually cheapness and simplicity at the expense of small gains in efficiency; otherwise the cost is not justified.

Simple and inexpensive collectors can be useful particularly for preheating domestic hot water or for swimming pools. Typical performance of a 2 m² panel (~3 m × 0.6 m) with 6 heating tubes is to raise the temperature of 100 L (20 gal) of water from 25 C to 40 C in a working day from 8:00 AM to 4:00 PM [1].

Case Study: The Chanterelle Inn, Nova Scotia

This structure was devised by its owners to be fossil fuel free. Accordingly as seen in Fig. 15-4, [2] the south-facing roof area of 47.5 m² was covered with 16 passive solar hot water heaters and 3 photovoltaic panels (see Sec. 15.4 below) to operate the pumps. The system contributes 12800 kW hr which is 50% of the annual hot water requirement. On days when the solar harvest is in excess of hot water needs (the winter season) the system contributes 12800 kW hr or 25% of the annual space heating requirement. In 2001 the system saved over Can$2500 and incurred operating costs of Can$100.[3] (see Problem 15-7).

Figure 15-4 Chanterelle Inn, Nova Scotia. A non–fossil–fuel building.

---

[1] Emin Yilmaz, *Instrumentation And Evaluation Of Commercial And Homemade Passive Solar Panels* Proceedings of the 2004 American Society for Engineering Education Annual Conference & Exposition (2004), CD–ROM, Session 1359.

[2] Courtesy: General Manager Ms. Earlene Busch

[3] Ministry of Natural Resources, Canada, *Renewable Energy in Action,* ISBN 0–662–31574–X

EXAMPLE 15-1

Using the performance data given above for a single, home-made solar collector panel, calculate how much energy is collected in a day, what is the oil-equivalent and, if the price of oil were $100/bbl, what is the saving on a sunny day relative to heating the water by electricity?

SOLUTION

The energy is given by $Cm\Delta T = (4189 \text{ J kg}^{-1}\ ^\circ\text{C}^{-1})(100 \text{ kg})(15\ ^\circ\text{C}) = 6.3\times10^6$ J.
One barrel (bbl) of oil supplies 5500 MJ so this requires $6.3/5500 = 1.1\times10^{-3}$ bbl.
Electricity is produced in an oil-fired plant at about 33% efficiency so the saving of oil is 1.1×10^{-3} bbl/0.33 = 3.3×10^{-3} bbl or 33 cents.

In many countries with long periods of clear sunny skies (Israel, Australia, Japan, Southern USA) domestic water and space-heating have been accomplished for years with solar collectors. Let us consider the possibilities for a location like Ontario. A typical single family home has a roof space sufficient for about 30 m^2 of south-facing collector. At 33% efficiency, this would provide 15 kW·h (1.5 kW·h/m^2 × 0.33×30 m^2) per day in the winter and 45 kW·h per day in the summer. At an average of 30 kW·h per day this is 11,000 kW·h per year. According to the Ontario Energy Ministry, a 2-story 1800 ft^2 house in Toronto uses 30,000 kW·h of energy for heating per year; existing homes cannot be heated solely by solar energy. The situation is even worse because in the winter when heat is needed, there is not as much solar energy available, as shown in part (c) of Example 15-2 below.

EXAMPLE 15-2

As a case study, one of the authors' homes in Guelph, Ontario has a floor area of 1800 ft^2 and uses 2739 m^3 of natural gas per year for heating and hot water. The consumption is distributed over the year as shown in Fig. 15-5 which is taken directly from the gas company's bill.
a) Estimate how much gas is used each year for space-heating.
b) What is the energy equivalent of this gas in kilowatt hours?
c) Compare the gas consumption in December with the energy that could be obtained from a solar collector of area 30 m^2.

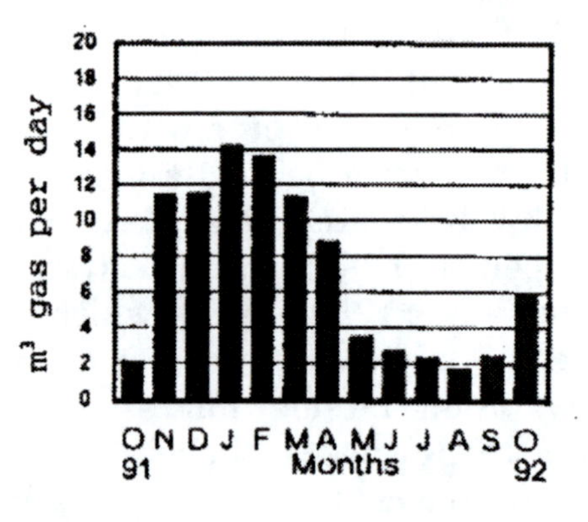

Figure 15–5 Yearly gas consumption for a 2–story home.

SOLUTION

(a) From the summer readings in Fig. 15-5 it can be estimated that hot water uses a steady 2 m^3 per day for a yearly consumption of 2 × 365 = 730 m^3. This leaves 2009 m^3 for space heating.

(b) The energy content of natural gas is given in Appendix III as 38×10^6 J/m^3. Therefore the energy equivalent of 2009 m^3 of natural gas is 2009 m^3 × 3.8×10^7 J/m^3 = 7.6×10^{10} J

To convert this to kilowatt hours we use the conversion factor from Section 1.5, 1 kW· = 3.6×10^6 J. Thus

$$7.6\times10^{10}\,\text{J}\times\frac{1\text{kW}\cdot\text{h}}{3.6\times10^{6}\,\text{J}}=2.1\times10^{4}\,\text{kW}\cdot\text{h}$$

(c) Looking at December we see that the gas consumption was 11.6 $m^3\times31$ = 360 m^3, which is equivalent to 3800 kW·h A 30 m^2 solar collector could supply at most 1.5kW·h/m^2/day. (This value is provided in the previous text material) Multiplying this by 30 m^2 and by 31 days gives about 1400 kW·h, which is 37% of the energy required in the house.

To some extent the inefficiency of solar heating can be improved if a method of storage can be implemented to save the heat from times of plenty to be used when needed. For single family dwellings this is usually not practical or economical for periods exceeding 24 hours, that is, daytime heat can be stored to use the following night, but not longer. Season-long storage is only practical in large buildings as discussed in Chapter 18. When houses do have heat storage it is usually in rock or water or heavy concrete walls designed for the purpose.

In most of the populated areas of North America the effectiveness and economic prospects of domestic space heating with solar energy are not great (See problems 15-4 and 15-5 for example). Some of the apparently successful *solar houses* are really *super-insulated* houses and it is the insulation and passive solar aspects that are worth enshrining in building codes, not active-solar aspects. The retro-fitting of existing houses is difficult, expensive and usually inefficient without massive reconstruction to incorporate super-insulation. After a flurry of activity in the 1970s when solar heating was somewhat oversold by its advocates, enthusiasm has considerably cooled on the part of the public and governments.

Simple domestic water-heating systems look rather more promising, since the need for hot water is year-round. Even here though there is little financial incentive unless the system is installed on a do-it-yourself basis.

Solar Ponds [4]

Large-area bodies of water absorb solar energy very efficiently but, because of convection, do not build up large temperature differences with depth. The energy is absorbed in the first metre or so below the surface and because of thermal expansion the water becomes less dense and starts to rise. This convective process results in the water being well mixed and of uniform temperature. However, in 1902 A. Kalecsinsky first reported an unusual phenomenon in the Medve Lake in Transylvania where the temperature of the water 1.3 m below the surface was 70 $^\circ$C higher than at the surface. The same phenomenon has been reported in small lakes in the United States and even Lake Vanda in Antarctica which, while frozen on the surface, had a temperature of 25 $^\circ$C at a depth of 65 m.

---

[4] H. Tabor, *Solar Ponds, in* Solar Energy, *27, pp 181–194, (1981).*

Of course something must be suppressing the natural convection of the water. This comes about because there is a varying concentration of salt dissolved in the water, with more salt at the bottom than at the top. The presence of the salt increases the density of the water so that it cannot become light enough to rise by thermal expansion alone. Such *solar ponds,* if made artificially, can support temperature gradients so steep that the bottom water is near the boiling point.

Many countries have shown an interest in solar ponds including the United States and Australia but nowhere have they been more developed than in Israel. Artificial ponds one to two metres deep with a blackened absorbing bottom and areas of thousands of square metres have been constructed which produce temperature differences of 85 to 90 °C. For power production the temperatures are still not very high so the thermodynamic efficiency is very low; the theoretical maximum is about 20% and half of this is a more likely performance. Nonetheless small solar pond power plants have been built. A 7000 m^2 pond at Ein Bokek on the shores of the Dead Sea provides a peak power of 150 kW of power. More realistically solar ponds have been used, in India and elsewhere, as a source of industrial process heat and to desalinate water and produce salt.

15.3 LARGE-SCALE DIRECT UTILIZATION OF SOLAR HEAT

The basic problem with solar energy is that it is so dilute; it has to be collected from a large area if it is to be useful as a substitute for our other large-scale energy systems, such as electric utilities. This generally precludes inhabited and agricultural areas since large tracts of land are unavailable or expensive. The Earth has many desert and semi-desert areas at low latitude, however, where the concentration of solar energy is at a maximum. Clearly solar plants, of whatever type, must be used to manufacture a high-value energy currency to ship to where the energy is needed; this means electricity or hydrogen.

Central-receiver Plants

In a solar "farm" a vast area is covered with steerable mirrors which collect the sunlight and direct it to a central receiver. This receiver is usually a steam boiler on a high tower in the centre of the mirror array, constructed so as to absorb the energy and transfer it to flowing water in steam tubes. In principle this concentration of light can produce steam at temperatures around 500°C, which compares to the best coal-fired plants; the high temperatures result in high efficiencies. Although the receiver may be complicated and expensive, the mirrors must be cheap since so many of them are required and they must be steerable. After much experimentation in the late 1970s and early 1980s a few small power plants were built.

The first was the Eurelios plant at Adrano in Sicily which develops one MWe under clear skies; it was opened in December 1980. The energy is collected by 182 heliostats (steerable mirrors) with a total area of 6400 m^2. The motor-drives track the sun to an accuracy of 0.2°; this high accuracy is necessary because the angular size of the Sun is only 0.5°. The receiver is a conical shell lined with water tubes on a tower 55 m high.

The steam at a pressure of 100 atmospheres is directed either to the turbines or to storage. When it opened, Eurelios produced power at twice the current cost and the situation has not improved since that time. Several other plants in the 1to 10 MWe range have been built since in the USA, Spain and Japan.[5]

Distributed Collector Plants

A promising type of solar electric plant is the distributed collector plant pioneered in California by the LUZ Corporation [6]. These plants are built in modules of 30 MWe output which each cover 6×10^5 m^2 of land. A single unit consists of a parabolic collector mirror 52 m long with an aperture of 2.5 m; 1600 of these units make one 30 MWe module. The mirrors concentrate the light on steel tubes encased in an evacuated glass tube with a concentration ratio of 61:1. Oil is pumped through the steel tubes where it is warmed from 245 °C to 310 °C and then produces steam at a temperature of 250 °C and a pressure of 38 atmospheres. This is a rather low temperature for operation of turbines and the thermodynamic efficiency would be low. The steam is further heated to 415 °C in a gas fired super-heater before being sent to the turbine. This is, then, a hybrid system; the solar collector serves as a pre-heater.

The installation cost was less than US$3000 per kWe but is the second highest of modern technologies. Of course, it again must be stressed that this is only feasible where sunlight is plentiful and land is cheap. In all, nine plants of increasing capacity up to 80 MWe were constructed deriving 75% of their output from sunlight. By the beginning of 1990 the total installed capacity was 354 MWe and production costs had fallen from 25 cents per kW·h to 8 cents, but the abundance and cheapness of gas forced the plant to close in 1992.

Solar Chimneys [7]

If hot air at ground level is exposed to the lower end of a tall chimney which has cooler, lower pressure air at the top, the hot air will rise, cooling off while converting thermal energy into kinetic energy. A turbine installed in the chimney can produce electricity from the vertical draft. A pilot plant of this type has been erected at Manzanares in Spain where a 195 m chimney generated 50 kW of electric power from 1986 to 1989.

Proponents of the scheme wish to erect a 1.5 km high chimney [8] in Northern Cape Province, South Africa to generate 200 MW at a cost of US$400 million. Estimates place the cost of the electricity at about 10 to 20% higher than for coal-fired power with a capital payback of 20 years.

---

[5] C.J. Weinberg and R.H. Williams, *Energy from the Sun*, Scientific American, Sept. (1990) pp 146–155.
[6] J.G. Ingersol, *The Energy Sourcebook,* (ed. R. Howes and A. Fainberg), The American Institute of Physics, (1991) p220.
[7] *Electric Dreams.* New Scientist 6 Mar. (1999) pp 30–33.
[8] This is 3 times higher than the world's highest free–standing structure–the CN Tower in Toronto, Canada.

15.4 DIRECT CONVERSION OF SOLAR ENERGY TO ELECTRICITY

Solar cells are based on semiconductors, of which silicon is the most widely used. Silicon is a material of valence 4 which forms a cubic crystal with each Si atom covalently bonded to four other silicon atoms. This is a very stable structure since the only electrons that are free are the few that are momentarily loosened by the vibration of the parent atoms on their sites; the material is a very good insulator at room temperature.

Si is made into a semiconductor by *doping*, which is the addition of a small impurity, say one part in 10^8, of an element of valence 3 or 5. For example, phosphorus with a valence of 5, has 5 outer electrons available for covalent bonding; it uses 4 of these to fit into the crystal lattice and its fifth electron is surplus. This extra electron is free to provide conduction in the lattice. Phosphorus is a *donor* atom and the semiconductor is called *n-type* because it has negative charge carriers. The substance is *not* negatively charged; it is just that all the electrons in the solid are not needed for the covalent bonding, and so a few are free to move in the presence of an applied electric field.

A different type of semiconductor can be made by doping with a valence –3 atom like boron. Boron is called an *acceptor* atom. This *p-type* semiconductor lacks one electron for covalent bonding at each boron atom; this gap is called a *hole*. As the atoms vibrate on the lattice the few free electrons can hop from one site to another; if a hole is filled with an electron a new hole is created at the site where the electron came from. It is as if the holes moved in the opposite direction, behaving as though they were positive charges. The holes will also move in the presence of an electric field.

Now suppose a slab of p-type silicon is placed in contact with a slab of n-type as in Fig. 15-6. At the junction the holes in the p-type will be filled with electrons from the n-type and this region will become rapidly depleted of both free electrons and free holes. The assembly as a whole is still electrically neutral. The depletion layer is thin since the depletion of holes in the p-type builds up an overall negative charge there while a positive charge is left behind in the n-type because of a loss of electrons. An electric field is created directed from the n- to the p-type. A state is reached where a hole, trying to enter the n-region, is unable to do so because of the force due to the electric field; a similar argument holds for electrons going the other way. The depletion region is very narrow (~10^{-4} cm); the voltage that appears across it is about 0.4 V.

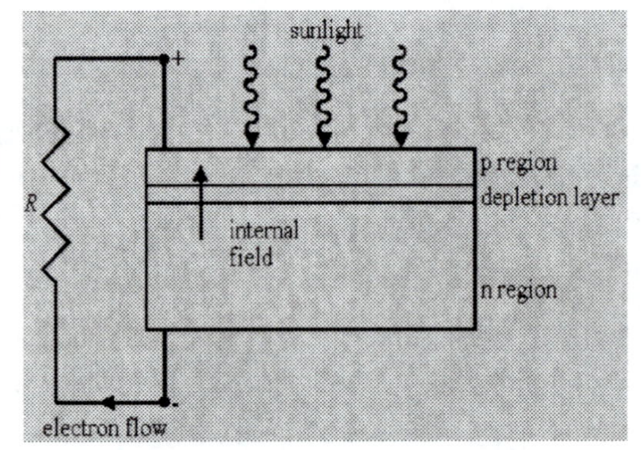

Figure 15-6 A Silicon Solar Cell.

A Si-photodiode, designed for use as a solar cell, is just such a p-n junction. One surface is doped with boron to about 0.002 mm depth; the remaining 1 mm is doped with arsenic to make it n-type. When the device is exposed to sunlight, a photon may transfer its energy to a bound electron, ejecting it from a valence bond and leaving it free to move as a conduction electron and, in the process, also creating a hole. If this happens in the depleted region the electric field will separate the electron-hole pair and the electron will flow around the external circuit to refill a hole on the p-side (See Fig. 15-6). The n- and p-layers act essentially as the terminals of a battery and the photoelectric interaction in the depleted layer is the analogue of the chemical processes in chemical batteries that release electric charge. A silicon solar cell will produce electron-hole pairs by absorption in the wavelength range 350 nm to 1100 nm, a range which covers the visible spectrum and some of the near infrared. This corresponds to photon energies of 3.5 eV to 1.1 eV.

EXAMPLE 17-2

Given that the maximum wavelength absorbed by a np-Si solar cell is 1100 nm, verify that the corresponding photon energy is 1.1 eV (as stated in the previous paragraph)

SOLUTION

The maximum absorbed wavelength is 1100 nm or 1.1×10^{-6} m.

The energy of a photon is given by Planck's equation: $E = hc/\lambda$, where h is Planck's constant, c is the speed of light and λ is the wavelength of the electromagnetic radiation.

$$\frac{(6.63\times10^{-34}\,\mathrm{J\cdot s})(3\times10^{8}\,\mathrm{m/s})}{1.1\times10^{-6}\,\mathrm{m}} = 1.81\times10^{-19}\,\mathrm{J}$$

$$= 1.81\times10^{-19}\,\mathrm{J}\times\frac{1\mathrm{eV}}{1.6\times10^{-19}\,\mathrm{J}} = 1.1\mathrm{eV}$$

Not all of the solar spectrum has photons of sufficient energy to release valence electrons, so a Si solar cell has a theoretical maximum efficiency of 29%. In practice the efficiency is much less than this for various reasons: Some of the light is reflected from the surfaces and some photons are absorbed in such a manner that they do not produce electron-hole pairs but only excite vibrations in the crystal lattice, simply producing heat. The cell also has electrical resistance so there is ohmic heating when a current is drawn from the cell. The practical efficiency is of the order of 12 to 16%. Much of modern research in this area is directed to improving the efficiency of solar cells. Efficiencies approaching the theoretical limit have been achieved in the laboratory but not yet in the field.

Figure 15-7 A Solar Cell–Powered Emergency Telephone. The solar cell array is at the top.

The difficulty of producing mono-crystalline silicon has led to the development of thin layers of amorphous silicon which is very efficient at absorbing light. A layer of amorphous silicon 1 μm thick absorbs as much light as a 50 μm layer of crystalline silicon. Amorphous layers have found considerable use in small power applications such as calculators but the silicon suffers damage with continued exposure to bright light and has not yet proven itself for large power applications.

In 1960 solar cells in large quantity could be manufactured for about $100 per peak watt; By 1980 that had fallen to $10 and today this is slightly better at about $4 [9] . If we take the best location in the American southwest with 300 W/m^2 the installation cost is $4 per watt or $4000 per kWe. These costs would have to fall by at least a factor of three to compete with conventional sources. In Canada where the solar flux is one half, the cost is worse by a factor of two. The cost of the generated power is likewise uneconomically high at 30 to 60 cents per kWe. Also, of all the renewable energy sources, the cost of delivered solar-generated power has been declining the most slowly.

In spite of this gloomy economic forecast several modest photovoltaic power plants have been built in the USA, West Germany, Switzerland, Finland, Austria and Britain with capacities ranging from 15 kWe to 6.5 MWe at Carissa Plains in Arizona; most have been abandoned or dismantled.

Photovoltaic power does, of course, have its uses. In regions where it is difficult or expensive to deliver power in other ways, photovoltaic power may be the answer. For example, telephone repeater amplifiers have been installed on poles in remote sunny areas to provide almost trouble free power. Another example is illustrated in Figure 15-7, which shows a common sight along highways in California - a solar-powered emergency call-box.

15.5 BIOMASS ENERGY

Biomass energy refers to the energy that is available from plant material such as wood, or animal matter such as manure. Wood is the most common source of biomass energy, but as we discuss in this section, gaseous and liquid fuels can also be produced from biomass. It is estimated that, with modern technology, biomass plants can be installed for about $1700/kW .

Wood

Around the world, fuel wood constitutes the single most common use of biomass energy. The forest-products industry frequently uses wood waste as an energy source, and residential wood-burning, even in North America, is not uncommon. Sweden produces about 1500 MW of electricity and Austria 500MW. Wood shares the positive features of all biomass: it is renewable, contains little sulphur (and hence produces little SO_2), and if trees are replaced at the same rate that they are harvested, there is no net contribution to atmospheric CO_2. However, wood-burning generates particulate matter (PM) and NO_X, and perhaps most importantly, it produces a carcinogenic hydrocarbon: benzo(a)pyrene (abbreviated B(a)P). About 25% of B(a)P emissions in the USA in the year 1975 were estimated to come from burning wood. (The single largest source was industrial coke ovens, contributing almost 40%.) If a large fraction of residential heating were converted to wood-burning, pollution due to B(a)P and PM would be a serious problem. In areas of the USA where many homes have wood-heating, the outdoor

[9] Energy Technology Perspectives 2006 OECD, International Energy Agency, #1960, p 130.

concentration of B(a)P has been measured at 10-100 ng/m^3, compared with only 0.7 ng/m^3 as the average concentration in US cities.[10] Unless B(a)P emissions can be controlled, large-scale burning of wood will not be acceptable as a major source of energy.
The precise meaning of "renewable" in the context of any forest use should be considered carefully. It does not mean, for example, that the primal rainforest of Canada's west coast will be regenerated in its original state. Rather, it means that new managed forests will replace it, with the type of trees selected for reasons that are in part economic; the overall ecology will be different.

Biofuels

Liquid and gaseous fuels can be produced from biomass. These fuels are produced by three methods, each discussed separately below: fermentation, anaerobic digestion, and extraction of natural plant oils.

Fermentation

Certain bacteria and yeasts are able to break down the sugars in plant material to produce alcohol which can be used as a motor fuel, or blended with gasoline to produce *gasohol* (~90% gasoline, 10% ethanol) to raise the octane rating of gasoline without the lead additives. Detractors of alcohol as a fuel point out that the production of alcohol is a very energy-intensive process. These evaluations of the total energy cycle are very difficult and nowhere are the opposing sides more divided than for biofuels and ethanol in particular. It would be a great mistake to pursue the use of gasohol fuel for automobiles if , in fact, there is more fossil fuel expended on its production than it displaces in transportation fuel. The issue is important as the US and the European Community are embarking on major expansions of ethanol fuel production.[11]

As the price of petroleum has risen in the years after 2000 more attention has been paid to using ethanol as a fuel additive or, indeed as in Brazil, as a total transportation fuel. The yearly world production of fuel ethanol is given in Fig. 15-8.

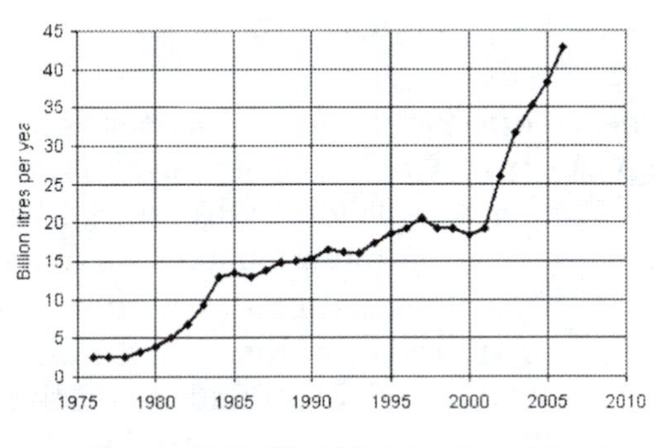

Figure 15-8 World Production of Fuel–Ethanol From 1975 to 2006.

It is not true however, that alcohol has negligible environmental consequences. In Brazil, where gasoline is very expensive, the government has encouraged a vast plan to replace a large fraction of gasoline consumption with alcohol. In the late 1970s 95% of cars produced in Brazil were run on pure ethanol. To do this, large tracts of the virgin rain forest have been cleared to grow cane sugar and cassava for fermentation. The large-scale clearing of the rain forest is an event of great global concern. In addition it has been discovered that although alcohol-burning reduces carbon monoxide by 50% and particulate emissions it produces more aldehydes, which react in the atmosphere to make peroxyacetyl nitrate, a compound that

[10] E. Calle and E. Zeighami, Ch. 3 in *Indoor Air Quality*, Ed. by P. Walsh, C. Dudney, and E. Copenhaver, CRC Press, Boca Raton, FL, (1984).
[11] *Ethanol burns more than it saves;* Toronto Globe and Mail, July 12, 2005, page B6.

stunts plant growth and causes eye irritation. Aldehydes also help increase ozone levels and thus urban smog. With the withdrawal of subsidies the price of alcohol rose at a time when petroleum prices were falling and in 2000 only 5% of cars produced in Brazil are pure alcohol vehicles. By 2005 this had risen again to 71%. Even so, all gasoline in Brazil is gasohol with 23% ethanol and it is the largest biofuel enterprise in the world.[12]

Some agricultural economists worry about large-scale development of bio-fuels because of their possible impact on the world food supply. At the present time there is not a world-wide food shortage in spite of the existence of widespread famine; the world's food supply is not effectively and equitably distributed. A very large excess of food can be, and is, grown and a small amount of this excess does find its way into the world's famine relief programs. Imagine what would be the effect on those countries with zero or small ability to pay for food if all of the surplus crops could be utilized to produce a high-value product like transportation fuel which only the wealthy countries could afford to purchase. The argument has been made that the famine problem would worsen.

It is certainly true that gasohol using alcohol derived from farm products is the only new liquid fuel to make any impression in the marketplace; gasohol is available in a few service stations in North America but again the oil glut of the late 1980s dampened enthusiasm. With the sudden increase in petroleum prices after 2000 there is greatly renewed interest in fuel-alcohol.

Anaerobic Digestion

In the absence of oxygen, certain bacteria break down organic matter to produce methane. On a small scale this technology has been used for years to provide cooking and heating gas for single families, particularly in the third world. Small digesters, usually using manure from a family's domestic animals, can supply a family with a clean source of cooking gas.

The production of *biogas* has been carried out on a somewhat larger scale particularly in agriculture. Considerable development was carried out during the energy crises of the 1970s to design low-cost digesters suitable for use on a large farm, especially those with large manure production such as dairy farms. The digesters are of two types:

- The "conventional" digester is usually a vertical tank into which preheated (35°C) manure is injected. The manure is stirred and maintained at this temperature, and biogas is produced and taken from the top of the tank.
- Plug-flow digesters are long concrete troughs having the manure continuously fed in at one end and removed at the other. The retention time varies from 14 days for cattle manure to 40 days for poultry. Typical figures for a 200-head dairy farm are:

Daily manure volume: 13 m^3
Volume of digester with water and gas storage: 360 m^3
Daily methane volume: 125 m^3

[12] Energy Technology Perspectives 2006, OECD, International Energy Agency, 1960, p 260.

If 30% of the methane is required to maintain the temperature of the digester then there is a net production of 88 m^3 per day with an energy content of 3.3×10^9 J. This is the equivalent of 95 L of gasoline. Because of its lack of purity and difficulty of storage in a small volume, biogas is unsuitable as a transportation fuel without further processing but it is an excellent heating fuel.

On a large scale the technology has not been much developed although there is currently some interest in utilizing large landfill sites. When garbage is dumped in a landfill and covered with earth, the covering keeps oxygen out and anaerobic digestion starts to break down the organic material. The evolved methane seeps out of the ground into the atmosphere and in the process adds to the greenhouse effect. If collecting wells are driven into the site the gas can be collected, cleaned and sold to gas distribution companies or used to generate electricity.

Several installations of this type are operating of which the one in the city of Waterloo, Ontario is representative. The landfill site was started in 1972 and is today half full with 6×10^6 tonnes of garbage. In 2001 a volume of 9.5×10^6 m^3 of methane was collected from 73 wells and was used to generate 28.6×10^6 kW·hr of electricity, enough to supply 2500 homes. It is estimated that landfill gas plants have an installation cost of $1400/kW.

The town of Linköping in Sweden is one of a small number of communities, mostly in Scandinavia, that have attempted to integrate the control of bio-waste and the use of bio-gas into the modern urban infrastructure. The city, which has a population of about 140000 uses the waste from two large abattoirs combined with agricultural waste to generate and purify methane. The plant, which is the largest in the world, processes 100000 t of waste annually to produce 3.6×10^6 m^3 of 95% pure methane with an energy content of 1.4×10^{14} J. This production is sufficient to fuel up to 50 municipal buses and a train running between Linköping and Vastervik. Studies have shown that urban populations over 100000 can implement and profit from this kind of infrastructure.[13]

Plant Oil

Oil doesn't come only from oil wells; the plant and animal kingdoms supply many useful and valuable oils. The nearest that plant oils have come to making an impact on the transportation fuel market is the case of *biodiesel*. This is a liquid fuel made from vegetable oils from various sources; the most common are soybean and rapeseed. The yield of oil from soybeans is 375 L/ha and for rapeseed 1000 L/ha.

Rape seed oil is extracted simply by crushing and yields one tonne of oil for three tonnes of seed. Although the oil can be used directly in place of diesel fuel, the glycerine in the oil clogs engines. To prevent this, the oil is mixed with methanol which, in the presence of a catalyst, precipitates out the glycerine leaving a clear clean fuel called "rape methyl

---

[13] JD Murphy, E McKeogh, G Kiely; "Technical/economic/environmental analysis of biogas utilisation,"*Applied Energy*, Volume 77, Issue 4, pp. 407–427, April 2004, also in: *Current Readings in Transport Economics*, Volume 1, No.4, 2004.
JD Murphy, E McKeogh; "Technical, economic and environmental analysis of energy production from municipal solid waste,"*Renewable Energy*, Volume 9 pp. 1043–1057, 2004 also in: *Current Readings in Transport Economics*, Volume 2, No.1, 2004

ester" or RME. A question that must always be asked about these new technologies is whether or not they actually yield more energy than that which must be invested to produce them. Farm machinery burns petroleum and agricultural fertilizer is largely made from petroleum. An exhaustive study in the USA[14] has analysed the total life cycle energy budget of several fuels and shows that there is considerable energy gain for biodiesel as shown in Table 15-1. Whether or not we can afford the agricultural land to grow crops that will displace a significant amount of petroleum is a matter of great debate.

Table 15-1: Life-cycle energy yield for common fuels.

FUEL	LIFE-CYCLE YIELD
Gasoline	0.80
Diesel	0.84
Ethanol	1.3
Biodiesel	3.2

EXERCISE

15-1. The total electrical generating capacity in Canada [15] in 2000 was 110 GW. If all this power were to be provided by wind-electric turbines each having 500-kW power output, how many such turbines would be required? (If the answer seems like a huge number of turbines, make an estimate of the number of automobiles in Canada; how does the number of turbines compare with the number of automobiles?)

PROBLEMS

15-2. In the Toronto region, solar flux varies from 46 W/m^2 in November to 253 W/m^2 in June; these figures are 24 hour averages. If you cover your roof with Si solar cells of 10% efficiency, how much power will you generate? Compare this to the consumption of your stove, TV etc. What is the oil equivalent of one week of energy from the cells?

15-3. Solar radiation could be used in a power tower for thermal dissociation of water to produce hydrogen. Taking a 75% collection efficiency into the boiler, calculate the H_2 produced daily per m^2 of collector in the Arizona desert. The dissociation energy of water is 2.4×10^5 J/mol. What amount of natural gas would 1 km^2 of collector replace?

15-4. Using Fig. 15-1 estimate the total solar radiation striking a horizontal plate at 44°N Lat. on March 20. Consider 1 m^2 of surface. How much water could be heated from 10°C to 40°C assuming that 50% of the Sun's energy striking the plate can be converted into heat?

15-5. The gas consumption profile for an older 6000 ft^2 house in Hamilton, Ontario is shown in the Fig. 15-9. Natural gas is used for space heating, hot water and cooking.

a) Estimate the yearly consumption of gas for space heating and the combined cooking and hot water.

---

[14] Sheehan, J; Camobreco, V; Duffield, J; Graboski, M; Shapouri, H, *Life cycle inventory of biodiesel and petroleum diesel for use in an urban bus*, NASA no. 19990032195

[15] *Electric Power in Canada 1990*, Energy, Mines and Resources Canada, Government of Canada, Ottawa, (1991) p. 46.

b) What energy is required to heat this house?
c) The house of Example 15-2 is 2 storeys plus a heated basement; the Hamilton house is 3 storeys plus a heated basement. What is the space heating requirement per m² of floor area? Which house is more efficient?
d) Estimate the size of the roof of this dwelling and determine what fraction of its December space heating requirement could be met by solar energy.

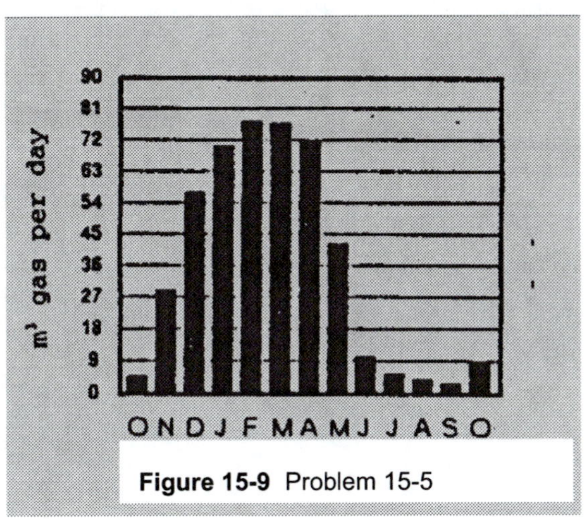

Figure 15-9 Problem 15-5

15-6 The data in Example 15-2 was for the year Oct. 1991 to Oct. 1992. In the intervening years the owner has made some improvements; new insulated vinyl cladding on the exterior, new furnace etc. The figure is the yearly gas usage graph from the bill showing data for July 2004 to July 2005. Analyze this data to see if there has been any reduction in consumption.

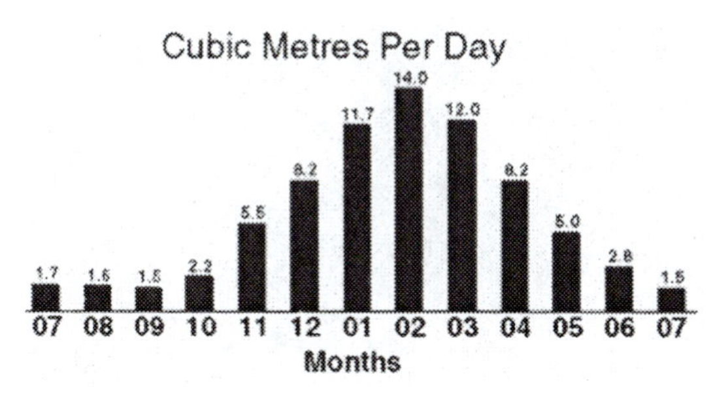

Figure 15-10 Problem 15-6

15-7 The installation cost of the Chanterelle Inn solar heating system was Can$36000 and the operating and maintenance costs are about $100 per year. With electricity costing 9.59 cents per kW hr, what is the theoretical payback time of the system excluding considerations of inflation?

ANSWERS

15-1. 2.20×10^5 = 220 thousand
15-2. (for 30 m²) June 760 W, 0.08 Barrel

15-3. 80 mole, 5×10^5 m³ of gas per day
15-4. Approx. 60 kg
15-5. **a.** 11500 m³, 2000 m³ per year, **b.** 1.2×10^5 kW·h per year, **d.** 70 m², about 7%
15-7 15 yr

CHAPTER 16

ENERGY USE IN THE RESIDENTIAL, COMMERCIAL AND INDUSTRIAL SECTORS.

In this chapter two of the three processes for the transfer of heat energy are described: conduction, and convection. They are illustrated by a consideration of the physical principles involved in the understanding of thermal insulation.

16.1 ENERGY TRANSFER

Energy transfer is an essential part of energy management. Heat must be transferred from a burner to water to run a steam turbine; it must be transferred out of an automobile engine to avoid overheating; we would like to minimize heat loss in our houses in the winter and, of course, energy is transferred from the Sun to the Earth.

Two of the three basic processes by which heat is transferred are easily understood on the premise that thermal energy is simply kinetic energy of the atoms, molecules, or free electrons composing the body. *Conduction* is the flow of heat, without the accompanying bulk movement of matter, from a hot to a cold region. Suppose one end of a copper bar is heated. Copper, being a metal, is composed of atoms arranged in a crystal lattice surrounded by a "sea" of electrons free to move (that is why metals conduct electric current). The thermal energy in the hot end is a result of the kinetic energy of the oscillations of the atoms at their sites, plus the kinetic energy of the randomly moving electrons. This kinetic energy is transferred from atom to atom by the atomic interactions in the crystal lattice or by collisions with the free electrons. In this way the kinetic energy at the hot end is shared with atoms and electrons further along the bar and the heat is thereby "conducted". In the case of metals, the free electrons are the most important factor in the transfer. It is for this reason that good conductors of electricity are usually also good conductors of heat. Insulating materials like wood contain very few free electrons and do not conduct either electricity or heat easily.

Convection is heat transfer in a fluid (gas or liquid) by actual macroscopic motion of atoms of the hot material into the colder regions. As a fluid in a vessel is heated from below, the lowest layer expands, and therefore its density decreases; as it becomes lighter it will move upwards to the top of the colder denser fluid above it. In older homes the old fashioned hot–air heating systems distributed heat entirely in this way; the air was heated in a furnace and, being less dense, rose up in ducts to the rooms above. Meanwhile the colder, heavier air near the floor flowed down into return grills and was re–heated in the furnace. In another design based on hot–water filled radiators rather than hot air, the hot radiators heated the air in the rooms by contact, and the air rose as a convection current, being replaced by cooler air from near the floor.

In many cases natural convection is too slow or has some other drawback. For example, the hot–air heating systems required large diameter pipes that occupied a lot of wall space. In such cases fans or pumps are used to make the process more efficient; such cases are called *forced convection.*

Neither of the above processes can explain the transfer of heat from the Sun to the Earth, since in empty space there can be no conduction or convection. This transfer is by *radiation*; any hot body emits a broad spectrum of electromagnetic waves which carry energy. Radiation will be discussed separately in Chapter 7.

16.2 CONDUCTION

In this section, heat conduction in solid materials (i.e., no convection) is examined quantitatively. Suppose a temperature difference $T_1 - T_2 = \Delta T$ is maintained between two parallel faces of area $\mathcal{A}$, of a slab of thickness Δx as shown in Fig. 16–1. The flux J of the heat–flow expressed as energy per unit area per unit time (J m^{-2} s^{-1}), or power per unit area (W/m^2) is given by $Q/(\mathcal{A}\,t)$, where Q is the heat transferred in a time interval t across area $\mathcal{A}$. The flux is proportional to the "temperature gradient" $\Delta T/\Delta x$ which drives the flow [1]:

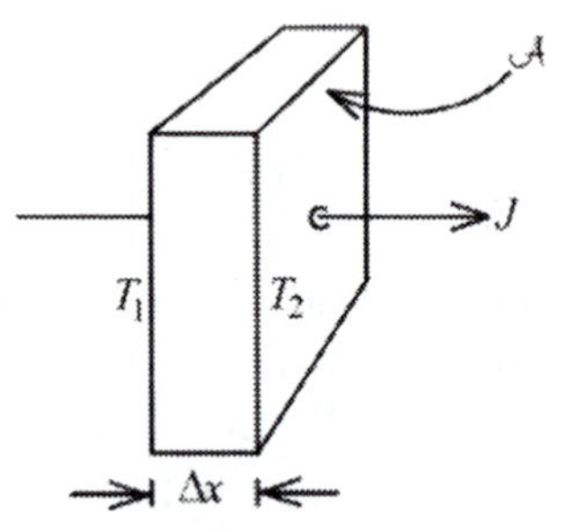

Figure 16–1 Thermal Flux (J) Due to aTemperature Difference $T_1 - T_2$ Across a Slab of Heat–Conducting Material

$$J = \frac{Q}{\mathcal{A}t} = k\frac{\Delta t}{\Delta x} \qquad \textbf{[16–1]}$$

where k is a constant of proportionality called, in this case, the *thermal conductivity.*

Equation [16–1] can be rewritten to give the heat transferred per unit time, Q/t (i.e., power P):

$$P = \frac{Q}{t} = k\mathcal{A}\frac{\Delta t}{\Delta x} \qquad \textbf{[16–2]}$$

Equation [16–2] accords with our intuition; we certainly expect the heat transferred per unit time, Q/t, to increase with the cross–section area, A, and with the temperature difference, ΔT, and to decrease with an increase in the thickness, Δx. Values of k for a few materials are given in Table 16–1.

---

[1] This is just one example of a whole class of phenomena where the gradient of some potential drives the flow of some measured quantity. Other examples: in electricity, Ohm's Law ($V = IR$) can also be written as $q/(At) = (1/\rho)(\Delta V/\Delta x)$ where ρ is the resistivity, ΔV is the voltage, and q is the electric charge. In diffusion, $n/(At) = D(\Delta C/\Delta x)$ is Fick's Law where D is the diffusion coefficient, n is the number of moles transferred and ΔC is the concentration difference of the diffusing molecules across the distance Δx.

Table 16–1: Thermal Conductivities of Several Materials

MATERIAL	K (W m^{-1} °C^{-1})
Vacuum	0
Copper	385
Plywood	0.12
Glass	0.8
Brick	1.3
Concrete	1
Air (non–convecting)	2.4 10 $^{-2}$
Argon (non–convecting)	1.6 10 $^{-2}$
Fibreglass	4 $^{-2}$

10

Two facts might be noted from the table: Firstly, the conductivity of the metal (copper) is about three orders of magnitude greater than for the insulating materials as expected from the arguments given in Sec. 16.1. Secondly, the conductivity of air is half that of fibreglass. Why then use fibreglass insulation? The reason is that air, and all gases, are very easily set into convection which greatly increases their ability to transport heat. One way to suppress convection is to divide the gas up into very small cells which, if less than a critical size (~1 mm), will not permit convective circulation. This is the function of fibreglass or foam insulation, or even feathers and fur; it allows us to profit from the great insulating property of stagnant air by confining the air in small pockets within the material.

Methods for measuring the thermal conductivity of conductors and insulators are outlined in Problems 16–4 and 16–5.

EXAMPLE 16–1

How many joules per day are transferred through a plywood sheet that is 4.0 ft. by 8.0 ft (1.22 m × 2.44 m) in size and 5.0 mm thick if the temperature on one side is –10°C and 20°C on the other?

SOLUTION

From Eq. [16–2], and using Table 16–1

$Q = kA(\Delta T/\Delta x)t$
$= (0.12\ \text{W m}^{-1}\,{}^{\circ}\text{C}^{-1})(1.22\ \text{m} \times 2.44\ \text{m}) \times [20^{\circ}\text{C} - (-10^{\circ}\text{C})](1/5.0\times10^{-3}\ \text{m})(24\ \text{hr}\times3600\ \text{s/hr})$
$= 1.9\times10^{8}\ \text{J}$

EXAMPLE 16–2

A 2.5 cm thick layer (1") of plywood is glued to a 10 cm (4") thick brick wall as shown in Fig. 16–2; the glue is of negligible thickness. The exposed face of the plywood is maintained at 30°C and that of the brick is 10°C. What power is passed across each square metre of this assembly and what is the temperature, T, of the plywood–brick interface?

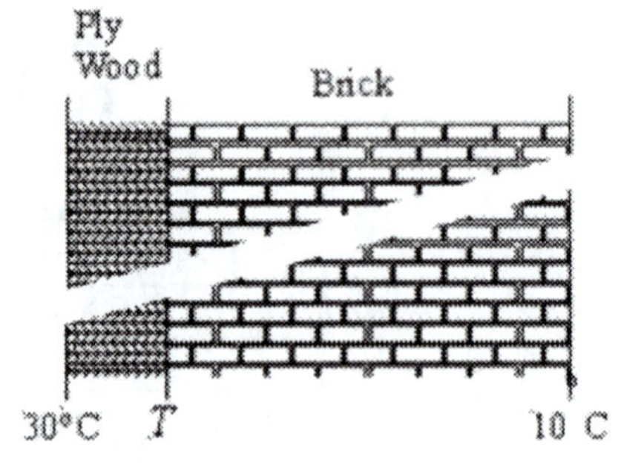

Figure 16–2 Example 16–2

SOLUTION

Rewriting Eq. [16–2] for the plywood and the brick, with ΔT on the left–hand side:

wood: $30 - T = (Q/t\mathcal{A})(\Delta x/k) = (Q/t\mathcal{A})(2.5\times10^{-2}/0.12)$
brick: $\underline{T - 10} = \underline{(Q/t\mathcal{A})(10\times10^{-2}/1.3)}$
adding: $20 = (Q/t\mathcal{A})(0.285)$

Therefore, $Q/t\mathcal{A} = 20/0.285 = 70$ W/m^2

Substitute in either of the above expressions and solve for $T = 15^\circ$C

16.3 THE CONDUCTIVE–CONVECTIVE LAYER

It might seem possible to calculate heat loss through walls, windows, etc. by direct substitution into Eq. [16–1] or [16–2]. However, in addition to the conduction process of Eq. [16–1], there is also a convection process – one which supplies heat on one side of a conductor and carries it away from the other. Obviously air will circulate around depending on the temperature gradients, but the thin layer of air right against the surface experiences a frictional drag and cannot follow the flow of the main body. Thus there is a transition air–film called the *conductive–convective layer* across which the heat transfer is dominated by the conduction of the air (near the solid surface) and gradually becomes dominated by convection further away from the solid surface. For this reason the air actually contributes substantially to the insulating value of a surface; in the case of a window, for example, it is the most important part!

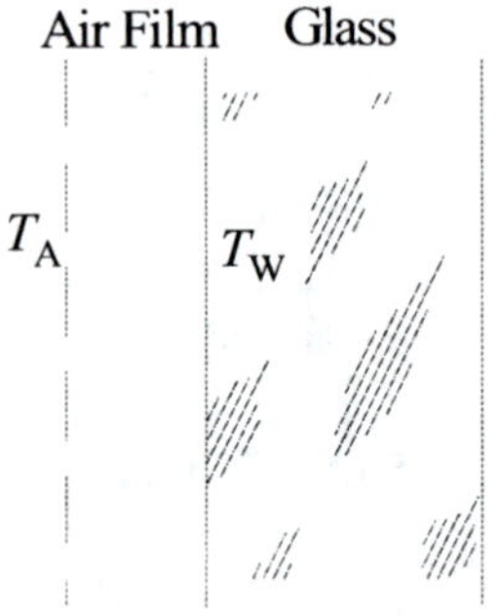

Figure 16–3 The Conductive–Convective Layer of Air on a Glass Window.

As shown in Fig. 16–3 it is assumed that this layer has a temperature drop $T_A – T_W = \Delta T$ across it, where T_A is the temperature of the air in the room and T_W is the temperature of the window surface. (Put your finger on a window in cold weather if you want to be convinced that the inner surface of the window is not at room temperature.)

Table 16–2: Conduction–convection Parameter in Air at Atmospheric Pressure (ΔT = temperature difference across conductive–convective layer)

	h (W m^{-2} °C^{-1})
Horizontal surface facing up	$2.5\,(\Delta T)^{1/4}$
Horizontal surface facing down	$1.3\,(\Delta T)^{1/4}$
Vertical surface	$1.8\,(\Delta T)^{1/4}$

The mathematical description of convection is very complex but fortunately it has been found that, for temperature differences which are not too great, this conductive–convective layer is well described by a simple empirical equation:

$$\frac{Q}{t} = h(T_a - T_w)A \qquad [16\text{–}3]$$

where h is a parameter described in Table 16–2.

The factor $(\Delta T)^{1/4}$ makes the solution of practical problems non–analytical and slightly awkward, but as will be seen in Example 16–3, a simple iterative method leads quickly to a solution.

EXAMPLE 16–3

Heat is transferred through a vertical window, glazed with a single pane of glass of dimensions 1.0 m × 1.0 m × 3.0 mm, from a room at 20°C to the outside at 0°C. The thermal conductivity of the glass is 0.80 W m^{-1} °C^{-1}. Assume that the outside and inside air is perfectly still, so that there is a conductive–convective layer on both sides of the glass as shown in Fig. 16–4. Determine the rate of heat loss through the window, and the temperature difference across the glass.

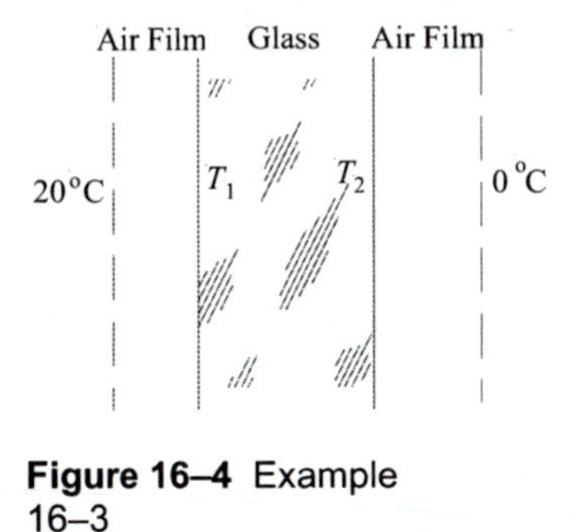

Figure 16–4 Example 16–3

SOLUTION

Rewriting Eq. [16–2] and [16–3] to give the temperature drops for the three layers we have:

$$20 - T_1 = \frac{Q}{t}\frac{1}{A}\frac{1}{h}$$

$$T_1 - T_2 = \frac{Q}{t}\frac{1}{A}\frac{\Delta x}{k}$$

$$T_2 - 0 = \frac{Q}{t}\frac{1}{A}\frac{1}{h}$$

$$20 = \frac{Q}{t}\frac{1}{A}\left(\frac{2}{h} + \frac{\Delta x}{k}\right)$$

Re–arranging to solve for Q/t:

$$\frac{Q}{t} = \frac{20A}{\frac{2}{h} + \frac{\Delta x}{k}}$$

Of course Q/t is the same in each equation since, when equilibrium is established, the heat which enters at one side must pass through every interface and be removed at the final surface.

It is necessary to make some reasonable assumptions–or even guesses– about T_1 and T_2 and then iterate the solution to an answer. Let's assume to start with that $T_1 = T_2 = 10^\circ C$ (i.e., we are assuming that the insulating properties of the thin glass are negligible (or $T_1 - T_2 \approx 0$) and that the two air layers are equivalent) [2]. Thus, ΔT for each air film is $10^\circ C$, and from Table 16–2, we have:

$$h = 1.8(10)^{1/4} = 3.2 \text{ W m}^{-2}\,^\circ\text{C}^{-1}$$

Calculating Q/t with this value of h gives:

$$\frac{Q}{t} = \frac{(20^\circ\text{C})(1.0\text{m}^2)}{\left[\dfrac{2}{3.2\text{W m}^{-2}\,^\circ\text{C}^{-1}} + \dfrac{0.0030\text{m}}{0.80\text{W m}^{-1}\,^\circ\text{C}^{-1}}\right]} = 32\text{ W}$$

Knowing the value of Q/t, we can now use our earlier expression for $T_1 - T_2$ to determine the temperature change across the glass:

$$T_1 - T_2 = \frac{Q}{t}\frac{1}{A}\frac{\Delta x}{k} = 32\text{W} \times \frac{1}{1.0\text{m}^2} \times \frac{0.0030\text{m}}{0.80\text{Wm}^{-1}\,^\circ\text{C}^{-1}} = 0.12^\circ\text{C}$$

instead of zero as we assumed. The calculation could be repeated with ΔT for each air film set equal to $\frac{1}{2}(20 - 0.12)^\circ C = 9.94^\circ C$, but it is hardly worth it, since our original guess of $\Delta T = 10^\circ C$ was very close to this value.

Example 16–3 points out several interesting things: The insulating property of a window is almost completely determined by the surface air films; without these films the heat flow would be

$$\frac{Q}{t} = 20A\frac{k}{\Delta x} = 53 \times 10^3\text{ W}$$

or about 170 times larger! The conductivity of the glass is almost irrelevant; if visibility were not a factor it could be a sheet of copper with almost no change in heat loss. The glass only serves to provide two surfaces to form two air films. It is now clear why windows lose so much heat on a windy day. The wind disrupts and effectively removes one layer, cutting the insulating property of the window in half. It is also clear that the function of storm windows is not to form a "dead air space" as is often stated, but to create two more surfaces to which air layers may adhere. A single glazed storm window cuts heat loss by a factor of two on a still day and three on a windy day over that of an unprotected window.

---

[2] One should always think about the problem carefully and make reasonable guesses. For instance it would not be reasonable to assume $T_1 < T_2$ in this example, but $T_1 > T_2$ would be reasonable; our choice of $T_1 = T_2$ is the simplest.

16.4 THERMAL RESISTANCE

Commercial insulation is usually specified by its R–value which stands for its *thermal resistance*, defined by

$$R = \frac{\Delta x}{k} \quad [16\text{–}4]$$

If this is substituted into Eq. [16–2], then

$$\frac{Q}{t} = \frac{\mathcal{A}\Delta T}{\mathrm{R}} \quad [16\text{–}5]$$

This equation can be rewritten as

$$\Delta T = \left(\frac{1}{\mathcal{A}}\right)\left(\frac{Q}{t}\right)R \quad [16\text{–}6]$$

which is similar in form to Ohm's Law in electricity, $V = IR$ (see footnote 1 p 16–2), so the (thermal) resistance is well named.

The numbers usually quoted for the R–value of insulation (e.g., R10) are based on an awkward and archaic set of units: Q in British thermal Units (Btu), t in hours, $\mathcal{A}$ in square feet, Δx in inches, and ΔT in °F. The relation between the thermal conductivity based on these units and the SI thermal conductivity is (see Exercise 16–1):

$$k(\text{Btu hr}^{-1}\text{ in }^{\circ}\text{F}^{-1}\text{ ft}^{-2}) = 6.93k(\text{W m}^{-1}\,^{\circ}\text{C}^{-1})$$

The R–values of several materials are given in Table 16–3 in British units ($\text{Btu}^{-1}\text{ ft}^2\cdot\text{hr }^{\circ}\text{F}$) for 1–inch thicknesses and in SI units ($\text{W}^{-1}\text{ m}^2\cdot{}^{\circ}\text{C}$) for 1–cm–thick samples.

EXAMPLE 16–4

Calculate the R–value of 1 inch of non–convecting air (i.e., the first line in Table 16–3).

SOLUTION

First, the value of k for air in British units is found using Table 16–1 and Eq. [16–7], and then Eq. [16–4] is used to calculate R:

$$k = 6.93 \times 0.024 = 0.17\text{ Btu hr}^{-1}\text{ in }^{\circ}\text{F}^{-1}\text{ ft}^{-2}$$
$$R = \Delta x/k = 1/0.17 = 6 \text{ (as shown in Table 16–3)}$$

Of course an R–value of 6 for one inch of air, as calculated in Example 16–4 above, can never be realized in practice because such a thick layer of air would be set into convection and the heat transfer would increase greatly. If the air volume is broken up into small cells by some insulating material, such that convection is not possible, then this high insulating property of air can be utilized. This is the purpose of fibre or foam insulations. As can be seen from Table 16–3, the R–value of such insulations actually approaches the ideal value of 6. In some cases it is even exceeded as for polyurethane foam. This is because the gas in the cells is not air but some other gas used in the manufacture of the foam that has a lower thermal conductivity even than air. The same effect operates in some sealed storm windows where the space between the double glazing is filled with the gas Argon; as can be seen from Table 16–1 this increases the insulating *R*–value by 50% (since $R \propto 1/k$).

Table 16–3: Thermal Resistance (*R*) Values

SUBSTANCE	R (1 IN.) $BTU^{-1}\ FT^2\ HR\ ^{\circ}F$	RSI (1 CM) $W^{-1}\ M^2\ ^{\circ}C$
Air (non–convecting)	6	0.42
Copper	3.7×10^{-4}	2.6×10^{-5}
Plywood	1.25	8.6×10^{-2}
Glass	0.18	1.3×10^{-2}
Brick	0.11	7.7×10^{-3}
Concrete	0.14	1.0×10^{-2}
Fibreglass	3.6	0.25
Polyurethane foam	6.3	0.43

The *R*–value of a conductive–convective layer is given by:

$$R = \frac{1}{h} \qquad [16\text{–}7]$$

where *h* is given in Table 16–2.

The *R*–values are particularly useful since, for multilayered structures they add, just as do series resistances in an electric circuit.

Let us redo Example 16–3 using the thermal resistances:

EXAMPLE 16–5

Re–visit Example 16–3 using thermal resistances.

SOLUTION

Again assume $T_1 = T_2 = 10^{\circ}C$
Therefore $h = 1.8\times(10)^{1/4} = 3.2\ W\ m^{-2}\ ^{\circ}C^{-1}$

Now calculate R values:
2 Air films: $2\times(1/3.2\ W\ m^{-2}\ ^{\circ}C^{-1}) = 0.625\ W^{-1}\ m^2\ ^{\circ}C$
Glass : $3\times10^{-3}\ m/0.80\ W\ m^{-1}\ ^{\circ}C^{-1} = 0.004\ W^{-1}\ m^2\ ^{\circ}C$
$R_{total} = 0.629\ W^{-1}\ m^2\ ^{\circ}C$

From Eq. [16–5]: $Q/t = A\Delta T/R = 1.0\ m^2(20\ ^{\circ}C)/0.629\ W^{-1}\ m^2\ ^{\circ}C = 32\ W$, as before.

EXAMPLE 16–6.

The wall of a room is constructed in two layers. In contact with outside still air at 0.0˚C is a 6" (15.2 cm) layer of concrete. A ½–inch (1.3 cm) layer of plywood is glued in close contact with the inside surface of the concrete. The inside temperature is 20.0˚C and the inside air is also still. Find the power transferred across 10 m^2 of the wall and the temperature at each surface. The situation is as illustrated in Fig. 16–5 with the air layers and the five relevant temperatures T_1, ...T_5.

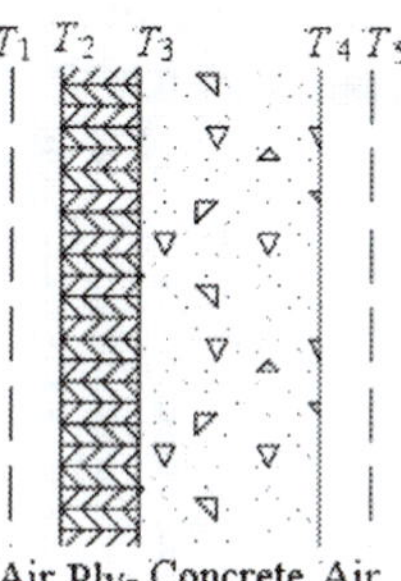

Figure 16–5
Example 16–6

SOLUTION

Assume for now that $T_2 = T_3 = T_4$. As we did in Examples 16–3 and 16–5 we are assuming, to begin with, that the air layers provide all the insulation.

Therefore, $T_1 - T_2 = T_4 - T_5 = 10^\circ\text{C}$ and
$h = 1.8 \times (10)^{1/4} = 3.2\ \text{W m}^{-2}\ {}^\circ\text{C}^{-1}$

Now calculate the R–value for each layer, using $R = 1/h$ for the air films and $R = \Delta x/k$ for the concrete and wood:

	RSI
2–Air films: $2 \times (1/(3.2\ \text{W m}^{-2}\ {}^\circ\text{C}^{-1}))$	= $0.62\ \text{W}^{-1}\ \text{m}^2\ {}^\circ\text{C}$
Concrete: $(15.2\times10^{-2}\ \text{m})/(1.0\ \text{W m}^{-1}\ {}^\circ\text{C}^{-1})$	= $0.15\ \text{W}^{-1}\ \text{m}^2\ {}^\circ\text{C}$
Wood: $(1.3\times10^{-2}\ \text{m})/(0.12\ \text{W m}^{-1}\ {}^\circ\text{C}^{-1})$	= $0.11\ \text{W}^{-1}\ \text{m}^2\ {}^\circ\text{C}$
***R* (total)**	= $0.88\ \text{W}^{-1}\ \text{m}^2\ {}^\circ\text{C}$

Use Eq. [16–5] to determine Q/t:

$$\frac{Q}{t} = \frac{\mathcal{A}\Delta T}{R} = \frac{(10\text{m}^2)(20^\circ\text{C})}{0.88\text{W}^{-1}\text{m}^2\ {}^\circ\text{C}} = 227\text{W}$$

Now that this approximate value for Q/t is known, Eq. [16–6] can be used to find better values for the ΔT's for the plywood, concrete, and air:

wood: $T_2 - T_3 = (1/\mathcal{A})(Q/\Delta t)R = (1/(10\ \text{m}^2))(227\ \text{W})(0.11\ \text{W}^{-1}\ \text{m}^2\ {}^\circ\text{C}) = 2.5^\circ\text{C}$

concrete: $T_3 - T_4 = (1/10)(227)(0.15) = 3.5^\circ\text{C}$

either air layer: $T_1 - T_2 = T_4 - T_5 = (1/10)(227)(0.31) = 7.0^\circ\text{C}$

Notice that the total temperature difference across all the layers (including the two air layers) adds up to 20.0˚C, as it should.

Now calculate T_4, T_3, and T_2:

$T_4 = T_5 + 7.0^\circ\text{C} = (0 + 7.0)^\circ\text{C} = 7.0^\circ\text{C}$

$T_3 = T_4 + 3.5^\circ\text{C} = 10.5^\circ\text{C}$
$T_2 = T_3 + 2.5^\circ\text{C} = 13.0^\circ\text{C}$

A second iteration using $T_1 - T_2 = 7.0^\circ\text{C}$ (instead of 10°C as we first assumed) gives:

$Q/t = 213$ W, which we round to 2.1×10^2 W. Then

$T_2 - T_3 = 2.3^\circ\text{C}$

$T_3 - T_4 = 3.3^\circ\text{C}$

$T_1 - T_2 = T_4 - T_5 = 7.2^\circ\text{C}$

$T_3 = 10.5^\circ\text{C}$ and $T_2 = 12.8^\circ\text{C}$

Notice that this second iteration produces relatively minor changes in the values and another iteration will produce no further significant change.

16.5 DEGREE–DAYS

In designing insulation for buildings, one has to remember that the outside temperature varies a great deal during the winter. For this reason, units called *heating degree–days* or HDD are used to give a measure of the insulation or heating requirements in various locations. The building industry in metric countries including Canada use, or are converting to, SI units but in the US the older British system of units is still in universal use. It is assumed that, if the average outside temperature (T_{av}) is over 18°C (65°F), no heat need be supplied to the building. When T_{av} is less than 18°C then for 1 day

$$\text{No. of Celsius degree–days} = 1 \text{ day} \times (18 - T_{av}) \qquad \textbf{[16–9]}$$

or

$$\text{No. of Fahrenheit degree–days} = 1 \text{ day} \times (65 - T_{av}) \qquad \textbf{[16–10]}$$

The sum of these quantities for every day on which T_{av} is less than 18°C (or 65°F) gives the total "degree–days" for the locality. The Celsius degree–day map for Canada is shown in Fig. 16– 6 and that for the United States in Fig. 16–7.

EXAMPLE 16–7

Suppose we have a house through whose walls, roof and windows we have calculated a heat–loss rate of 4000 W (4000 J/s) on a day when the temperature is 21°C inside and 6°C outside. What is the annual heating requirement of this house?

SOLUTION

Per degree temperature difference and per day, the loss is given by:

$$\text{Loss} = \frac{4000\frac{\text{J}}{\text{s}} \times \frac{3600\text{s}}{1\text{h}} \times \frac{24\text{hr}}{1\text{day}}}{15^\circ\text{C}}$$

$$= 2.3 \times 10^7 \text{ J per day per degree temperature difference}$$

This is then the heat loss per degree–day. In the city of Toronto, the annual heating energy requirement for this house would be about

$$(4000 \text{ degree–days}) \times (2.3\times10^{7} \text{ J/(degree–day)}) = 9\times10^{10} \text{ J}$$

(using Fig. 16–6 and knowing that Toronto lies near the 4000 contour to the lower right).

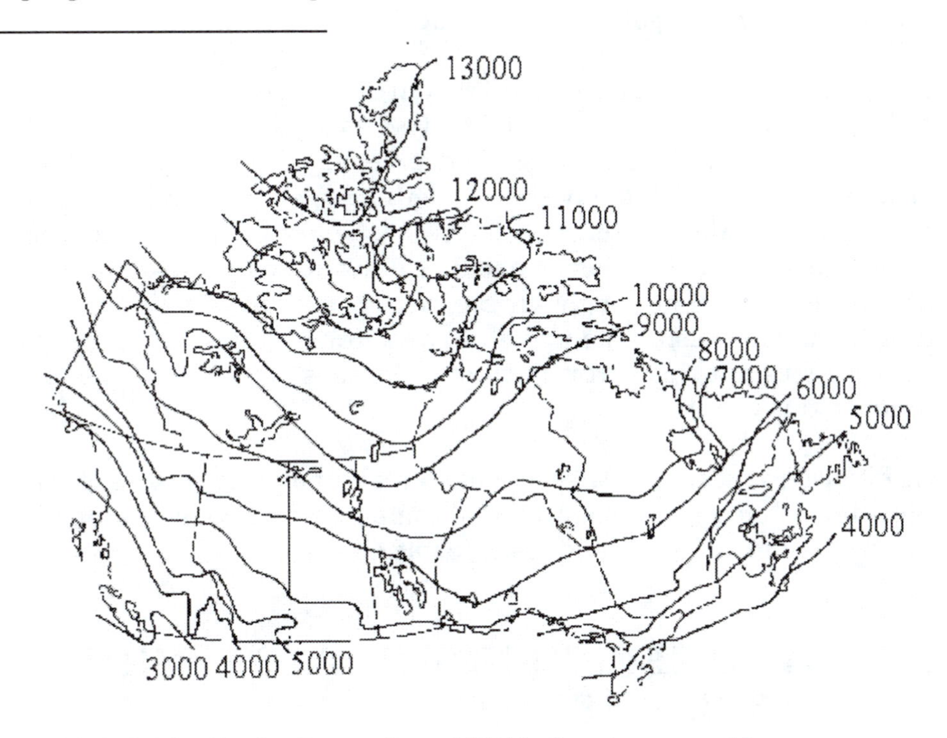

Figure 16–6 Celsius Heating Degree Days (HDD) in Canada averaged for one year.

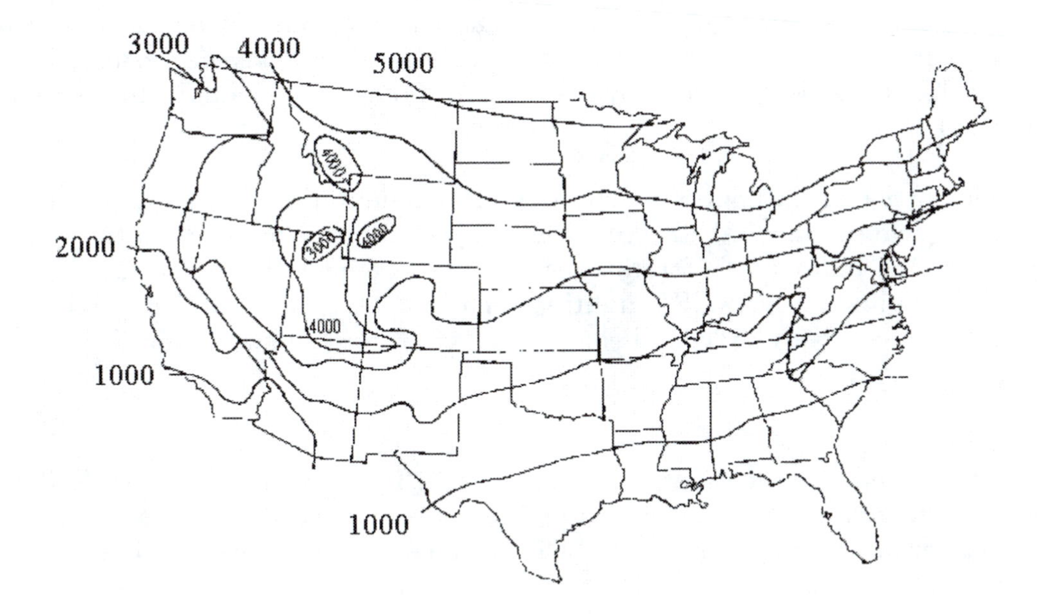

Figure 16–7 Celsius Heating Degree Days (HDD) in USA averaged for one year.

16.6 Conservation or Efficiency?

In the previous two chapters, we have been discussing new technologies, such as coal gasification, that will help make better use of present energy sources, and alternative sources of energy – the wind and sun, for example – that can make an important contribution to the energy supply where appropriate.

In the following sections we explore another "source" of energy, but not one that we can mine from the ground or capture in sunlight. The source is commonly called "energy conservation" but is probably better described as "energy efficiency". If we reduce our energy consumption by a kW·h today then we will have an extra kW·h available tomorrow, in effect providing a "source" of 1 kW·h of energy. "Energy conservation" is achieved by increasing the efficiency of end–use devices, so that less energy is required to achieve the same level of heating, lighting, transportation, etc. (Energy "conservation" could also mean "doing without," in other words, cutting back on heating, lighting, etc., but this curtailment of services is not what we mean by energy "conservation" in this chapter.)

The topic is divided into three broad sections, corresponding to sectors of society that use large amounts of energy: residential and commercial buildings (which consume 38% of world energy), industry (38%), and transportation (24%).

16.7 IMPROVED EFFICIENCY IN THE RESIDENTIAL AND COMMERCIAL SECTOR

Residential and commercial buildings include houses, apartment buildings, offices, stores, hotels, schools, and government edifices such as post offices, courthouses, etc. Many businesses and households have been conserving energy over the past three decades. In Canada from 1973 to 1987, the average household energy use declined [3] by about 32%, and energy consumption per m^2 of commercial floor space declined by almost 18%. Some of this decrease was aided by government grants, for example, to upgrade home insulation.

A considerable portion of the energy consumption in residential buildings goes into lighting. In a typical commercial structure, lighting consumes about 40% of the electricity used, plus another 10% for air conditioning to remove the unwanted heat generated by the lights [4]. In the USA, lighting (in all sectors, not just residential and commercial) consumes about 25% of all electricity, 20% directly plus 5% for air conditioning [2].

Lighting

Estimates of the potential reduction[3] in energy used for lighting in commercial buildings range from 55% to 85%. One obvious way to reduce energy for lighting is to change from incandescent bulbs, which convert only 5% of the electrical energy into light, to

[3] The State of Canada's Environment, Government of Canada, Ottawa, (1996).

[4] A. Fickett, C. Gellings, and A. Lovins, Scientific American **263**, (1990) No. 3 (Sept.), p. 66.

fluorescent bulbs which have a conversion efficiency of 20%. There are also the *metal-halogen* lamps, more often used as automotive headlights and to illuminate public spaces. They have an improved lifetime and produce light similar to that from a very high temperature blackbody i.e. bluish-white. Finally there has been the development of the compact fluorescent lamp which has the ballast[5], in the base of the lamp so that it can physically replace an incandescent lamp. Some properties of common household lamps are given in Table 16 – 4

Table 16–4: Properties of Household Lamps.

	INCANDES-CENT	**HALOGEN**	**LINEAR FLUORES-CENT TUBE**	**COMPACT FLUORES-CENT LAMP (CFL) (MAGNETIC BALLAST)**	**COMPACT FLUORES-CENT LAMP (CFL) (ELECTRONIC BALLAST)**
Light output in lumens[6]. per watt	8 - 20	15 - 25	20 - 90	40 – 65	60 – 70
Purchase Cost	$3	$5	$9	$5	$9
Lamp Life	750 –1500 hr	2,000 – 4,000 hr	10,000-20,000 hr	6,000 - 15,000 hr	6000 – 15000 hr

Modern fluorescent lamps have phosphors that can provide a wide range of "warm" and "cool" colours with higher efficiency than previous phosphors, and electronic ballasts have been introduced with a resulting increase in efficiency [7] of about 20%. The electronic ballast permits switching of the light at a frequency of up to 20 kHz instead of the 60 Hz of the magnetic ballast producing the further gain in efficiency. In addition the high frequency switching allows the lamps to operate on dimmers with a further potential saving of electricity.

EXAMPLE 16-8

A standard 100 W incandescent bulb costs $1 and has an expected lifetime of 750 hr when operated at 120 V. A house contains 10 of these bulbs. If electricity is 10 cents per kW hr what is the total cost to operate these bulbs for their rated lifetime?

SOLUTION

The cost of the bulbs is 10 × $1 = $10. While running, these lamps draw a power of 1 kW. Therefore they use 750 kW hr in their rated lifetime. The cost of the electricity for their rated lifetime is $0.10 × 750 = $75. The total cost is $75 + 10 = $85.
Now try Problem 16-13.

---

[5]The "ballast" in a fluorescent fixture provides a high voltage to initiate the ionization of the mercury vapour in the tube, and also limits the current for safe operation.
[6]A lumen is the SI unit of luminous flux, which is radiant flux adjusted for the response of the human eye.
[7]D. Hafemeister and L. Wall, *Energy Conservation*, in "The Energy Sourcebook", Amer. Inst. of Physics, New York, (1991), p. 455.

In addition to upgrading lighting fixtures, there are a number of other methods to reduce energy consumption for lighting. Using existing daylight from windows or skylights can decrease lighting requirements, but attention has to be paid to the heating effect of direct sunlight, which is a potential advantage in cold climates, but a problem in warm climates. Even in cold regions of the world, a large building in winter often requires air conditioning instead of heating because of the large amount of heat generated by people and equipment. Interior or exterior devices to diffuse the daylight can be used to reduce heating and glare produced by direct sunlight. It has been estimated [8] that the energy requirements of a typical commercial building can be reduced by about 10% with the appropriate use of "daylighting." Another energy–saving step is to reduce the overall light level in a room, but provide compact fluorescent desk lamps to illuminate work areas. As well, glass partitions in the upper part of walls between offices and rooms allow light to pass from room to room, and light–coloured furniture, carpets, and wall–coverings reflect light and hence reduce lighting requirements. Occupancy sensors can be used to turn off lights automatically when rooms are not in use. A number of these ideas, such as conversion from incandescent bulbs to fluorescent bulbs and turning off lights when not needed, can be applied also in houses and apartments.

Heating, Ventilating, and Air Conditioning (HVAC)

Another area of potentially large reduction in energy consumption is the heating, ventilating, and air conditioning system in a building. Heating and air conditioning needs can clearly be reduced if insulation with a high R–value is used during construction or renovations, and if caulking around windows and weather stripping around doors are installed and maintained. Computer–control of the air condition in individual rooms, based on automatic monitoring of temperature, humidity, etc., can lead to greater efficiency. Proper maintenance, such as cleaning the coils in air conditioners and replacement of air filters in ventilating systems, also decreases energy consumption.

The use of high–efficiency motors in HVAC systems can play a significant role. In Canada, 50% of the electricity used in the commercial sector is consumed by electric motors [9], and in industry, the figure is 75%. In the USA, more than half of all electricity generated is used by electric motors[10], and in industry they consume 65% to 70% of the electricity. High–efficiency motors have designs that reduce the magnetic, resistive, and mechanical losses to less than half the levels of a decade ago. For maximum effect, a high–efficiency motor is combined with an electronic variable–speed drive, which controls the speed of the motor. In a typical older motor installation, which might control a ventilation fan, for example, the motor operates at full speed all the time, and the amount of ventilation is controlled by a valve or damper that restricts the air flow. (This is somewhat like driving with one foot on the gas and the other on the brake.) Improved motor and drive systems use about half the electricity of older systems [11].

---

[8]Ibid, p. 451.

[9]Toronto Globe and Mail, Energy Management Business Report, (1992) Oct. 20, p. 3.

[10]Ref. 4, p. 67.

[11]Ref. 4, p. 68.

Electrical needs for air conditioning can be reduced by use of thermal storage of cold water or ice, produced overnight when exterior temperatures are lower and hence the efficiency of air conditioning is greater. The cold water or ice can then be used during the daytime to supplement regular air conditioning. As well, overnight use of an air conditioner is usually less expensive because of lower commercial electricity rates for off–peak hours. Bell Canada in Ottawa has installed a state–of–the–art cool storage system, and will save $80000 in energy costs annually[12]. Combined with other energy–efficiency improvements, including lighting upgrades, Bell will be saving $250000 each year. The capital cost was $1.5 million, giving a payback time of about six years, which was shortened to three years by a financial contribution from Ontario Hydro.

Installation of energy–efficient modern windows can reduce heating needs, and wise placement of windows can also decrease electric–lighting requirements as previously mentioned. It has been estimated that about 5% of the energy used in the USA is lost through windows[13]. A single–glazed window has an R–value (Section 5.4) of about 1, and double–glazed windows have $R \approx 2$. Glass is a very strong absorber of the infrared (IR) radiation emitted by the interior of a building. A thin film of an IR–reflecting material (called a low–emissivity [14], or low–E material) can be used to coat the inside surface of a window to reflect the radiant heat back into the building. Whereas glass absorbs most of the IR, a low–E window absorbs only 10%, reflecting 90%. [15] A low–E double–glazed window has an R–value of about 3. Adding a second low–E film and filling the space between the double glazing with a poorly conducting gas such as argon increases R to about 4. Thermal resistance values as high as R–6 to R–10 can be achieved by filling the space between a double–glazed low–E window with a transparent, poorly conducting material such as aerogel, which is a very open skeleton of tiny glass particles (5%) and air (95%).

In warm climates, and in large buildings in cold climates, heat gained through windows is more of a problem than heat loss. For large office buildings, it is common practice to place reflective coatings on the exterior of windows to transmit visible light, but reflect solar IR radiation. Still under development are photosensitive windows that darken when the ambient light levels get too high.

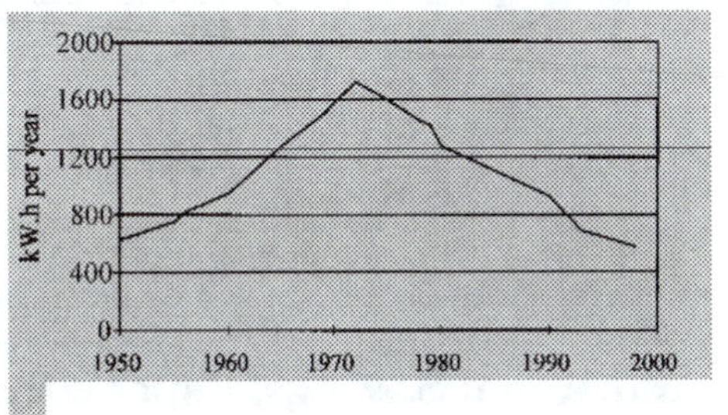

Figure 16–8 Energy usage by refrigerators in the USA increased from 1950 to 1972 and has declined since then.

Appliances and Office Equipment

Efficiency improvements in lighting and motors have been matched by various appliances. For example, current refrigerators and freezers now consume about half the energy of the models from 1972. Between 1950 and 1972, energy consumption by new refrigerators in the USA[16] increased from

[12]Ref. 9

[13]Ref. 7, p 451..

[14]The IR–reflective materials are poor absorbers of IR and also poor emitters of IR — hence the phrase "low emissivity."

[15]A.H. Rosenfeld, T.M. Kaarsberg, J. Romm. *Technologies to Reduce Carbon Dioxide Emissions in the Next Decade,* Physics Today Nov. (2000) p 31.

[16]Ref. 7, pp. 456–457.

approximately 600 kW·h per year to almost 1800 kW·h per year (Fig. 16–8), because of increased size (from about 0.25 m^3 to 0.6 m^3), a decrease of 40% in actual efficiency, and the addition of features such as automatic defrosting. The "energy crisis" of 1973 led to the establishment of tighter standards on refrigerator efficiency by the California and US Governments, resulting in a continuing decline in electricity usage by refrigerators[17]. This has been accomplished by improving the insulation (both in thickness and quality), using more efficient motors and compressors, and increasing the area of the heat exchangers.

Other devices can show large efficiency improvements: Efficient televisions can consume 75% less electricity than their inefficient counterparts[18], photocopiers can save 90%, and computers 95%. Clearly a consumer who wishes to reduce energy consumption can do so by shopping wisely.

Commercial Sector Consumption

In the commercial sector, a useful way to look at energy consumption is the energy consumed per year per square meter of floor space (sometimes called, quite incorrectly, "energy intensity"). Table 16–5 gives the average energy use per year per square metre in the retail sector. The value for food retailers is more than twice the others reflecting the heavy use of refrigeration, most of which is operating inefficiently. This large inefficiency is a result of the industry's reluctance to abandon open freezers on the grounds that closed ones are more expensive and that the public dislikes them. The cheapest enclosure method is the plastic strip curtain but, unless mandated, supermarket chains have largely been reluctant to use them.

Table 16–5: Retail and Shopping Centres: Yearly Energy Use

STORES, SUPERMARKETS AND MALLS	ENERGY USE
Non–Food Retailers	0.9 GJ/m^2
Non–Food Big Box	1.1 GJ/m^2
Food Retailers	2.8 GJ/m^2
Enclosed Shopping Malls	1.4 GJ/m^2
Strip Malls	1.2 GJ/m^2

Adapted from Natural Resources Canada publication, ISBN 0–662–35666–7, *Energy Innovators Initiative Retail and Shopping Centre Sectors; Saving Energy Dollars in Stores, Supermarkets and Malls* (2003)

Additional Savings, Especially for Houses

Owners of houses can take advantage of many of the energy–saving strategies mentioned above, such as upgrading insulation and using low–E windows, and there are many other ways to effect energy savings, some involving very small cost. In cold climates, deciduous trees can be planted on the south and west sides of a house to provide summer shade, but when the leaves are absent in the winter, the sun provides light and warmth. Coniferous trees can be planted on the north side of a house to block winter winds, and ideally, this side of a house should have fewer windows to reduce

[17]Ref. 16; and Ref. 2, p. 74.
[18]Ref. 4, p. 68.

heat loss. Turning down the furnace thermostat a few degrees at night and when no one is in the house results in a saving of heat energy of about 10–15%. The temperature of the water heater can also be turned down from 60°C (140 °F) to 49°C (120°F) to reduce heat loss from the tank and hot–water pipes. With modern detergents, using cold water for the laundry is possible.

A detailed study [19] on a typical home in Northern California showed that its total energy consumption could be reduced from about 70 MW·h to 30 MW·h per year, with only a modest investment. Energy–reducing steps included upgrading ceiling insulation from R–11 to R–19 and wall insulation from R–0 to R–11, installing a fluorescent light fixture in the kitchen, using a low–flow showerhead, and decreasing the hot–water temperature. The payback time is about three years.

16.8 DISTRICT HEATING

In Sweden, Denmark, and many Eastern European cities, the waste heat from electrical power stations is used to heat surrounding communities. This district heating has the obvious advantage of using thermal energy that would otherwise be discarded. In addition, controlling pollution from one central plant is easier than trying to control individual furnaces. In at least one Swedish city, there is even enough waste heat to melt snow from the roads, thereby saving on snow removal costs. District heating is easier to implement in European cities than in North America because of the denser housing patterns. In addition, some North Americans would undoubtedly object to their lack of choice in home heating if district heating were mandatory, as it is in some European cities.

About 70% of the population of Russia get their domestic heat from district heating and the largest single system is that in St. Petersburg supplying 2.4×10^2 PJ per year The largest 23 systems in the world together supply 8×10^2 PJ per year. The heat comes from a variety of processes such as geothermal heat in Reykjavik, Iceland and cogeneration in Russia. Biomass in the form of agricultural waste (wood–chips, seed–husks and straw) supply district heat in some villages in Austria. Where garbage is incinerated the heat is always sold as district heat or as process heat for industry.

16.9 CONSERVATION IN INDUSTRY

Industry can avail itself of all the energy–conserving devices discussed for residential and commercial buildings. In particular, appreciable savings are possible from the use of high–efficiency motors, improved lighting, and upgrading insulation and HVAC systems. At least as important, however, are improvements in industrial processes. For example, the introduction of a new steel–making procedure in the 1960s reduced the energy requirements by 50% relative to the open–hearth methods that were previously used. The newer technique involves blowing oxygen through the molten

[19] A. Rosenfeld and A. Meier, *Energy Demand,* Physics Vade Mecum, H. Anderson, Ed., American Institute of Physics, New York, (1981), p. 170.

metal to burn much of the carbon that is present; the heat thus generated provides energy to remove other impurities in the form of slag. In the older open–hearth process, all the heat was provided from outside the steel. Another general area of process improvement is the use of computers to monitor and control complex manufacturing procedures, reducing waste and inefficiency.

From 1973 to 1987, Canadian industry reduced its energy consumption[20] per unit of industrial output by approximately 6%. In the USA from 1971 to 1986, total industrial energy use declined by about 1% per year, while output increased by roughly 2% per year.[21] However, it has been estimated that about 40% of the US energy reduction is not due to conservation measures and process improvements, but rather to a shift toward the manufacturing of products that happen to require less energy per unit of output [22].

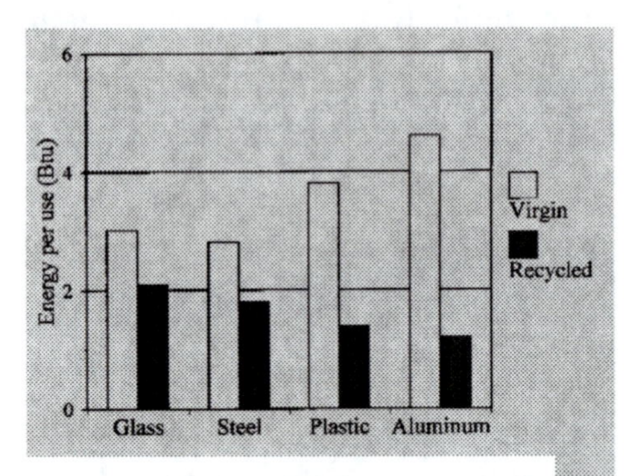

Figure 16–10 Producing a consumer container (glass bottle, steel can, etc.) from recycled material requires less energy than producing it from virgin material.

Industries are usually careful to recycle materials, because the energy needed to manufacture a product from virgin materials is typically about double that required when recycled materials are used. Recycling consumer products also saves energy; Figure 16–10 shows the energy required [23] to produce a consumer container (glass bottle, etc.) from various materials, either virgin or recycled. Aluminium clearly offers large savings in recycling. Of course, even recycling consumes more energy than using refillable containers.

16.10 COGENERATION

Industries often are in need of heat for various industrial processes, as well as electricity for machinery, lighting, etc. Since most electricity is generated by thermal plants, which produce a great deal of waste heat, why not supply industry with both electricity and heat? This cogeneration of electricity and heat can occur in two ways: an electric power plant can sell waste heat to industry (or the heat could be used for district heating, as described in Section 16.8), or an industry that now produces heat for itself can convert to generation of both electricity and heat, possibly selling excess electricity to a local utility.

However, there are barriers to cogeneration. If an electric plant is to provide heat to industry, the plant and industry must be physically close together. As well, the times when an industry requires the most heat might not correspond to times when electricity generation is high. Industries are often hampered from considering cogeneration

---

[20]Ref. 1.
[21]M. Ross and D. Steinmeyer, Scientific American **263**, (1990) No. 3 (Sept.), p. 89.
[22]Ref. 21, p. 94B.
[23]Ref. 21, p. 96.

themselves because they consider the payback time for the electrical generating system to be too long. In addition, if surplus electricity is to be sold to a utility, the industry will likely not be paid a good price for it (because the surplus would probably not be supplied on a regular basis) and the industry is apt to be subject to stricter environmental laws regarding the generation of the electricity.

Although cogeneration is clearly a good idea in principle, the barriers have prevented much development of this concept.

EXERCISES

16-1. Show that: k(Btu hr^{-1} in °F^{-1} ft^{-2}) = 6.93 k (W m^{-1} °C^{-1})
using: 1 Btu = 1054.8 J   1 inch = 2.54×10^{-2} m   1 ft^2 = 0.0929 m^2 9 F° = 5 C°

[Hint: You will probably discover that 1 Btu hr^{-1} in °F^{-1} ft^{-2} = 0.144 W m^{-1} °C^{-1}, and wonder how to get from 0.144 to 6.93. Here is an analogy: You know that 1 cm = 0.01 m; suppose that the length l of some object is measured both in cm and in m. How is the numerical value of l in units of cm related to l in units of m? So, how is the numerical value of k in Btu hr^{-1} in °F^{-1} ft^{-2} related to k in W m^{-1} °C^{-1}?]

16-2. Determine the heat energy lost in 1.0 h through a slab of styrofoam that is 3.0 cm thick, and that has a size of 2.0 m × 1.5 m. The temperature difference between the two sides of the styrofoam is 12°C. The thermal conductivity of styrofoam is approximately 0.01 W m^{-1} °C^{-1}. Neglect the effect of air films on the surface of the styrofoam.

16-3. There are several algorithms to convert from °C to °F (or vice versa). The easiest is as follows:
Convert °C to °F: Add 40 to °C**,** then multiply by 9/5, then subtract 40.
e.g., Convert 20°C to °F: ①20 + 40 = 60. ②60×(9/5) = 108. ③108 – 40 = 68°F
Convert °F to °C: Add 40 to °F**,** then multiply by 5/9, then subtract 40.
This algorithm is based on the following facts:
1 Water freezes at 0°C and 32°F
2 Water boils at 100°C and 212°F
From this you should be able to explain the algorithm. (Hint: at what temperature do the two scales have the same reading?)

16-4. Make a numerical estimate of the total amount of energy that you use in a typical day. Include transportation, heating, lighting, appliances, etc. For each category, how could your consumption be reduced (without significantly affecting your lifestyle), and how much of an effect would such reductions have on your total consumption?

PROBLEMS

16-5. An apparatus to measure the thermal conductivity of a metal (called a "conductometer") consists of a round copper bar A heated by a steam jacket B at one end and cooled by coils C carrying water at the other. The heat conducted along the copper is removed by the water (specific heat 4186

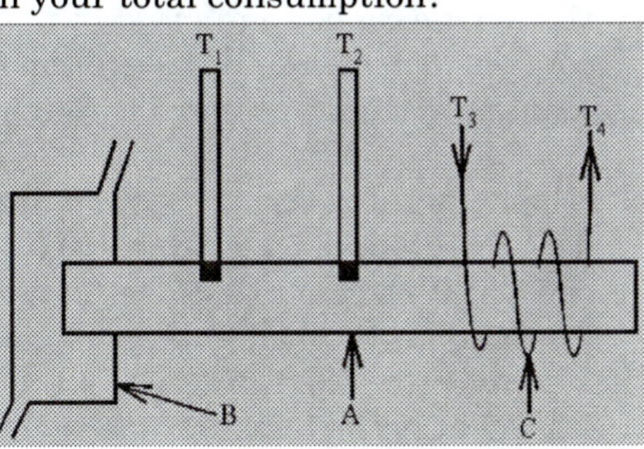

Figure 16–11 Problem 16–5. The "Conductometer".

J kg^{-1} °C^{-1}). Two thermometers T_1 and T_2 are inserted in small holes in the copper 2.0 cm apart. The temperature difference $T_1 - T_2$ is 10°C when water flows through the tubing at 50 cm^3 per minute and enters at a temperature of T_3 = 10.0°C and leaves at a temperature T_4 of 27.3°C. The diameter of the bar is 2.0 cm. Find the thermal conductivity of copper.

16-6. To measure the thermal conductivity of an insulator, thin samples are usually used as in the method of Fitch shown in Fig. 16–12. A heat reservoir at constant temperature T_1 (e.g. a metal pot of boiling water) is placed on the insulating layer (thickness, x, and conductivity, k) which in turn is placed on a metal block (mass, M, cross–section area, A, and specific heat, C) at time t = 0. In a time interval dt an amount of heat dQ flows across the insulator, raising the temperature of the block by an amount dT.

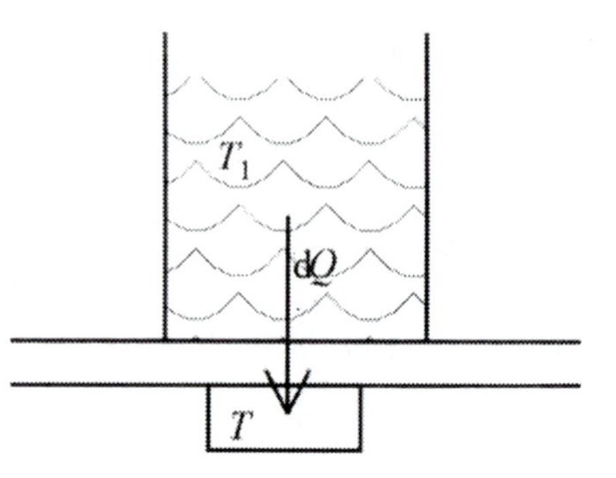

Figure 16–12 Problem 16–6. The "Fitch Conductometer".

a) Show that:

$$dt = \frac{CMx}{kA}\frac{dT}{(T_1 - T)}$$

b) Show that a graph of ln($T_1 - T$) vs. t gives a straight line whose slope is $-kA/(CMx)$ from which the conductivity k can be determined.

16-7. The door of a refrigerator is 1.5 m high, 0.80 m wide and 6.0 cm thick. Its thermal conductivity is 0.21 W m^{-1} °C^{-1}. The inner temperature is at 0°C and the outer is at 27°C. (a) What is the heat loss per hour through the door neglecting convection effects? (b) The air is usually still inside a refrigerator so there will be a conductive–convective layer on the inside but in a busy kitchen such a layer will be absent on the outside. What is the heat loss per hour in this case and what is the temperature of the inside surface of the door? (c) At night the kitchen is also still; what is the heat loss per hour now, and what are the temperatures of the inside and outside surfaces?

16-8. A beer cooler, made of plastic foam, is in the form of a box. The inside dimensions of the box are 50 cm × 30 cm × 30 cm and the walls are 5.0 cm thick. The plastic foam has a thermal conductivity of 0.040 W m^{-2} °C^{-1}. Ice is placed inside the cooler to keep it at 0°C inside. How much ice melts each hour if the outside temperature is 30°C?

16-9. Two copper plates 0.50 cm thick have a 0.10 mm sheet of glass sandwiched between them. One copper plate is kept in contact with flowing ice water and the other is in contact with steam. What are the temperatures of the two copper–glass interfaces, and what power is transferred through a 10 cm × 10 cm area?

16-10. A home has a concrete basement floor 15 × 15 m square and 10 cm thick. On a winter day, the ground below is at –10.0°C while the basement recreation room is at 20.0°C.
(a) Calculate the rate of loss of heat, remembering to include the air film.
(b) A do–it–yourself enthusiast glues a plywood floor of thickness 1.0 cm onto the concrete. What fraction of the heat loss is eliminated?
(c) What is the temperature at the interface of the plywood and concrete?

16-11 The total electrical consumption in Canada[24] in 1990 was 466 TW·h, of which roughly 20% was used for lighting. If conservation measures were to reduce energy consumption for lighting by 50%, what would be the power saving in megawatts, and what could be the reduction in the number of 1000–MWe plants in operation? (Assume that the power consumed by lighting is constant throughout each day of the year.)

16-12. Suppose that an electric pump is being used to drive a fluid through a pipe. If the pump is run at a slower speed, so that the speed of the fluid is reduced by a factor of two, by what factor is the power requirement of the pump reduced? Hint: consider a quantity of fluid (having mass m) to be at rest prior to passing through the pump, and travelling with speed v when leaving the pump a short time later.

16-13. The lamps of Example 16-8 are run at a reduced voltage of 110 V to increase their rated lifetime by a factor of 3 and achieve a saving in operating cost. a) What is the power now drawn by each of these lamps? b) Using the data of Table 16-4, how many lamps are required to maintain the illumination? 100 W bulbs operate at the high end of their light output. c) Lowering the voltage to 110V increases their rated lifetime by a factor 3 to 2250 hr. What is the cost to operate this array for their rated lifetime? d) Is this operation cheaper or more expensive than that in Example 16-8?

ANSWERS

16-2. 4×10^4 J
16-5. 3.8×10^2 W m^{-1} °C^{-1}
16-7. **a.** 4.1×10^5 J, **b.** 2.0×10^5 J, 13.5 °C, **c.** 1.3×10^5 J, 9.5 °C, 17.5 °C
16-8. 0.20 kg
16-9. 9°C, 91°C, 6.6×10^3 W
16-10. **a.** 2.3×10^4 W, **b.** 24%, **c.** –2.2°C
16-11. 5.3×10^3 MW; 5 power plants
16-12. 8
16-13. a) 84 W, b) 12, c) $238.80 d)Less expensive by $16.20

[24]"Electric Power in Canada" 1990, Energy, Mines and Resources Canada, Government of Canada, Ottawa, (1991), p. 29.

CHAPTER 17

THE AUTOMOBILE AND TRANSPORTATION

One of the largest sectors for the consumption of energy is transportation. Transportation vehicles may be roughly divided into two classes:

- bulk- (or mass-) transportation such as ships, trains and most air-transportation
- unit (or near unit) transportation such as automobiles, trucks and buses.

A large portion of the first category involves fossil fuel consumption. Even electric trains get their electricity mostly from fossil fuels and simply have the advantage of central rather than distributed burning. Central generation of electricity assures higher fuel efficiency and lower per-unit pollution problems. In a very few countries like France the fossil fuels for electricity generation have been substantially replaced by non-fossil sources, e.g., nuclear.

The unit transportation is virtually entirely driven by fossil fuels and that almost entirely from petroleum and its by-products. A few non-fossil sources are being developed such as alcohol and rapeseed oil but, as yet, their impact on petroleum use is negligible. In 2003, 95% of transport was driven by petroleum. Taken with the petroleum-driven bulk-transportation, it uses by far the largest fraction of petroleum, which is also our most vulnerable energy supply.

17.1 THE OTTO CYCLE

There are several types of petroleum-fueled engines in use in transportation - e.g., gasoline 4-stroke, gasoline 2-stroke, diesel, turbine, etc., but the ubiquitous one is the gasoline 4-stroke engine operating in the *Otto cycle*[1]. The others differ in detail but not in principle so the discussion of this chapter will focus on the automobile using a 4-stroke, Otto-cycle gasoline engine.

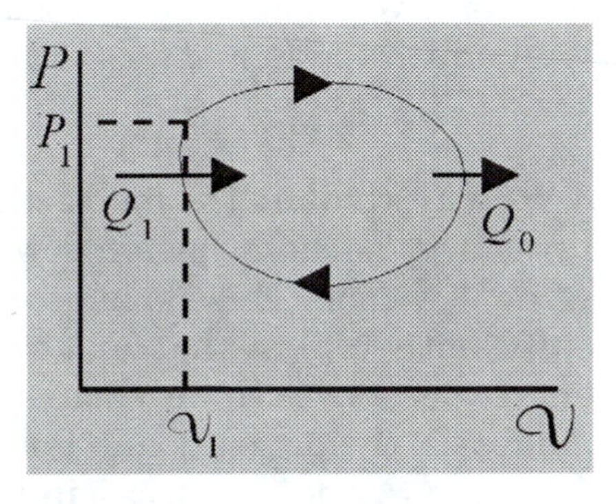

Figure 17–1 Thermodynamic (Heat–engine) Cycle

Like all heat engines (Section 4.6) the Otto cycle takes a working fluid (air-fuel mixture) through a cycle involving pressure P and volume $\mathcal{V}$ (see Fig. 17-1), taking in heat Q_1 and expelling heat Q_0 ($Q_1 > Q_0$) in such a way that the cycle is closed (i.e., it returns to its starting value of $P_1, \mathcal{V}_1$) and its enclosed area is not zero. In such a case the area of the cycle is equal to the mechanical work done

[1] Named after Nicolaus Otto who patented the first 4–stroke gasoline engine in 1867.

$$W = \int P \cdot d\mathcal{V} = Q_1 - Q_0 \qquad [17\text{-}1]$$

In the Otto cycle this is achieved by the performance of 4 strokes (2 up, 2 down) of a piston in a closed cylinder with the appropriate operation of valves to admit the air-fuel mixture and to exhaust the burnt air-fuel mixture as shown in Fig. 17-2.

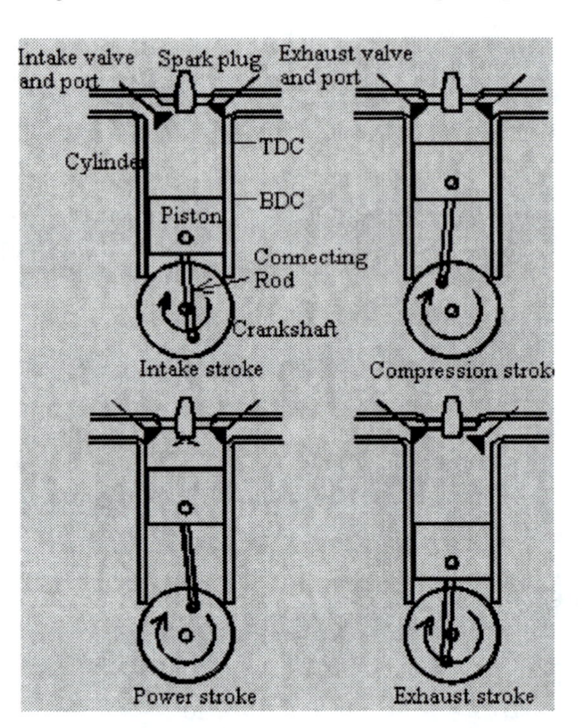

Figure 17–2 The 4-stroke Otto Cycle

1. **Intake stroke:** Starting with the piston at top-dead-center (TDC) with the intake valve open, the piston descends to bottom-dead-center (BDC) filling the cylinder with the air-fuel mixture at pressure P_1 and temperature T_1. ($1 \Rightarrow 2$ in Fig. 17-3)

2. **Compression stroke:** The intake valve closes and the piston goes from BDC to TDC compressing the mixture. This takes place so quickly that we may assume that the compression is *adiabatic,* i.e., no heat enters or leaves during the compression. As a result the pressure rises to P_3 and the temperature to T_2. [2] ($2 \Rightarrow 3$ in Fig. 17-3)

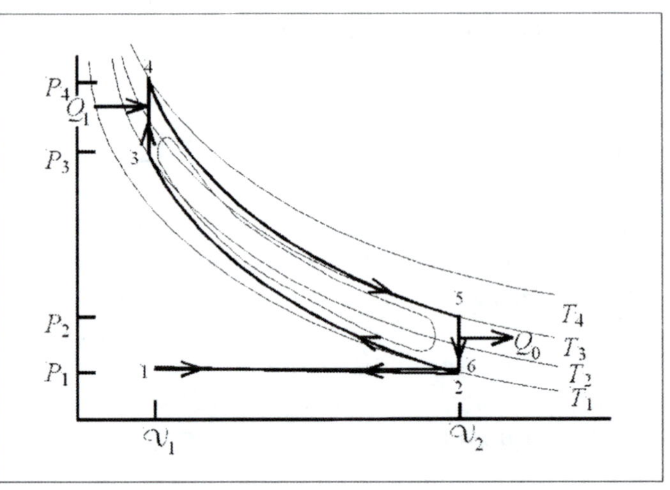

Figure 17–3 The 4-stroke Otto Cycle

3. The mixture is ignited by a spark and the energy input Q_1 takes place so quickly that the piston movement is negligible. This results in an *isochoric* (constant volume) rise in pressure and temperature to P_4 and T_4. ($3 \Rightarrow 4$ in Fig. 17-3)

4. **Power stroke:** The increased pressure pushes the piston down doing work $W = \int F \cdot dx$. This again occurs sufficiently rapidly that heat neither enters nor leaves so the expansion is adiabatic, the pressure and temperature falling to P_2 and T_3 . ($4 \Rightarrow 5$ in Fig. 17-3)

5. The exhaust valve opens and an isochoric expansion occurs to P_1, T_1. ($5 \Rightarrow 6$ in Fig. 17-3)

---

[2] The compression ratio P_3/P_1 must be kept low enough so that the temperature does not rise so high as to ignite the air–fuel mixture; ignition is done at a carefully controlled time by a spark. The compression ratio in gasoline engines is usually below 10. If such *pre–ignition* takes place it is called 'knocking' or 'pinging' and can damage the engine. Diesel engines operate with compression ratios of about 18 and use the temperature rise of compression itself to ignite the fuel.

6. **Exhaust stroke:** With the exhaust valve open the piston returns to TDC and expels the remaining burnt gases and returns the cycle to its starting point. ($6 \Rightarrow 1$ in Fig. 17-3)

In the portion of the cycle $5 \Rightarrow 6 \Rightarrow 1$ an amount of energy Q_0 ($Q_1 > Q_0$) is expelled into the external environment.

Note that in the portion of the cycle $6 \Rightarrow 1 \Rightarrow 2$ there is ideally no enclosed area so no work is done by, or on the system. The only area is in the portion $2 \Rightarrow 3 \Rightarrow 4 \Rightarrow 5 \Rightarrow 6$.

The net result of the cycle is to convert an amount of heat $Q_1 - Q_0$ into work done on the piston. The *thermal efficiency* is defined in Chapter 4:

$$\eta = \frac{W}{Q_1} = \frac{Q_1 - Q_0}{Q_1} = 1 - \frac{Q_0}{Q_1} \qquad \textbf{[17-2]}$$

It can further be shown that for an ideal gas (a reasonable approximation at these temperatures) η depends only on the compression ratio $\mathcal{V}_2/\mathcal{V}_1$. For a compression ratio of 10, η is about 60%. In other words, approximately one third of the fuel energy is lost in the exhaust due to the thermodynamic cycle alone. This is <u>not</u> avoidable loss like conductive and radiative heat loss; this is the unavoidable thermodynamic loss as a result of finite T, P, and $\mathcal{V}$.

Of course the real situation is that none of the branches of this cycle are perfectly adiabatic or isochoric, and the cylinder walls, piston, etc., conduct heat away continuously. The real cycle might be more like the rounded contour shown inside the ideal cycle in Fig. 17-3. These losses take away another portion of the fuel energy about equal to the thermodynamic losses and the thermal efficiency of a real engine is of the order of 40%.

The engine of a mid-sized automobile (mass = 1500 kg) traveling at 90 km/hr (56 mph) might use energy from the fuel at a rate of 70 kW. This means that fuel is consumed at a rate of 8.8 L/100 km (32 mi/Imp gal, 27 mi/US gal; see Problem 17-2). The 70 kW is called the '*fuel input power*' (P_{fi}) and is given by

$$P_{fi} = qve \qquad \textbf{[17-3]}$$

where q is the heating value of the fuel in J/L, v is the vehicle's speed in km/s, and e is the fuel consumption in L/km.

As seen above, fully 60% of the P_{fi} is immediately lost as heat to the environment leaving 40% or 28 kW to run the automobile. This latter is called the *indicated power* P_i.[3]

---

[3] The word 'indicated' comes from the early days of steam power. Shortly after James Watt began to manufacture steam engines his craftsmen made mechanical contrivances that drew a chart with a pencil in which the displacement of the piston and the steam pressure controlled the position of the pencil in two directions at right angles. In other words the *P–V* cycle was drawn. The devices were called 'indicators'.

If the numerator and denominator in Eq. [17-2] are divided by the time then the thermal efficiency may be rewritten as

$$\eta = (\frac{W}{t})(\frac{t}{Q_I}) = \frac{P_i}{P_{fi}} \qquad \textbf{[17-4]}$$

Increasing Engine Thermal Efficiency

To increase the thermal efficiency, the area of the closed cycle in Fig. 17-3 must be increased. One way to do this is to increase the compression ratio $\mathcal{V}_2/\mathcal{V}_1$. Compression ratios around 10 are vast improvements over early engines; they have come about because of the ability of gasoline formulators to blend fuels with a high *octane rating*, that is, they will not pre-ignite due to the heating by compression. Very little further improvement can be foreseen in this area.

Alternatively, to increase the area of the cycle in the direction of the pressure, entails operating between a greater temperature difference. T_1 is at, or near, environmental temperatures so it is T_4 that must be increased. Increasing T_4 will result in the engine structure itself becoming hotter and engine temperatures are today running near the limit that modern lubricants can withstand. It is possible to envisage an engine that would not involve petroleum-based lubricants and might be made of ceramic-based materials and run red-hot. Such an engine would indeed have a higher thermal efficiency but would come from a far different technology than is available at the present time. Higher T_4 is aided by minimizing the transfer of heat from the hot gas mixture to the cylinder walls and piston head. This is achieved by making the piston diameter to piston travel (bore/stroke) ratio equal to 1 and thus minimizing the internal surface area for a given volume. In addition, higher engine speeds mean that the residency time of the hot gases in the engine is reduced and less heat is transferred to the engine structure. Modern engines run at very high speeds compared with those of a half-century ago.

Why, then, should engines not be made to run even faster to improve their efficiency? One important reason is the time it takes to introduce the air-fuel mixture (and expel the exhaust). When the intake valve opens the mixture must be accelerated from zero speed and introduced into the cylinder before the valve closes a very short time later (see Problem 17-3). Modern engines, when running at their optimum speed are already starved for fuel and do not get the optimum amount (running "lean"). This is the reason for introducing *superchargers* on high performance engines. These are blowers that increase the pressure in the intake manifold to increase the amount of air-fuel that enters when the intake valve opens.

The intake manifold is also an acoustically resonant cavity. Its length is adjusted so that at optimum engine speed the acoustic resonance in the manifold produces a pressure maximum at the intake valve to aid fuel injection. Of course this condition can only exist at one engine speed or its harmonic multiples.

Another reason that engine speeds cannot be continuously increased is the ever-increasing friction losses between pistons and cylinder walls with increasing speed. In fact this effect is so important that optimum engine speeds have been decreased from ~3500 rpm to ~2500 rpm in recent designs.

Many other small improvements in engine carburation and ignition as well as fuel blending have raised the thermal efficiency to the 35% to 40% level of today. Little improvement in this figure can be anticipated in the near future. Further improvement in the overall efficiency of automobiles must be sought elsewhere. It is important to note however, that improvements in the utilization of the 40% which is useful work will result in an almost threefold saving in fuel. For example: The power left to run our sample automobile is 28 kW. If the required power could be reduced to 27 kW by some small improvement (more efficient electrical ignition, better lubrication etc.) the savings in fuel amounts not to 1 kW but 2.5 kW (1 kW/0.4) because of the thermal efficiency. The fuel input power would be reduced to 67.5 kW from 70 kW, a fuel saving of 3.6%.

Because of various internal losses (e.g., friction), the power that is effective in the form of forces on the top of the pistons (P_i) is not that delivered to the output crankshaft of the engine; the latter is called the *brake power*, P_b. This name arises from a device called a 'Prony brake' once used to measure it (see Problem 17-5). The *mechanical efficiency* η_m is defined as

$$\eta_m = \frac{P_b}{P_i} \qquad \textbf{[17-5]}$$

Multiplying this by Eq. [17-4] gives:

$$\eta \eta_m = \frac{P_b}{P_{fi}} \qquad \textbf{[17-6]}$$

The power lost internally, $P_i - P_b$, is mostly from the friction between the piston and cylinder wall, but there are other contributions as well, e.g., bearing and gear friction in the crankshaft and camshaft, power losses in operating the fuel, oil, and water pumps, and the electrical power consumed by the ignition system. All losses in components required to operate the engine must be included, so the fan, fan belt, and battery charging losses must also be included.

Major improvements in these losses have taken place in the last three decades. The switch from the DC electrical generator (60% efficient) to the AC alternator (80% efficient) is an example. Similarly the change from rigid fan blades always driven at full engine speed whether or not the engine needed fan-cooling, to flexible (or variable-pitch) blades or even electrically driven fans switched on only when needed are an important example. The indicated power of about 28 kW in our example car is thus reduced to a brake power of 14.5 kW, i.e., η_m is approximately 52% ($70 \times 0.4 \times 0.52 = 14.5$). Accessories such as lights, wipers, etc. have their power extracted at this point and may take 1.5 kW leaving 13 kW delivered to the transmission.

There are further losses however before the power is delivered to the drive wheels. The drive-train, consisting of the transmission (automatic or manual gear-box) and differential, use up about 2.5 kW. The *drive-train efficiency* η_d can be written as

$$\eta_d = \frac{P_r}{P_b} \qquad \textbf{[17-7]}$$

where P_r is the road power.

Combining Eq. [17-6] and [17-7]:

$$\eta_{fi}\eta_m\eta_d = \frac{P_r}{P_{fi}} \quad \text{[17-8]}$$

Of the original 70 kW, only about 10.5 kW remains to actually propel the automobile. A summary of the preceding power budget is shown in Fig. 17-4.

The road power is also ultimately converted to environmental heat through rolling friction (heating the road and tires), braking friction (heating the brake disks, drums and shoes), heating the air through viscous flow of the air over the car, and heating the damping medium in the shock absorbers. Ultimately all of the 70 kW is converted into environmental heat.

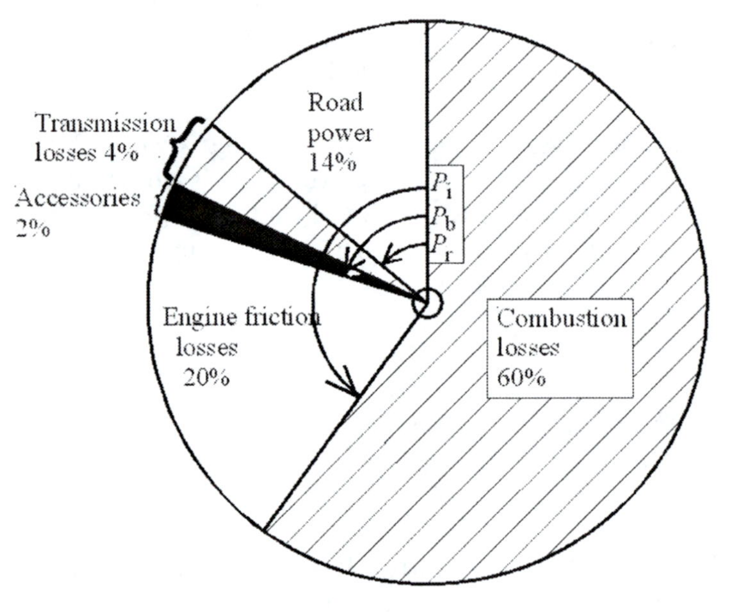

Figure 17–3 Power Budget of an Automobile

Improving efficiency in any of the drive train, accessories, internal friction and propulsion friction losses is the area where further improvements in automobile efficiency are to be anticipated. This is because of the effect that for every kW saved in this area about 2.5 kW of fuel are saved. For example, even a traditional speedometer driven by a cable takes power from the engine. If it is replaced by a solid-state electronic sensor and display that requires negligible power then almost three times the fuel power is saved. One of the areas that have been addressed with considerable success is the aerodynamic loss. This has been reduced considerably by wind-tunnel studies and applications of principles long known to the aircraft industry.

EXAMPLE 17-1.

An automobile of P_{fi} = 70 kW consumes fuel at a rate of 8.8 L/100 km at 90 km/hr. A roof rack is added which increases the aerodynamic drag and thus the road power at 90 km/hr increases by 15%. What is the increased road power and fuel consumption required to sustain this extra load?

SOLUTION

Taking data from Fig. 17-4:

Road power increase = (0.15)(0.14)(70 kW) = 1.5 kW

Because the road power has increased, there is an increase in fuel consumption, which will increase losses in combustion, engine friction, and transmission (but not accessories). The ratio of these losses to road power is (60 + 20 + 4)/14 = 84/14 (from Fig. 17-4). Hence the fuel input power will increase by 1.5 kW × 84/14 = 9 kW.

Increased fuel consumption = (8.8L/100 km) × 9/70 = 1.1 L/100km

---

One accessory which consumes considerable power is the automobile air conditioner which can be rated at 3 to 4 kW; this is not always a wasteful accessory. Certainly at lower speeds, if opening the windows provides equivalent comfort in hot weather, then the air conditioner is wasteful. At highway speeds, however, open windows can create so much air turbulence that the automobile will actually operate more efficiently with the windows closed and the air conditioner on (see Problems 17-8, 9, 10).

The gasoline-powered automobile has been developed almost to the stage where few further improvements will be significant. With modern pollution controls the modern engine is a marvel of clean burning; only very marginal improvements will be made. If further large gains are to be made in this area it will be to change the fuel entirely, e.g., to hydrogen or to transfer the burning of the fuel to another locality where further gains in environmental impact are possible. The electric car is an example of the latter. The burning of coal to generate electricity is one of the areas where the greatest gains can be made in applying new technology. Of course adopting other forms of electric generation such as nuclear and solar energy would have an even greater impact on atmospheric cleanliness.

17.2 FUEL COMBUSTION AND ENGINE EMISSION CONTROL

Combustion chemistry

The combustion process in the automobile engine is not simple; there is more than just the conversion of hydrocarbons and oxygen into carbon dioxide and water with the release of heat. There are other complex reactions taking place and they depend on a variety of conditions in the combustion chamber. The result is that, in addition to CO_2 and H_2O, in the exhaust there are other components, most of which have some detrimental effect on humans and the environment. The most important of these are:

- Carbon monoxide (CO) is very detrimental to human health. In 1970, just before the imposition of exhaust standards, automobiles were responsible for 50% of the total CO emissions in the US.
- Unburned gaseous hydrocarbons (HC) are known carcinogens. In 1970 automobiles were responsible for 20% of the total HC emissions in the US.
- Nitrogen oxides (NO, NO_2, etc) usually referred to as NO_x . NO is essential in the complex chemical cycle that produces photochemical smog that plagues urban areas. In 1970 automobiles were responsible for 20% of the total NO_x emissions in the US.

It has been the task of emission control devices to reduce the 1970 concentrations of these pollutants. By 2000 they had been reduced to: CO - 45%, HC - 14% and NO_x - 16% in spite of a growth of the automobile fleet at a rate of 1.8% per annum. Emission control had eliminated 96% of CO and HC and 76% of NO_x in engine exhaust largely due to the effectiveness of *catalytic converters.*

These unwanted exhaust contaminants are present because the air/fuel mixture is not always stoichiometrically correct and the two are not completely mixed at the moment of ignition. The correct stoichiometric ratio air/fuel is 14.7/1 by weight. At this mixture the combustion is almost ideal; the production of CO_2 is a maximum and CO emissions are low. For "leaner" mixtures (air/fuel > 14.7/1) CO_2 falls slightly, CO remains low but an excess of O_2 remains in the exhaust. For "richer" mixtures (air/fuel < 14.7/1) CO rises rapidly and the exhaust is free of excess O_2. HC is high for rich mixtures and decreases as the mixture gets leaner as would be expected. The production of NO_x is a reaction of the atmospheric nitrogen with oxygen in the combustion chamber. Its rate of production increases with increasing temperature and is a maximum at the lean mixture of air/fuel $\approx$16/1.

Removal of all these pollutants is a challenge since HC and CO require an oxidizing environment to produce CO_2 and H_2O, but NO_x requires a reducing environment to be changed back to N_2.

The Catalytic Converter

A converter has to carry out three chemical reactions on the exhaust products and each reaction by itself would require far different conditions from the others for high efficiency. In general the reactions are:

- Reduction of nitrogen oxides to nitrogen and oxygen: $2NO_x \rightarrow xO_2 + N_2$
- Oxidation of carbon monoxide to carbon dioxide: $2CO + O_2 \rightarrow 2CO_2$
- Oxidation of unburned hydrocarbons (HC) to carbon dioxide and water: $C_xH_y + nO_2 \rightarrow xCO_2 + mH_2O$

The necessary complex chemistry takes place in a unit containing the appropriate catalysts called a "catalytic converter." The first catalysts used were platinum (Pt) and palladium (Pd) in a ratio 5/2 coated on, or embedded in, an inert substrate. In the earliest converters, introduced in 1975, the substrate was a bed of pellets of alumina (Al_2O_3). In the later *dual bed converter* a second catalytic bed was added to convert NO_x back into N_2; they are still commonly used on diesel engines. The *three way catalytic converter,* introduced in 1981, consists of a monolithic, extruded structure of "corderite" ($2MgO.2Al_2O_3.5SiO_2$) in the form of a honeycomb of about 60 channels per cm^2. The dilute catalyst is "washed" onto the surface of these channels and sintered into place. At temperatures between 250 and 600°C the exhaust gases pass over the catalyst and the appropriate reactions are induced. A more recent third generation of converters have rhodium (Rh) and a large array of trace elements and compounds to accomplish almost complete conversion of pollutants. For proper action the engine must be kept operating in a very narrow range about the value air/fuel = 14.7/1. This requires the use of a sensitive O_2 detector placed in the exhaust just before the converter and a computer to use the information from the sensor to adjust the air/fuel ratio. A serious problem with these converters is that they are inefficient at low temperatures and therefore pass a lot of pollutants while the engine and the converter are warming up. This aspect is particularly severe then for short-trip city use.

As a new generation of vehicles, with even stricter emission standards is developed, new types of catalysts are under investigation. Much higher standards, however, will require a conversion of a substantial part of the automobile fleet to entirely new kinds of motive power. Possible candidates are discussed in the next three sections.

17.3 THE ELECTRIC AUTOMOBILE

The petroleum crises of the 1970s prompted a great deal of activity in the United States and elsewhere to develop a practical electric automobile. In the 1990s an additional reason emerged to urge the development of the electric vehicle: air pollution. Some jurisdictions, notably the State of California in the USA have recognized that automobile exhaust is the major contributor to bad air quality in large cities. As a result they passed legislation to require a certain fraction (e.g., 2%) of automobile sales to be "zero-emission" vehicles in the early years of this century.[4] The increasing cost of crude oil in the first decade of the 21st century has renewed interest in alternative fuels for cars.

Electrically powered vehicles have been available for many years but only in a form that requires low acceleration, relatively low speed and sufficient off-duty time to recharge the batteries. The electric milk-van is common in England and other countries, and electrically powered warehouse trucks are common everywhere. An automobile, however, needs reasonable acceleration, short-term high power for emergency manoeuvring, and a cruising range of over 100 km between recharges.

The requirements for a viable electric-car are threefold:

- High torque, high speed electric motors of high efficiency.
- Motor control circuitry of high reliability.
- Batteries with a high power and energy density.

With the technology available in the 1970s none of these three requirements could be met. Since that time continued research has resulted in great improvements in the first two. DC motor efficiencies are now very high with reduced friction losses and efficient cooling. Motor control circuits that had to be made from discrete components can now be made with high reliability using integrated circuits. These control circuits can employ "regenerative braking" in which the electric motor is converted to dynamo operation where the kinetic energy of the moving car is used to recharge the battery. Only the battery remains a problem and, unfortunately, little has happened to improve them

[4] California and other jurisdictions have since put these regulations on hold because of the automotive industry's inability to comply at present. Doubtless, however, they indicate an increasing pressure that will be exerted in the near future.

since the 1970s. As shown in the next two sections almost all batteries fail to match the internal combustion engine in energy and power density. A few exotic systems involving expensive or hazardous materials (or both) are a possibility but only the lead-acid battery is sufficiently cheap and reliable to consider at present.

Production electric cars currently are small, slow and suitable only for neighbourhood use. The most widespread model is the Indian-built Reva sold as the G-Wiz in the UK for about $20,000 (See Problem 17-15).

With major improvements in two of the three areas, however, some manufacturers believe that a marketable electric automobile is possible. Using light composite materials for the body, advanced aerodynamic design, high efficiency motors, low-rolling-friction tires, but still lead-acid batteries, electric automobiles could have an acceleration of 0 to 95 km/hr in 8 to 10 seconds and a range of 150 km on a single charge. They will probably sell in the range of $20,000 to $30,000.[5] Unfortunately for the all-electric car, by 2005 attention had shifted to the hybrid vehicle led by developments in the Japanese and German automobile industries; this is discussed in Section 17.5.

The gasoline automobile has reached the stage where major improvements in efficiency will be very difficult to achieve; the electric automobile offers an advantage. In a recent study the cost was traced right back to the well-head and included all transportation and manufacturing costs; the electric automobile uses only 60 to 75% of the energy of its gasoline counterpart even when using electricity generated from fossil fuels.

A further difficulty with the electric car is connected with the problem of re-fuelling. There is a large infrastructure in place to distribute gasoline and refuel cars quickly and cheaply. Such an infrastructure does not exist for battery recharging or replacement. There will undoubtedly be great consumer resistance to owning a vehicle that must stand idle at home for long periods while it is recharged at a relatively slow rate. We are used to completely refuelling our cars in about 5 minutes, not 8 hours. In addition to this it is not clear that the electric utility infrastructure is presently able to support a major transfer of transportation energy to electricity. British Columbia Hydro in Canada has investigated the problem [6] and concluded that electric-car owners will require an additional electric service at least equal to that of an electric clothes dryer and that for most homes this could add an additional $35,000 to the cost of the car. Furthermore no city has the electrical distribution system that could handle a major transfer of transportation energy to electricity nor has any utility in North America the excess generating power. Charging overnight is also not the answer as utilities rely on their "base-load" generation to carry the night demands. The base load is supplied by those systems which are difficult to shut down and start up like hydro and nuclear. The timescale required to increase the base load is of the order of decades and the investment is very high. Massively increased electric transport in the short term, at least in North America, would require the "peak load" generation to be used at night. This type of generation is expensive relative to the base load and often polluting as well.

---

[5] In 1996 General Motors introduced the EV–1 electric automobile based on its earlier sports model, the "Impact". It has since been withdrawn.
[6] R. Williamson, *Electric Cars Fuel Utility Headache.* Toronto Globe and Mail, Aug. 20, (1996).

Large-scale electric transport can not be introduced on a short timescale; a fact that is not often understood by its more vocal advocates.

17.4 CAPACITY OF STORAGE CELLS

The physics and chemistry of batteries and storage cells was discussed in Section 8.7. Clearly there is a vast difference between a useful battery to power a flashlight and one required to power an automobile. To be useful in transportation a storage cell must contain a large quantity of energy and be able to deliver it at a variable rate which at times can be rather large. The longer a cell can supply current the better and this is determined by the amount of materials used as well as the type. The objective is to store as much energy in the cell as possible.

Peukert's Law.

It is an experimental fact that the rate at which current is drawn from a storage cell affects the amount of energy available; the higher the rate the less energy. How the battery capacity C (Amp · hr), the rated discharge time R (hr), the actual discharge time T (hr) and the actual current drawn I (Amp) are related is given by a relation known as Peukert's Law. The performance of storage cells is usually quoted as a specified capacity in ampere ·hours (amp ·hr) at a rated current for a rated time. For example a common rating for a car battery might be 100 amp ·hr at a rated current of 5 amp for a rated time of 20 hr. Such a battery then will deliver 5 amp for (100 Amp · hr) /(5 Amp) = 20 hr.

For how long will this battery deliver 10 amperes? The simple answer would be 100/10 = 10 hr, exactly half the time. That, however, is not correct. Peukert's law says these quantities are related by:

$$T = \left(\frac{C}{I}\right)^n \frac{1}{R^{n-1}} \qquad \textbf{[17-9]}$$

The dimensionless constant n is known as *Peukert's constant* and is usually between 1.1 and 1.5. This law assures that the time, during which currents larger than the rated current can be drawn from a battery, will always be less than that calculated by a simple ratio of the currents.

EXAMPLE 17-2.

An automobile battery is rated at 100 amp ·hr for 20 hr at a rated current of 5 amp. For how long can this battery be expected to deliver 10 amp? Peukert's constant for this battery is 1.3.

SOLUTION

In this problem C = 100 amp ·hr, R = 20 hr and I = 10 amp.

$T = (100/10)^{1.3}(1/20^{0.3}) = 19.95 \times 0.407 = 8.1$ hr

---

Figure 17-5 shows the discharge behaviour of the battery of Example 17-2 for a range of discharge currents and values of the Peukert constant.

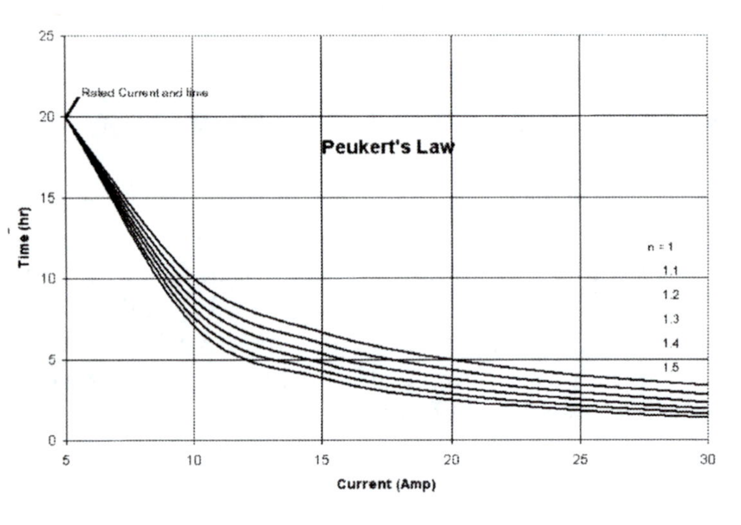

Figure 17–4 Peukert's Law Applied to the Battery of Example 17–2

The Ragone Plot

Storage cells contain a lot of metal and can be quite heavy, as anyone who has lifted a car battery knows. Lighter batteries are very desirable but difficult to achieve; satellites don't use lead-acid cells, for example, because they simply are too heavy. A more meaningful characteristic than energy capacity is *specific energy capacity*, that is, the energy stored per kilogram of mass, usually measured in W·h/kg.

There is another consideration, and that is the rate of delivery of the stored energy. In many applications, an electric-car for example, a cell that stores large amounts of energy in a small mass would be useless if that energy can only be taken out very slowly. In other words, we also want a large *specific power,* i.e., power per unit mass, usually expressed as W/kg.

How are these two related? Here is a case where the internal resistance is very important (see Sec. 7.7). If the energy is extracted very rapidly, i.e., the current I is large, then we will sacrifice much of the stored energy as heat I^2r in the internal resistance This is in addition to the unavailability of some of the battery capacity as the discharge rate is increased according to Peukert's Law. If we draw a small current we will not waste so much energy but we will not be able to develop as much power. In general the characteristics of a cell are pictured in Fig. 17-5 which is known as a Ragone [7]plot. It shows that the useful energy is large at small power and vice versa.

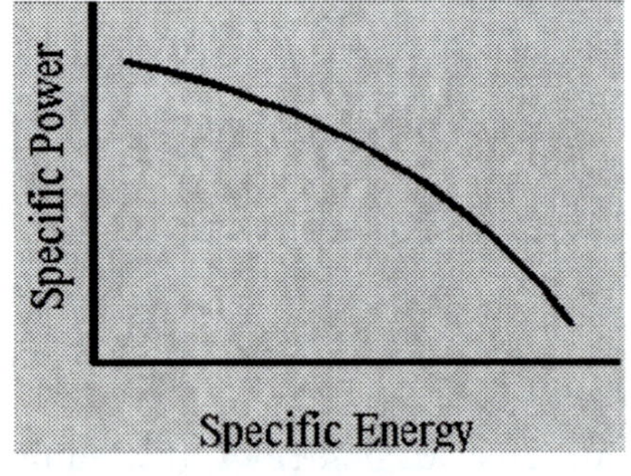

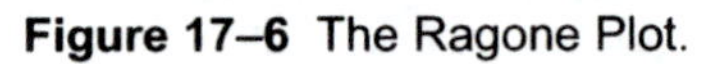
Figure 17–6 The Ragone Plot.

In an automobile we require large bursts of power for acceleration and large specific energy if we wish to make long trips on one charge. The Ragone plot shows that, these two requirements are incompatible, a fundamental fact that has plagued the development of the electric-car and limited it, in the past, to slow moving vehicles like milk-vans and indoor tractors.

[7] Pronounced "Ra–go–nee".

There is a lot of active research on storage cells [8] but progress is slow. Figure 17-7 is a Ragone plot describing various types of storage cells. A study of the plot will show how storage cells fall far short of the internal combustion engine in performance. Consider as an example a 1000 kg electric-car with a range of 80 km on a charge, in which the batteries account for 30-35% of the vehicle weight. The required specific energy is about 40 W·h/kg and the mean specific power is about 16 W/kg, with peaks to 80 W/kg. The Ragone plot shows that lead-acid (Pb/PbO_2) is unable to satisfy the peak power requirement, although it is suitable for steady driving. The Zn, Ni-Metal hydride and Li-based batteries qualify but they are all, at present, in the development stage and require improvements as regards safety, cost and reliability.

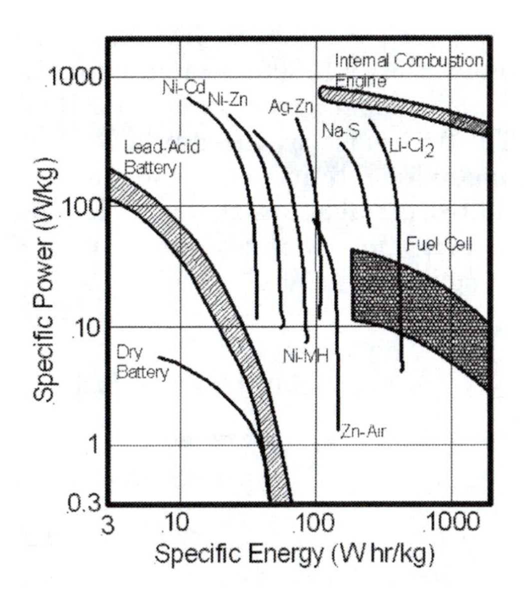

Figure 17–7 Ragone Plot for Storage Cells Compared to the Internal Combustion Engine and Fuel Cells.

17.5 THE HYBRID VEHICLE

One of the interesting possibilities that has come to pass is the *hybrid vehicle* which has a conventional engine an electric power source as well as an electric drive for cruising when high power is not needed. The idea, in fact, is not new: a hybrid vehicle was produced and sold by Ferdinand Porsche in 1898.

In general hybrids can be of two types:

- A series hybrid has an engine which drives a generator which charges batteries. The batteries drive electric motors as in a standard electric-car.
- A parallel hybrid has two drive trains in parallel; An electric power source and electric motor, and a standard engine. Each is connected to the drive shaft by a clutch and they may be run independently or in parallel.

Such a vehicle requires far less weight in the electric system than a pure electric vehicle since the electric drive is only used for low-power cruising; acceleration and emergency power is provided by the boost of the standard engine. From the Ragone plot (Fig. 17-7) it is seen that batteries now become a possibility (although not Pb/PbO_2). The battery being chosen by some manufacturers is the Ni/Metal hydride.
Further, a vehicle using AC motors to drive the wheels in electric operation can switch the motors to a generator mode while braking (so-called "dynamic braking") and convert the KE of the car to electrical energy to recharge the batteries. The specific power and energy requirements of the batteries are now much less severe. Modern solid-state electronics make this type of "inversion" from the battery DC to AC feasible at high efficiency. Automobiles of this type with fuel economies of between 3 and 4 L/100 km

[8] S.R. Ovshinsky, M.A. Fetcenko and J. Ross. *A Nickel Metal Hydride Battery for Electric Vehicles.* Science, Vol. 260, (1993) pp 176–181.

have been available in Japan for a few years and became available in the USA in small quantities (< 10000) beginning in the year 2000. Their price is about 20% greater than a comparable internal combustion driven vehicle.
The first mass-produced vehicle of this type is the Toyota Prius first introduced in Japan in 1997 and worldwide in 2001. The economy of the vehicle is somewhat controversial but by the European method of calculation the Prius gets 4.3L/100 km or 66 mpg (Imp). Some of the characteristics of the power plant of the Prius 4-door sedan are given in Table 17-1.

Table 17-1 Characteristics of a Hybrid Automobile Power-plant.

Body style		4 Door Sedan
Year of First sales		2000
Battery	Modules	38
	Cells per module	6
	Total cells	228
	Volts per cell	1.2
	Total volts (nominal)	273.6
	Capacity amp hours	6.5
	Capacity Watt hours	1778.4
	Weight kg	50
Gasoline Engine	Power kW	52
	Max rpm	4500
Electric Motor	Power kW	33
Combined	Power kW	73

17.6 THE FUEL CELL VEHICLE

A more attractive option for an electric or a hybrid vehicle may be to use a fuel cell for the electric drive as discussed in Chapter 7. With the proper chemical regenerator to extract hydrogen from the fuel, the fuel-cell and the standard engine can use the same fuel, e.g., gasoline or natural gas. The Ragone plot (Fig. 17-7) shows the region for full-power anticipated in the near future for an automotive fuel cell. A large development effort is going into this option as the development of practical storage batteries for a purely electric car has proven almost intractable. Even so, the power units are presently still rather large and they will probably find their first commercial use in buses. The cost for a fuel cell drive system is about $2000 per kW.[9] A leader in this technology is Ballard Power Systems of Canada. They have a number of demonstration projects underway with buses in Australia, Europe and California. Although reporting trouble-free performance there has not yet been a commitment from any authority to adopt them on a scale that would support commercial production.

[9] *Energy Technology Perspectives 2006*, OECD, International Energy Agency, 2006, p 145.

17.7 CONSERVATION IN TRANSPORTATION

> "That the automobile has reached the limit of its development is suggested by the fact that during the past year no improvements of a radical nature have been introduced."
>
> — *Scientific American*, 1909

Perhaps there were no significant changes in automobiles in 1909, but in the past few decades there have been a great many improvements in the energy-efficiency of automobiles. In Canada [10], the number of automobiles increased from 8.9 million in the year 1975 to 13.2 million in 1990, but during that period the total consumption of gasoline decreased from 3.45×10^7 m^3 to 3.39×10^7 m^3. In 1975, the average car in Ontario required 18.6 L of gasoline to travel 100 km in combined city and highway driving; by 1985, the comparable figure was 9.3 L. Since that time the figure has increased by about 5% as the sales of light trucks and SUVs have increased.[11]

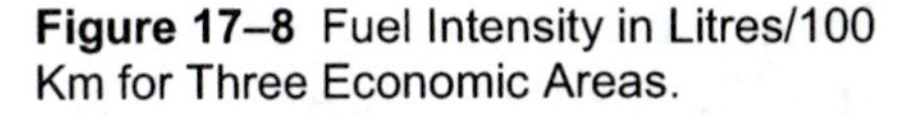

Figure 17–8 Fuel Intensity in Litres/100 Km for Three Economic Areas.

It is not necessary for automobiles to consume that much gasoline. It is the North American preference for large and heavy vehicles that keeps the average consumption high. Figure 17-8 [12] shows the average "fuel intensity" in Litres/100 km for the automobile fleet in three different economic areas: Canada and USA, U.K. and Japan from 1985 to 2002. The curve for the U.K. is typical of the fleet in Europe and that for Japan is typical of the rest of the developed nations where fuel is much more expensive. The increasing price of petroleum is having a modest effect on increasing the efficiency in Canada and the USA but not enough to enjoy the economies of the rest of the world.

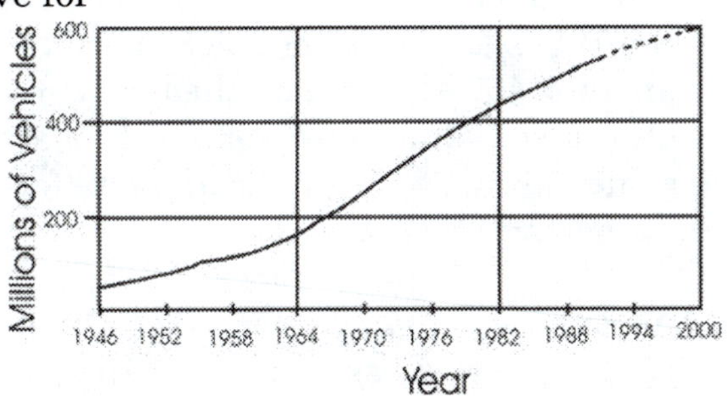

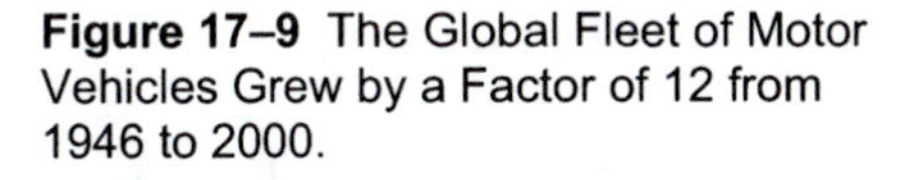

Figure 17–9 The Global Fleet of Motor Vehicles Grew by a Factor of 12 from 1946 to 2000.

Road vehicles account for about half of the world's oil consumption each year, and the number of registered vehicles is steadily growing [13] (Fig. 17-9). Many people in industrialized countries have become heavily dependent on the use of automobiles, and the layout of cities is often arranged for the convenience of automobile-drivers. In the USA in 2001, 92% of all households owned one or more cars, and in 1983, 73% of journeys to work in Canada were made by car[14]. Many people travel a

[10] *Toronto Globe and Mail*, January 25, 1992, p. B1–B4.

[11] A.H. Rosenfeld, T.M. Kaarsberg, J. Romm. *Technologies to Reduce Carbon Dioxide Emissions in the Next Decade.* Physics Today, Nov. (2000), p 37

[12] Adapted from *Energy Technology Perspectives 2006*, OECD, International Energy Agency, 1960, p 261.

[13] D. Bleviss and P. Walzer, Scientific American **263**, (1990) No. 3 (Sept.), p. 109.

[14] *The State of Canada's Environment*, Government of Canada, Ottawa, (1991), pp 13–21.

great distance to work; in Canada, about 13% of commuters [15] had a one-way commuting distance greater than 25 km in 2001.

In the USA in 1988, about 27% of the energy consumed was devoted to transportation[16]. A breakdown of the transportation energy [17] used in 2003 in the US is provided in Figure 17-10. Notice that automobiles consumed over 60% of the energy in the transportation sector, and cars and trucks together accounted for 85% of the consumption. Aircraft took up most of the remainder.

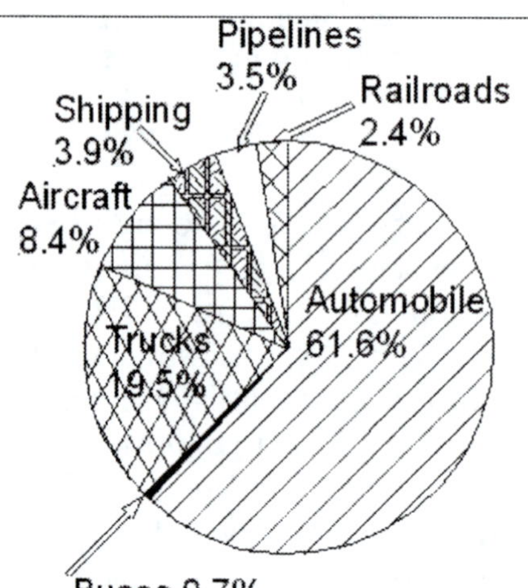

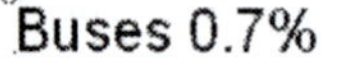

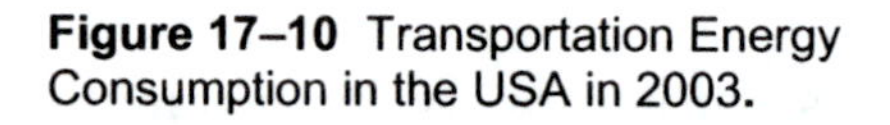

Figure 17–10 Transportation Energy Consumption in the USA in 2003.

Increasing Fuel Efficiency

As a result of the energy crisis after 1973 and a growing awareness of the pollution problems resulting from the use of motor vehicles, the US and California Governments passed laws requiring a gradual increase in fuel efficiency of automobiles and light trucks. Much of the resulting improvement has been due to a reduction in weight. A typical mid-size car in the early 1970s had a large, iron engine, rear-wheel drive that required a heavy drive shaft and differential gearbox, and a metal chassis. A comparable car today has a smaller, aluminium engine, front-wheel drive, and no chassis. Many of the parts that used to be metallic — such as the dashboard and bumpers — are now made of plastic, thus reducing the weight. A typical car 20 years ago had a mass of about 1800 kg; in the 1980s, the mass had decreased to about 1400 kg, and in 1991 it was anticipated that another 400 kg could be shed [18] by replacing steel and cast-iron with high-strength low-alloy steel, aluminum, composite plastic, and magnesium. Although a few models have achieved this[19] the public appetite for larger and heavier cars has kept most small cars in the 1500 kg region.[20]

Automotive engines now tend to be smaller and more fuel-efficient than their gas-guzzling predecessors in the 1970s, and on average, more North Americans are driving smaller cars. (However, in comparison with typical car sizes in Europe, North American cars are still larger.)

Other changes that have led to a decrease in fuel consumption include:

- body shapes that are more aerodynamic, thus reducing air friction;
- the use of radial tires, which can have over 30% less rolling resistance than bias-ply tires[21], producing fuel savings of 15% on the highway, and 10% overall;

---

[15]Canada Census Data 2001
[16]A. Chachich, "Energy and Transportation," in *The Energy Sourcebook*, American Institute of Physics, New York, (1991), p. 358.
[17]Stacey C. Davis, Susan W. Diegel; *Transportation Energy Data Book, Ed 25;* U.S. Dept. of Energy (2006)
[18]Ref. 16, p. 370.
[19] The 2006 Toyota Corolla has a curb weight of about 1100 kg.
[20] The Volkswagen Jetta and the Nissan Altima are about 1500 kg.
[21]ibid, p. 386.

- electronic engine control, which regulates air intake, exhaust recirculation, fuel injection, spark timing, and idle speed; fuel savings of up to 10% can result;
- computerized traffic control in cities, thus reducing time spent waiting at lights and accelerating and decelerating.

Future changes in automobiles will probably include further reduction in weight, even better aerodynamics, improvements in lubricants, and use of a continuously variable transmission that permits the engine to work at its most efficient speed almost all the time.

Table 17-2 Energy Consumption in Urban Transportation

Transportation Mode	Typical Energy Used ($MJ{\cdot}person^{-1}{\cdot}km^{-1}$)
Automobile (1 person)	8
Automobile (5 persons)	2
Commuter Rail	0.8
Bus	0.6
Walking	0.2
Bicycle	0.1

Thus far in this section we have been discussing automobiles (and light trucks), which consume a large fraction of the energy devoted to transportation. However, fuel consumed by transportation could be reduced significantly if people would use mass transit instead of automobiles. Table 17-2 shows the energy used per person·km for various forms of urban transportation[22]. Many urban travellers drive alone in their cars, but this is the most energy-inefficient way to travel. Unfortunately, car-drivers are unlikely to use more efficient types of transit unless there is considerable intervention by governments through subsidies, introduction of special traffic lanes for buses, or development of cities in which access to the downtown areas by private automobiles is limited.

Notice in Table 17-2 that the most efficient mode of transportation, even better than walking, is the bicycle. In the Netherlands, a large fraction of commuters use bicycles on an extensive network of bicycle paths that even include separate bicycle traffic lights.

For intercity transportation, the energy used per person·km decreases. Buses offer the most energy-efficient transportation, followed by railroads, automobiles, and aeroplanes. A typical aeroplane consumes about six times as much energy per person·km as does a bus.

Long-Haul Trucks

No fuel-efficiency standards have been mandated for large trucks, and hence progress in this area has been driven only by cost savings to operators. Various energy-saving (and hence cost-saving) measures undertaken so far include installation of air deflectors above cabs to reduce aerodynamic drag (by 15-25%), design of new cabs with rounded

[22]Ref. 15, pp. 13–20.

corners, angled windshields, etc., and use of radial tires and diesel engines (which are more energy-efficient than their gasoline counterparts). As time passes, we can expect to see engine improvements such as electronic engine control, and further aerodynamic changes such as skirts around tractor-trailers.

Although trucks carry most of the freight in North America, ships, pipelines (for oil and gas), and railroads offer more energy-efficient ways of moving goods [23] (Ch. 5, Table 5-1). Only aeroplanes have a greater energy cost for freight transport than do trucks.

Aeroplanes

Aeroplanes have also undergone improvements in design to decrease their fuel consumption. Engines have become more efficient with the use of lighter-weight materials such as graphite fibres in epoxy resin which are being utilized in new aircraft in place of aluminium. Lighter carpets and seats are being installed, and less paint is being applied (thus reducing weight). Operational factors are also important: changes in airport design, airspace routing, and flight scheduling can all make improvements in fuel efficiency. In the future, it is expected that new wing designs that create less turbulence and hence save fuel will come into common use, and that there may be a return to propeller-driven aircraft (having better fuel efficiency) for short-haul routes.

EXERCISE

17-1. In Table 17-2, the energy used in urban transportation is given in units of $MJ{\cdot}person^{-1}{\cdot}km^{-1}$. In the USA, another common unit for this quantity would be $Btu{\cdot}person^{-1}{\cdot}mi^{-1}$. Convert the energy consumption given in Table 17-2 for an automobile containing 5 people to $Btu{\cdot}person^{-1}{\cdot}mi^{-1}$.
(1 mi = 1.609 km)

PROBLEMS

17-2. The heating value of gasoline is $3.2{\times}10^7$ J/L. Show that a mid-sized automobile, traveling at 90 km/hr and with a fuel input power of 70 kW, consumes fuel at a rate of 8.8 L/100 km or 32 mi/Imp gal or 27 mi/US gal. (1 L = 0.2200 Imp gal = 0.2642 US gal).

17-3. Consider a 4-stroke engine turning at 2500 rpm. If the intake valve is open for one full stroke (it is actually open for far less), how much time is there for the air-fuel mixture to enter the cylinder?

17-4. As a model of the intake manifold of an engine consider it to be a long straight cylinder 25 cm long with one end closed by the intake valve. It is filled with air-fuel mixture at one half atmosphere pressure ($0.5{\times}10^5$ Pa) and of density 0.8 kg/m^3. When the valve opens it exposes one end of the column of gas to zero pressure and the pressure difference accelerates the gas. If the intake valve is open for 1 ms, how far can the column move before the valve closes? (Assume the gas experiences constant acceleration.)

[23] R. Hemphill, *Energy Conservation in the Transportation Sector*, in "Energy Conservation and Public Policy", J. Sawhill, ed., Prentice–Hall, Englewood Cliffs, N.J., (1979), p. 92.

17-5. The 'Prony brake' is a simple device used to measure the output power (or brake power) of an engine. It consists of a flat pulley of radius r attached to the output shaft of the engine with a half turn of a flat strap wrapped around the pulley as shown in the Fig. 17-11. The tension on both sides of the strap is measured, for example, with two spring balances fixed at their upper ends. These tensions are a result of friction between the strap and the pulley and are T_1 and T_2 ($T_1 > T_2$). The pulley turns with a constant angular velocity ω driven by a constant torque τ supplied by the engine. Clearly, since ω is a constant, the torque τ is equal to the friction torque τ_f. Show that the power output of the engine is given by

$$P_b = (T_1 - T_2)r\omega$$

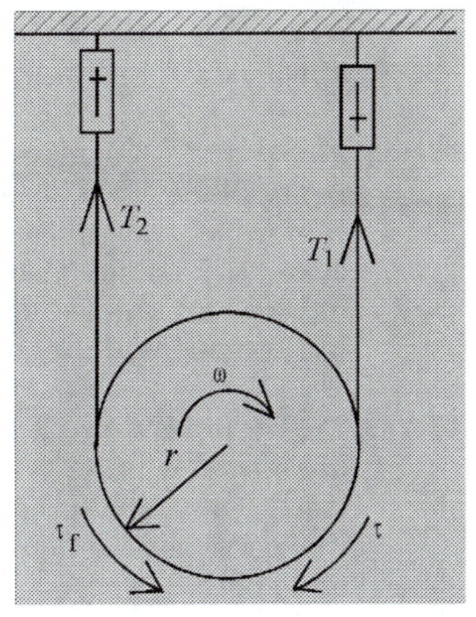

Figure 17–11 Problem 17–5

17-6. The pulley of a Prony brake as described in Problem 17-5 is 10 cm in diameter and attached to the shaft of an electric motor. The spring balances record tensions of 500.0 and 490.1 N. A strobe light shows that the motor is turning at 3600 rpm. What is the brake power of the motor in watts and in horsepower?

17-7. When a 1500 kg automobile is towed slowly on a level smooth road it requires a force of 220 N to move it at a constant speed. Wind tunnel tests show that the aerodynamic drag is equal to the rolling friction at 65 km/hr and varies as the square of the speed. What power is delivered to the wheels (road power) by the engine at 65 and 120 km/hr?

17-8. Problem 17-7 assumes that the rolling friction is constant at all speeds. This is not exactly true as there is some aerodynamic lift that reduces the apparent weight of the automobile as the speed increases. At 65 km/hr the lift is 300 N and at 120 km/hr it is 900 N. In the light of this further information re-solve Problem 17-7. Assume that rolling friction is proportional to the normal force.

17-9. When the windows of an automobile are opened, the road power increases because of the increased air turbulence which increases with speed. By how much must the road power be increased before it is equally economical to operate a 3 kW air conditioner? Air conditioners derive their power at the crankshaft output of the engine.

17-10. An automobile as described in Problem 17-2 is equipped with an air conditioner of 3 kW capacity. While driving at 90 km/hr the air conditioner is switched on. What is the new fuel consumption rate if the speed is maintained?

17-11. An automobile hobbyist fits an air conditioner to his automobile by driving it from a home-made coupling to the driven axle of the automobile rather than the engine crankshaft. Ignoring the fact that the unit would run at variable speed, would it be as efficient as the case presented in Problem 17-10? What would be the fuel consumption at 90 km/hr?

17-12. a) When you refuel your auto you can transfer about 80 L of gasoline from the pump into the gas tank in about 2 minutes. What power does this represent?
b) If you owned an electric car with the same range as the automobile in a) it would need periodic refueling (i.e., battery recharging) from an electricity source. A heavy-duty household circuit such as the type that runs an electric stove can supply 30 A at a voltage of 220 V. How long would it take to refuel your electric car, remembering that a gasoline

automobile operates at an efficiency of 20% whereas the electric car is nearly 100% efficient? Assume that the charging is done at an efficiency of 75%.

17-13 The Duracell Co. gives the technical specifications for all their batteries. For the MX1500 AA Alkaline Manganese Dioxide battery the performance is given as a group of graphs from which one can infer that the rating of the cell is a capacity of 3.2 Amp·hr at a rated current of 5.6 mA. Cells of this type have a Peukert constant of 1.1.
a) How long will the cell deliver this current?
b) How long will the cell deliver a current of 11.2 mA? 56 mA?

17-14. Figure 17-12 shows a semilog plot of the number of registered motor vehicles in the world vs. time. The graph has essentially two straight-line portions, one from the year 1946 to 1976, and another (with a smaller slope) from 1976 to 1988. (Use a straight-edge to confirm this.)
(a) Determine the average annual percentage growth rate between 1946 and 1976, and between 1976 and 1988.
(b) If the growth rate remains constant at the 1976-88 value, how many motor vehicles will there be in the year 2030?
(c) Assume that the average mass of a motor vehicle is 1400 kg. If the growth rate were to remain constant at the 1976-88 value, when would the total mass of motor vehicles equal the mass of the earth (5.98×10^{24} kg)?

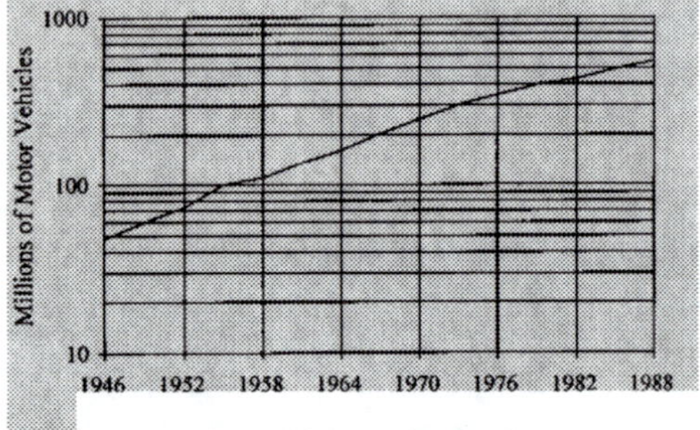

Figure 17–12 Problem 17–14

17-15. Let us examine some of the data published for the Indian-built Reva electric car.
a) Use Fig. 17-3 to determine how much power is required if a 70 kW engine is removed and replaced with an electric motor and batteries of the same weight.
b) Assume that the electric motor and its controller operate with 95% efficiency and 20% of the kinetic energy of the car is recovered by "dynamic braking". Further assume that these figures are for the car operating at a speed of 65 km/h. If the range of the car is to be 65 km between rechargings, what capacity in A·h (ampere-hours) must the battery have? The Reva's battery is 48 V and it is claimed that it has a capacity of 200 A·h.

ANSWERS

17-1 610 BTU·person^{-1}·mi^{-1}
17-3. 6 ms
17-4. 13 cm
17-6. 186 W, 0.25 hp
17-7. 7.9 kW, 32.3 kW
17-8. 7.8 kW, 32.2 kW
17-9. 2.4 kW
17-10. 10.6 L/100 km
17-11. No, 11.3 L/100 km
17-12. **a.** 23 MW **b.** 31 hr!
17-13. **a.** 570 hr *b*. 270 hr, 45 hr
17-14, **a.** 6.6% and 3.6%, **b.** 2.5 billion, **c.** in the year 2810 (approx.)
17-15. 11.8 kW, 195 A h

CHAPTER 18

NEW ENERGY TECHNOLOGIES

In an effort to meet the world's ever-increasing demand for energy at reasonable cost, it will be necessary to modify and improve the technologies associated with energy sources and their uses. Here we will examine some of the steps that are being taken to upgrade present sources and technologies in cleanliness, efficiency or both.

18.1 ENERGY CURRENCY

Many of the sources of energy are available to us at times and in places that are inconvenient or impractical. For example:

- Water flows over Niagara Falls continuously, but during the night-time hours there is a reduced demand for electricity; the water however continues to flow.
- The Sun shines for a longer time during summer days, but more energy is needed in temperate climates in the winter when the days are short.
- There are more sunny hours over deserts — where few people live — than over more populated agricultural areas.
- Populous areas have developed for historical and commercial reasons with little regard to the proximity of energy sources. Energy, or its raw materials, must usually be brought to people, rather than the reverse.

As a general rule it is more economical and ecologically sound to deliver energy than to deliver the raw material from which it is extracted. For example, the transportation of coal by rail is a greater strain on the economy and the environment than the delivery of electricity by high-voltage transmission lines. In the past, at a time when energy was cheap and apparently plentiful, the method chosen for the energy distribution infrastructure was not the best for later times. For example, most coal is delivered long distances (even inter-continentally) and little electricity is generated on-site.

One of the problems is that there is not enough variety in *energy "currencies"* [1] compared to energy sources. In fact, there is only one currency in widespread use and that is electricity. As the world's supply of petroleum decreases, there could develop a crisis in transportation because of lack of a convenient currency. The use of electricity in transportation as we now practise it is difficult. Of course we could all abandon our cars and trucks and revert to electric rail transport, but if there were another currency that could replace gasoline, the present transportation infrastructure could continue. One currency that probably will be developed is *hydrogen*.

---

[1] Energy currencies are discussed in Section 1.1.

18.2 THE HYDROGEN "CURRENCY"

Hydrogen is *not* a source of energy; there are no hydrogen mines or hydrogen wells. Hydrogen must be manufactured with the expenditure of energy, but by manufacturing hydrogen, a fuel is created that has almost no further environmental impact. This is because the energy is recovered from the hydrogen via the chemical reaction

$$H_2 + O \rightarrow H_2O \qquad [18\text{-}1]$$

The waste product is simply water, and in some cases even oxygen is not used up since it can be created in the manufacture of hydrogen in the first place. As we saw in Section 7.7, hydrogen is an ideal fuel for fuel-cells which can produce electricity directly via chemical reactions and are thus free from thermodynamic efficiency limitations. As a result, much effort has gone into research on methods of producing hydrogen from fossil fuels. There are several methods under investigation, such as steam reformation, thermal cracking, and partial oxidation.

Steam Reformation

Steam and fuel gas (e.g., methane) are combined in a reactor at temperatures between 750 and 900°C where the reaction

$$CH_4 + H_2O \rightarrow CO + 3H_2 \qquad [18\text{-}2]$$

takes place. Reaction of the product gases with water converts the CO to CO_2 and the production of more H_2. Removal of the CO_2 results in hydrogen gas of 95-99% purity.

Thermal Cracking

In the presence of an appropriate catalyst, and at high temperature, it is possible to separate the carbon and hydrogen in methane according to the equation

$$CH_4 \rightarrow C + 2H_2 \qquad [18\text{-}3]$$

At a temperature of 1200°C, methane from natural gas is blown through a bed (fluidized bed, Sec. 18.5) of catalyst pellets (7% Ni on Alumina). The carbon is deposited as coke on the pellets, which must be continuously removed and regenerated in another fluidized bed with high-temperature air. The energy for the process is supplied by the burning of the coke. The process can produce 95% pure hydrogen.

Partial Oxidation

In the presence of steam and oxygen and with an appropriate catalyst, virtually any hydrocarbon (not only natural gas) can be converted into hydrogen and carbon monoxide. (The CO is also converted with water to CO_2 and H_2.) The ability of this process to convert coal and heavy oils to the cleaner fuel has made it an attractive choice for development.

Of course, none of this does anything to reduce CO_2 in the atmosphere; so long as energy is extracted from hydrocarbons — by whatever method — CO_2 will be released. However, in principle, hydrogen can be manufactured from any energy source if the temperature is high enough. For some time there was enthusiasm for the production of hydrogen using the heat of nuclear reactors. Even if any future reactors are built, that possibility seems less likely as the trend is away from very high-temperature reactors (such as breeders). A further possibility is the production of hydrogen from the very high temperature of the receivers in large solar power plants.

It is possible to produce hydrogen by the electrolysis of water according to the equation

$$2H_2O + \text{energy} \rightarrow 2H_2 + O_2 \qquad \textbf{[18-4]}$$

The energy is provided by passing an electric current through the water, which has been made electrically conducting by the addition of a small amount of acid or a salt. Each electrolysis cell requires a DC voltage of about 2 V, so a battery of 100 cells in series can be operated from a 200-V DC source. The highest purity hydrogen is made in this way.

This process does not constitute a source of energy; it is simply an exchange of one energy currency (electricity) for another. This trade has a considerable cost since electrolysis cells operate at an efficiency of 50-60%; the energy is lost as resistive heating of the water. It could be of importance in the future as a way of utilizing excess electrical energy production to make a fuel that is portable and suitable for transportation.

Storage of Hydrogen

One obvious way to store hydrogen is as a gas at high pressure using the well established technologies that exist for handling gases at pressures up to 150 times atmospheric pressure. This technique, however, makes it a poor source for transportation fuel. The energy density of gasoline is 35×10^6 J/L, whereas the 6600 L of hydrogen gas (pressurized to 44 L at 150 atm.) in a standard industrial gas cylinder has only 69×10^6 J of energy, the equivalent of only 2L of gasoline! Clearly it is impractical to operate an automobile using hydrogen in cylinders. (Even this comparison neglects the energy cost of carrying around the cylinder which has a mass of about 40 kg and the energy required to compress the hydrogen into the cylinder in the first place.)

Hydrogen as a portable fuel will probably not be practical unless it is carried as a low-temperature liquid in a light, well-insulated tank. However, here there is a large energy price to pay as it takes about one third of the energy recoverable from the hydrogen when it is burned to liquefy it in the first place. In addition, liquid hydrogen has a very low latent heat of vaporization, so the containers must be very well insulated indeed or the boil-off rate becomes very large. Liquid hydrogen can be used as a fuel in the space-program rockets because the tanks are topped up until just before launch and it is all burned up in the first five minutes; even the storage tanks don't have to be very well insulated under those operating circumstances. A similar situation could exist in the context of take-off of large aircraft. A scenario can be envisaged where liquid hydrogen powers the take-off, following which the aircraft reverts to a different fuel for cruising. Liquid hydrogen, while dangerous, is not as dangerous as gasoline and even has some safety advantages. In the case of a leak or spill it rises quickly and disperses unlike gasoline and propane which, being heavy, do not disperse well.

Hydrogen can also be stored within certain metals where it forms a hydride compound that, in some cases, can be made to release its hydrogen by the application of a modest amount of heat. Aluminum and titanium hydrides are examples that have been investigated extensively. One of the virtues of metal hydride storage is that the metal packs the hydrogen in an even smaller volume than for liquid hydrogen. This is because the liquid is a molecular liquid and the molecules of H_2 occupy a large volume. As a hydride the hydrogen is stored as atoms which have a much smaller volume. The downside of metal hydride storage is that the hydrides are heavy. Table 18-1 summarizes some of these properties for the fuel equivalent of 80 L of gasoline, a typical automobile load. The energy content of 80 L or 64 kg of gasoline is 2.8×10^9 J. An examination of the table shows why gasoline is such an ideal portable fuel; it has remarkable properties of concentrated energy in a reasonable weight and volume along with relatively user-friendly handling properties.

Table 18-1: Properties of Stored Hydrogen Equivalent to 80 L of Gasoline

Name	%H	Mass (kg)	Volume (L)
Gasoline	(N/A)	64	80
Liquid H_2	100	21	300
Lithium Hydride	13	166	200
Calcium Hydride	5	440	230
Titanium Hydride	4	520	133
Iron Titanium Hydride	2	1123	205

The Hydrogen-Oxygen Fuel Cell

The operation of the H_2-O_2 fuel cell was described in Section 7.7. The perfection of such a fuel cell with its inherent high efficiency raises the possibility of a new kind of energy distribution infrastructure. The concept that has attracted most attention provides a method of storing the energy generated in off-peak times and using it efficiently later. Off-peak electrical supply may be used in several ways, as for instance to electrolyze water, or to produce hydrogen from any hydrocarbon, in particular coal, as described above. The hydrogen can be easily stored and delivered by pipeline to localized fuel cells where it is used to generate electricity. From a transmission point of view this is very attractive; an underground gas pipeline leaves the surface land free of unsightly transmission lines and is cheaper in capital cost. The land area required is 1% of that required for conventional hydro-storage systems (see Section 18.4).

Since the 1970s, when there was intense interest in the "hydrogen economy," there has been a waning of interest with the precipitous decline in the world price of oil. The price-rise of the late 1990s has seen some new interest in the development of hydrogen.

18.3 SYNTHETIC FUELS

Coal can be used as a feedstock to produce methane from which, in principle, almost any organic fuel can be made. The process starts with the gasification reactions described in the previous section which produce carbon monoxide and hydrogen. In the presence of a suitable catalyst the following reaction can be made to occur: [2]

$$CO + 3H_2 \rightarrow CH_4 + H_2 \qquad \textbf{[18-5]}$$

The methane can be used directly as a fuel or as the feedstock in further catalyzed reactions even to the manufacture of gasoline.

Synthetic gasoline was made in this way in Germany during the Second World War and is currently made in South Africa where it supplies 30-40% of the national liquid fuel requirement. It has never been able to compete, however, with petroleum-derived gasoline and indeed no synthetic fuel yet competes economically with natural fuels. Estimates of the delivered price of synthetic fuels are always more than that of normal gasoline (at least by North American standards), sometimes by a factor of two. In addition the conversion process is inefficient requiring much heat that results in 40% more CO_2 being released to the atmosphere than for direct use of the original fuel. The various chemical processes use large amounts of water which implies a further environmental cost and energy cost to clean up.

As a result of all these factors, research and development of synthetic fuels has not been attractive. Companies which embarked enthusiastically on these programs after the oil crises of the 1970s have since retrenched considerably or withdrawn completely from activity.

18.4 ENERGY STORAGE

Energy needs to be stored for a great variety of reasons, some of which have already been discussed. The most obvious one is to make it portable; this is usually the reason for the existence of batteries of all types. (The readily available stored energy in a lead-acid battery makes it easy to start an automobile and begin to utilize the energy stored in the gasoline.) We must always realize, though, that the battery is not an energy source; it is simply a temporary energy repository. There is, however, another major reason for storing energy: the energy source might be available only in a continuous supply while our energy needs vary with the time-of-day or the season. Rather than have the energy wasted we attempt to find ways to store it for recovery when the demand rises.

[2] This the "Fischer-Tropsch" process discovered by coal researchers F. Fischer and H. Tropsch in 1923.

Portability

Many methods have been devised to store energy and make it portable; indeed most of the transportation technology, with the exception of electric trains, relies on this technology in some form or other. So far, transportation has relied on rendering portable the source fuels such as coal and oil, perhaps after some industrial processing to forms like gasoline or jet fuel. Little progress has been made in the widespread use of stored energy currencies such as hydrogen, electricity, or fuel-cell fuels. As we have seen in previous sections, these technologies may be on the verge of development in the first decade of the 21st century; this is almost certainly the case for fuel cells.

Large-scale Storage

The water flowing over Niagara Falls does so continuously and most of it is used to generate electricity. The demand for electricity in the Ontario-New York area is variable, as has been discussed in Chapter 9. If the demand falls below that which can be supplied by the river flow, the energy will be wasted; this is particularly true at night. If large holding ponds are constructed at the top of the fall then, during periods of low demand the water can be accumulated in the ponds. This water can be called upon during peak-load periods to augment the normal flow. In this way the energy can be stored very efficiently. At the large tidal power station at La Rance in France described in Section 14.9, where the tidal flow must be utilized, or wasted, the off-peak generation is used to run electric pumps which pump the water back to the head side of the dam to be used when needed. In effect the surplus electric energy is stored as gravitational potential energy.

For this purpose it would not be a good idea to store the energy as some form of fuel, e.g., hydrogen. When this fuel is burned it will then be subject to the usual thermodynamic efficiency loss which does not apply if a thermal cycle is not involved.

If the electricity demand is met by a wide-ranging mix of energy sources then this technology is not so important since, as discussed in Chapter 9, sources which are difficult to restart or which produce energy at awkward times can be included in the base-load, the peak being supplied by more costly but more flexible technologies.

Small-scale Heat Storage

A very large fraction of the world's energy supply is used to provide heat in the winter and cooling in the summer to our houses and workplaces. The ideal situation would be to have buildings so well insulated that their heat content would be a constant; heat could neither enter nor leave the building. In fact, if such a situation were possible it would be necessary to remove heat constantly since each occupant would be adding metabolic heat at a rate of at least 100 W. In very large buildings this load of metabolic heat is a significant part of the energy budget of the building. Its effect is usually to make the cooling energy requirement in the summer greater than the heating in the winter.

The next best situation would be to save the heat from the air conditioners in the summer by storing it in some way to be used in the winter. Many schemes of this type have been designed and carried out in buildings ranging from the family home to large office complexes. The problem always is: what medium is cheap enough and has a sufficiently high specific heat capacity to serve as a storage medium? With cost being so important the simplest answer has usually been to use rock or water.

Another method of heat storage that has attracted attention and some development is to utilize the latent heat of fusion of some substance that has a freezing point near the temperature of our normal living environment. A small scale example of this is the phase-change hand-warmers that can be purchased in any sporting goods shop. They consist of a plastic bag filled with a concentrated solution of sodium acetate $NaC_2H_3O_2{\cdot}3H_2O$. This substance melts at 58°C but can be supercooled to room temperature (20°C) without solidifying. When heat is required the contents are given a slight mechanical shock and the crystallization begins, releasing the latent heat and warming the bag to 58°C. It can be regenerated by boiling the bag in water.

Table 18-2: Properties of Water, Rock and Glauber's Salt

Substance	Density (kg/m^3)	Specific Heat (J·kg^{-1}·°C^{-1})	Latent Heat of Fusion (J/kg)
Water	1000	4190	3.35×10^5
Rock (Granite)	2600	~300	N.A.
Glauber's Salt	1464	N.A	2.15×10^5

For heat storage in buildings the substance that has attracted the most attention is Glauber's Salt (sodium sulphate decahydrate $Na_2SO_4{\cdot}10H_2O$) which melts at 30.5°C and has a latent heat of fusion of 1.64×10^5 J/kg. In Table 18-2 the relevant properties of water, rock and Glauber's salt are given. The largest latent heat of fusion is that of water but its low melting temperature (0°C) makes it impractical for this use.

Table 18-3: Storage Capacity and Cost Relative to Water

Substance	Volume (m^3)	ΔT (°C)	Energy (J)	Mass (kg)	Cost Relative to Water
Water	1	20	8.3×10^7	1000	1
Rock (granite)	5.3	20	8.3×10^7	13800	4
Glauber's Salt	0.27	0	8.3×10^7	390	3

If the relative storage capacities and costs of these three are evaluated based on 1 m^3 of water the values shown in Table 18-3 are obtained. Examination of this table would indicate that the poorest choice is rock and yet that is the technology that has been best developed. Glauber's salt is very attractive and a flurry of research in the 1970s solved many of the application problems such as dehydration of a fraction of the salt with every cycle. However, as with many of the attractive technologies investigated in the 1970s, interest waned with the oil "glut" of the 1980s and 90s.

18.5 CLEANER COAL-BURNING

The one energy technology that produces the greatest number of environmental problems is the burning of fossil fuels, particularly coal. The production of carbon dioxide, of course, cannot be overcome by any improvement of technology short of the abandonment of fossil-fuel burning, and burning fossil fuels has many negative environmental impacts other than producing CO_2 (see Chapter 5).

When coal is burned on a large scale to produce electricity, it is first crushed to a powder and injected into the burner much as a liquid fuel would be handled. Indeed many such plants can burn powdered coal or oil with the same equipment. The powdered coal requires a large-volume reaction vessel in order that the high-temperature gases can have sufficient surface to deposit their heat in tubes which produce steam. Such furnaces are physically quite large and cannot be cheaply built at a factory and moved to the site; they must be built on-site. The temperature of combustion is very high (~1700°C) with a resulting high production of nitrogen oxides. The sulfur content of the coal must be removed from the flue gas in the form of SO_2. Improved furnace design should address several problems:

1. Remove the sulfur content before or during the burning process to eliminate expensive scrubbers.
2. Lower the combustion temperature to decrease the production of nitrogen oxides.
3. Improve the efficiency of heat transfer to the water tubes to make the furnace physically smaller.

Two new coal-burning technologies that address these concerns are *fluidized bed combustion* and *integrated gasification combined cycle furnaces.*

Fluidized Bed Combustion

The greatest mass in a fluidized bed furnace (Fig. 18-1) at any time is a charge of limestone or dolomite pellets which are kept in violent agitation by upward jets of hot air. The turbulent motion of the pellets is much like that of a boiling liquid, whence the name "fluidized bed." The powdered coal is fed into this hot "boiling" mass where it reacts with the oxygen in the air to release its energy. The bulk of the energy is transferred to the pellets where they, in turn, transfer it to the steam tubes, not in the furnace walls, but mostly in the bed itself. The hot gases also give up their heat to tubes on surfaces higher in the furnace. The heat transfer in the bed is much more efficient and so the furnace can be significantly smaller. In addition the combustion temperature is lower (~900°C) so there is much less production of nitrogen oxides.

The sulfur in the coal reacts with the limestone ($CaCO_3$) to form calcium sulphate ($CaSO_4$) as a crust on the stone. Between 150 and 300 kg of limestone are required to remove 90% of the sulfur from a tonne of coal. Of course this means that limestone must be mined as well as coal and transported to the plant site with the attendant increase in environmental impact and strain on the transportation systems. The crusting of the pellets renders them inactive and so they must be constantly removed and replaced.

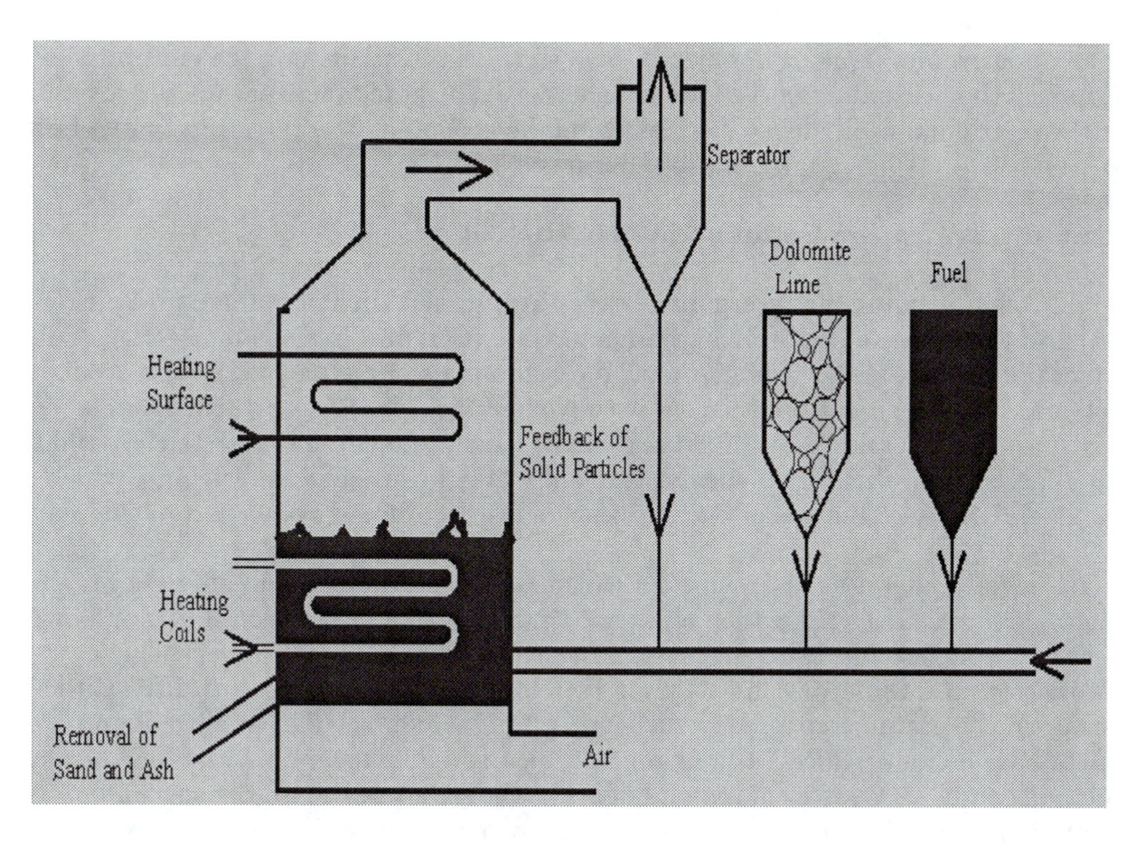

Figure 18-1 A Fluidized Bed Furnace (*Adapted from: F.D. Friedrich, Canmet Report 79-39*)

Some research has been carried out on processes to remove the crust but generally the pellets are discarded with the coal ash.

A further elaboration of the method is the "Pressurized Fluidized Bed." Here the whole reactor is pressurized to about 10 atm (1000 kPa). This permits more compact units for a given thermal output along with more efficient combustion of the fuel and removal of sulfur. The price paid is in a greater complexity of equipment to maintain the required pressure. The idea is, however, attractive as the high-pressure hot exhaust gases can, after cleaning, be fed directly to a turbine without generating steam.

In many cases the fluidized bed furnace is the most economical way to cope with the sulfur content of coal and the technology is now sufficiently well developed that many new installations or re-fittings are of this type. There is, however, another environmental price to pay. The formation of $CaSO_4$ is via the reaction

$$CaCO_3 + 3O + S \rightarrow CaSO_4 + CO_2 \qquad \textbf{[18-6]}$$

In other words, more atmospheric carbon dioxide is produced. One of the attractions of the fluidized bed process is its ability to handle high-sulfur coal which other furnaces cannot. But removing more sulfur means producing more carbon dioxide. In the 1970s when acid rain was causing concern, fluidized bed combustion was hailed by many, including environmental organizations, as the desired technology. In the 1980s as attention shifted to atmospheric CO_2 and global warming, the technology is now seen by many as undesirable. This is an object lesson in the shifting fashions in environmental concerns. In the present example, one is faced with the stark facts of burning fossil fuels: either the environmentally dirty business must be accepted, or alternatives must be found (and they may be equally harsh if not more so).

Integrated Gasification Combined Cycle (IGCC)

In an IGCC coal-burning plant, crushed coal is first mixed with hot steam and air (or oxygen) and the mixture produces carbon monoxide (CO) and hydrogen gas (H_2). This portion of the process is the "gasification;" if oxygen is used instead of air, the products are CO_2 and H_2. Any ash from the coal is converted to a glassy slag during the gasification. The CO and H_2 are burned in a gas turbine to generate electricity, and then the hot exhaust from this turbine is used to boil water to run a steam turbine, which also generates electricity. This use of two turbines is the "combined cycle" part of the plant.

There are several advantages to IGCC. Current IGCC plants have an efficiency for electricity generation of 45%, which is higher than that of a typical coal-burning power plant, and it is hoped that this will reach 60% as the technology improves. The pollutants from IGCC plants come out in forms that can be captured with little difficulty. The sulfur in the coal is emitted as hydrogen sulfide, which is easily removed, and instead of NO_X being produced, the nitrogen typically exists as ammonia which also can be captured easily. If O_2 is used instead of air in the gasifier, the highly concentrated CO_2 that is produced can be captured straightforwardly. Once the CO_2 has been captured, it could be sequestered in deep geological strata or in the ocean, and considerable research is underway to investigate these possibilities, as discussed in the following section.

IGCC technology is still in its infancy. As of September 2006, there are only two commercial-size coal-based IGCC plants in the United Sates and two in Europe.[3]

18.6 Carbon Dioxide Sequestration

The continuing increase in CO_2 concentration in the atmosphere and the serious climate-change problems that are resulting have been discussed in Chapter 6. One of the possible methods of reducing the rate of increase in CO_2 concentration is carbon sequestration, which refers to the capture and secure storage of CO_2 that would otherwise be emitted to, or remain in, the atmosphere. One general way in which CO_2 could be stored is by the enhancement of the uptake of CO_2 by natural sinks such as vegetation and soils. Another approach is to capture CO_2 from, for example, fossil-fuel-burning electrical plants and inject it into geologic strata or deep in the ocean.

Sequestration by Natural Sinks

Plants, soils and the ocean already absorb a great deal of CO_2. The process of photosynthesis in plants and the development of humus-rich soil removes about 2.8

[3] www.clean-energy.us

gigatonnes (Gt = 10^{12} kilograms) of carbon from the atmosphere annually.[4] Absorption by the ocean removes another 2 Gt of carbon per year, giving a total natural removal of 4.8 Gt C/year. Anthropogenic emissions are about 8 Gt C/year, and so natural sinks are removing about 60% of these at present.

There are several options for increasing the absorption and retention of carbon by plants and soils. Conversion of marginal agricultural lands back to natural systems would increase the amount of carbon in the soil. (When natural ecosystems such as forests are converted into agricultural areas such as cropland or grazing land, the organic carbon in the soil decreases by 50-80%.) The vegetative and soil carbon pool could also be enhanced by choosing fast-growing plant species with deep root systems, using no-till farming with incorporation of meadows in the planting cycle, restoring eroded and degraded soils, and using manure and other biosolids to improve nutrient content in the soil. The suggested changes in farming practices would have the added benefits of helping to increase crop output and reduce soil erosion and river sedimentation. How quickly these practices might be encouraged and adopted remains to be seen.

Capture and Geologic or Oceanic Storage of Carbon Dioxide

The largest potential sources of CO_2 for capture are fossil-fuel electric power plants, since they are normally built as large centralized units and emit huge quantities of CO_2, approximately one third of the CO_2 emissions worldwide.[5] There are three general categories of capture processes: flue gas separation, oxyfuel combustion, and precombustion separation. All of these alternatives are more expensive than using fossil fuels to generate electricity without capturing CO_2, and cost has been a major barrier to implementation.

Flue gas separation involves absorption of CO_2 by a liquid solvent in which the CO_2 becomes chemically bonded. Afterward, heating the solvent releases the CO_2 which can then be stored or used for various industrial processes such as urea production, carbonation of beverages, etc. Flue gas separation is currently employed at only about a dozen facilities in the world.[5]

In oxyfuel combustion, the fossil fuel is burned in pure oxygen or air that has been oxygen-enriched. The resulting combustion products are mainly CO_2 and H_2O, and the H_2O can easily be condensed, leaving the CO_2 readily available for storage. Of course, the oxygen must initially be separated from the nitrogen in the air, and this process consumes about 15% of a power plant's electricity output.[6] Pilot projects indicate that existing coal plants could be retrofitted straightforwardly for oxyfuel combustion.

Precombustion separation is the type of process carried out at IGCC plants (Section 18.5). The coal is first gasified, the CO_2 is separated, and then the H_2 undergoes combustion.

---

[4] R. Lal, Carbon Sequestration, Terrestrial, in *Encyclopedia of Energy*, Editor C.J. Cleveland, Elsevier Academic Press, New York, 2004, Vol. 1, p. 290

[5] H. Herzog & D. Golomb, Carbon Capture and Storage from Fossil Fuel Use, in *Encyclopedia of Energy*, Editor C.J. Cleveland, Elsevier Academic Press, New York, 2004, Vol. 1, p. 278

[6] Ref. 5, p. 279

Once the CO_2 has been captured by any of the three procedures described above, it could be stored. There are several important criteria concerning storage: it must be secure for hundreds or thousands of years, its cost should not be prohibitively large, and environmental impact and the risk of accidents must be small. A number of possible types of sites are being proposed. One option is to use depleted oil and gas reservoirs and, indeed, injection of CO_2 into such reservoirs has been practiced for many years, especially in Western Canada. The purpose of these injections has been to dispose of "acid gas," which is a mixture of CO_2, H_2S, and other gaseous byproducts of oil and gas exploration and refining. CO_2 has typically been the largest component (approximately 90%) of the injected gases.

Captured CO_2 is also already being used for enhanced oil recovery, especially in the US, where CO_2 has been injected under pressure into oil fields to displace oil that has remained in porous rock after normal production of crude oil. This displaced oil can then be pumped easily to the surface.

Other possible sites for CO_2 storage are unminable coal seams, where the CO_2 injection can enhance the recovery of methane that is in the coal bed. As well, saline pools deep underground or undersea are potential locations for storing CO_2.

All of the options mentioned are being considered for long-term storage of CO_2. Areas of research and study that are underway include the investigation of the possibility of leaks and their environmental and public health impact, the potential for slow migration of CO_2 from storage sites, and possible increases in seismicity.

By far, the largest potential sink for CO_2 is the ocean. Already it holds about 40000 Gt C, compared with only 750 Gt C in the atmosphere and 2200 Gt C in the terrestrial biosphere.[7] Two principal methods are under study for possible injection of CO_2 into the ocean. One technique involves piping liquid CO_2 to a depth of 1500-3000 m for dispersal. The liquid CO_2 at this depth is less dense than seawater and will rise, eventually being absorbed before it reaches the surface. The second method is to inject the CO_2 at a depth below 3000 m, where the CO_2 is denser than seawater and will form a deep "lake." A major concern about either of these approaches is the effect of the CO_2 injections on marine organisms. Of particular worry is the resulting change in pH of the water, which is predicted by some researchers to produce a large increase in mortality rates of organisms in the injection region.[8] Duration of exposure and methods of dispersal are important considerations in the ongoing research.

At this time, it is not clear whether it will be wise to adopt CO_2 capture and storage to make an impact on the climate-change problem, and if so, how quickly the technologies can be executed. Environmental concerns as well as cost considerations will undoubtedly mean a slow and measured approach to any implementation.

---

[7] Ibid, p. 282

[8] T. Flannery, *The Weather Makers*, HarperCollins, Toronto 2005, p. 251

PROBLEMS

18-1. Consider a standard cylinder of hydrogen gas compressed to a volume of 44 L at a pressure of 150 atmospheres.
(a) What is the volume of this gas if it is released to the atmosphere? Assume hydrogen is an ideal gas.
(b) The density of liquid hydrogen is 71 g/L and gaseous hydrogen at 1 atm and 20°C has a density of 0.084 g/L. What volume of liquid hydrogen can be made from one cylinder?
(c) The energy content of liquid hydrogen is 120×10^6 J/kg. What is the energy content of 1 L of liquid hydrogen?
(d) The energy content of gasoline is 35×10^6 J/L. What is the gasoline equivalent of one cylinder of gaseous hydrogen?

18-2. An automobile fuel tank holds 80 L of gasoline.
(a) How large a tank would be required to hold enough liquid hydrogen to drive the car the same distance as the gasoline?
(b) What would be the mass of each load of fuel (excluding their containers)? The specific gravity of gasoline is 0.8.

18-3. The work W done in compressing a gas isothermally (constant temperature) from volume V_1 to V_2 is given by

$$W = \int_{V_1}^{V_2} P\,\mathrm{d}V$$

If the gas is ideal, then $PV = nRT$, where n is the number of moles and R is the gas constant (8.31 $\mathrm{J\cdot mol^{-1}\cdot K^{-1}}$).
(a) Show that $W = nRT\,\ell n(V_2/V_1)$.
(b) How much energy is required to fill a standard gas cylinder with hydrogen? (See problem 18-1 for relevant data).
(c) What percentage is this of its energy content?

18-4. Instead of pressurizing the 6600 L of hydrogen from a gas cylinder it is desired to cool it and liquefy it. The specific heat of hydrogen gas is 14×10^3 J/kg and the latent heat of vaporization of the liquid is 32×10^3 J/L; the liquefaction temperature of hydrogen is 20 K.
(a) What energy must be extracted from the gas to liquefy it?
(b) What percentage is this of its energy content?

18-5. Using the data of Table 18-2 derive the data of Table 18-3 (except for the cost).

ANSWERS

18-1. (a) 6600 L (b) 7.8 L (c) 8.5×10^6 J/L (d) 1.9 L
18-2. (a) 330 L (b) gasoline 64 kg, H_2 23 kg
18-3. (b) 3.7×10^6 J (c) 6%
18-4. (a) 2.4×10^6 J (b) 4%

EPILOGUE

It is extremely difficult to predict what lies ahead in the energy field. The direction that will be followed depends on political and economic situations, prices of various forms of energy, new developments in technology, and level of concern about the environment, especially as regards global warming.

Much of the interest in energy conservation and alternative sources of energy in the late 1970s and early 80s was the result of an increase in the price of crude oil between 1973 and 1982 from $3 (US) per barrel to $35 (in money-of-the-day)[1]. Most of this increase occurred in two distinct steps, each precipitated by political events: in late 1973 there was the Yom Kippur war between Egypt and Israel, and the Arab nations who then controlled the Organization of Petroleum Exporting Countries (OPEC) embargoed oil exports to the USA and other Western countries in order to put indirect pressure on Israel to settle territorial disputes. The price of crude oil roughly quadrupled, gasoline was in short supply (producing long lines of cars at gasoline pumps in the US), and an atmosphere of crisis surrounded the energy industry. Some people argue that it was not a coincidence that this embargo and the huge price increases occurred shortly after US domestic oil production had begun to decline after peaking in 1970. In 1979, a revolution in Iran led to temporary suspension of Iranian oil production and roughly a tripling of oil prices, with gasoline rationing in parts of the USA. Interest in alternative technologies grew dramatically, but the Three Mile Island (1979) and Chernobyl (1986) nuclear accidents brought the growth of the nuclear alternative to a halt in western countries.

In early 1986, oil prices plummeted to about $15 a barrel. Why this huge drop? To quote a newspaper of the day [2]: "The Organization of Petroleum Exporting Countries said late last year that it wanted a 'fair market share' of the world oil market, even if it meant price cutting and the risk of a price war. OPEC, which was seeing its market share cut dramatically as North Sea output increased, urged non-OPEC producers to cut production." As a result of the low price of oil, interest in alternative sources such as solar energy and wind energy waned quickly.

Only now in the early years of the 21st century has there been a rekindling of interest in these alternative technologies, driven this time by both oil supply issues and by the Kyoto Agreement. The price of oil has soared from the $20/ Bbl that prevailed through most of the 1990s to over $60 as this edition of our book went to press in 2006; this has produced huge price increases for gasoline, heating oil, aircraft fuel, etc. In addition, the price has become very unstable, reflecting both the difficulty of meeting increasing demand, and the influence of world politics and of natural disasters such as Hurricane Katrina on the US Gulf Coast in August 2005. There is once again an increasing awareness of the pressing need to investigate other energy sources. Hubbert's picture of both oil and gas supplies has gained vastly increased acceptance, and even some oil companies are diversifying to new energy types. The Kyoto Agreement is part of the reason for the dramatic expansion of wind energy in Europe, and for the strides made

---

[1] *BP Statistical Review of World Energy*, June 1985, p. 13, and July 1989, p. 14.

[2] *Globe and Mail*, January 23, 1986.

there in using energy more efficiently, far more so than in North America. Wind power is on the verge of being economically competitive, although it is unsuited to baseload provision; solar power has a long way to go. It is probably safe to predict that geothermal, tidal and wave power systems will benefit local situations but will have a negligible impact on the global appetite for energy in the first half of the 21st century.

A renaissance of nuclear power is widely predicted, in part because of its lack of greenhouse gas emissions, but the issue of waste storage remains a concern and the issue of nuclear material getting into the wrong hands in a new age of global terrorism attracts much attention. Different countries have very different views: France continues with its highly successful nuclear program. Germany decided to phase out its reactors but is now hesitating. Britain let its nuclear expertise run down, and is now faced with difficult choices as the end of North Sea oil becomes discernible. Several Asian countries are forging ahead with extensive nuclear programs.

A major factor that was not visible when we first published this book was the development of the economies of two emerging giants – India and China. Their use of coal and imported oil is growing at incredible rates, increasing the draw-down of global oil supplies. There is the potential here to undo the greenhouse gas reductions that are made in more developed countries. But why should anybody expect poor people in these countries to forego the chance for the living standards that they deserve?

In 2005, the G8 leaders and their energy ministers asked the International Energy Agency to advise them on future energy strategies. The IEA Report[3] projects an alarming situation if nothing changes: despite expected improvements in energy use, a shift to coal for electricity generation and as a source of liquid fuel for transport will increase the atmospheric carbon concentration by a factor 2.5 between 2005 and 2050. To avert this situation and return carbon concentration to its 2005 value would require accelerated development of a wide range of technologies; any individual technology on its own affords no chance of attaining that goal. The IEA describes various scenarios which have different balances among their component technologies. Massive improvements in the efficiency of buildings, industry and transport are urgently needed. New technologies to capture and store CO_2 from coal burning are fundamental. Advances to a new generation of nuclear power, and the accelerated development of wind and solar energy are also needed. Finally, the use of bio-fuels and hydrogen fuel cells in transportation is crucial.

We have discussed all these technologies in this book. But now a global collab-oration is required, involving both developed and developing countries. Whether this is possible in the current political climate is a big question. Whether politicians are ever capable of the required long-term commitments is quite another matter.

However, whichever energy futures are followed by individual regions of the world, one thing remains certain: as long as human population continues to grow, eventually there will simply be too many of us to extract energy from the limited resources available without doing irreversible environmental damage.

[3] *Energy Technology Perspectives*, International Energy Agency, 2006

APPENDIX I

WEB-BASED RESOURCES

The advent of the World Wide Web (WWW) has made available many excellent and even entertaining resources to aid in the learning of science in an interactive and dynamic way. The difficulty of including the addresses of these resources in a published book is that the sites are often ephemeral. By the time the book is published, many sites active during the writing period will have disappeared or been altered, and new ones will have appeared.

We have chosen, therefore, to list only one site which will be maintained in an active state by the authors and the Department of Physics at the University of Guelph.

http://www.physics.uoguelph.ca/energy/

On this site will be found links to excellent tutorials suitable for study using this book. The available material includes several *remedial tutorials* such as one on UNIT CONVERSIONS and interactive activities suitable for many of the chapters in the book. At the site they are grouped by chapter after the list of remedial tutorials, and include a brief description of the activity and what the student may expect to see. Many of the activities are "applets", which is a term applied to interactive applications which can be accessed and manipulated on the Web. These applets introduce the student to a new form of tutorial and are highly recommended as a strong reinforcement of many concepts.

Very importantly, a list of the inevitable ERRATA will be posted on this site as soon as they are found and reported. The authors welcome such reports, suggestions, comments or new web-based tutorials. They can be reached at the e-mail link on the web-site or at:

energy@physics.uoguelph.ca

APPENDIX II

SEMI-LOGARITHMIC GRAPH PAPER

Notice that this paper has normal spacings on the x-axis, but that equal distances on the vertical axis correspond to *multiplication* by a given number. For example, the distances between 1 and 2, 2 and 4, and 4 and 8 are all equal, corresponding to a multiplication by two. On the linear x-axis, equal distances indicate *addition* by a given number.

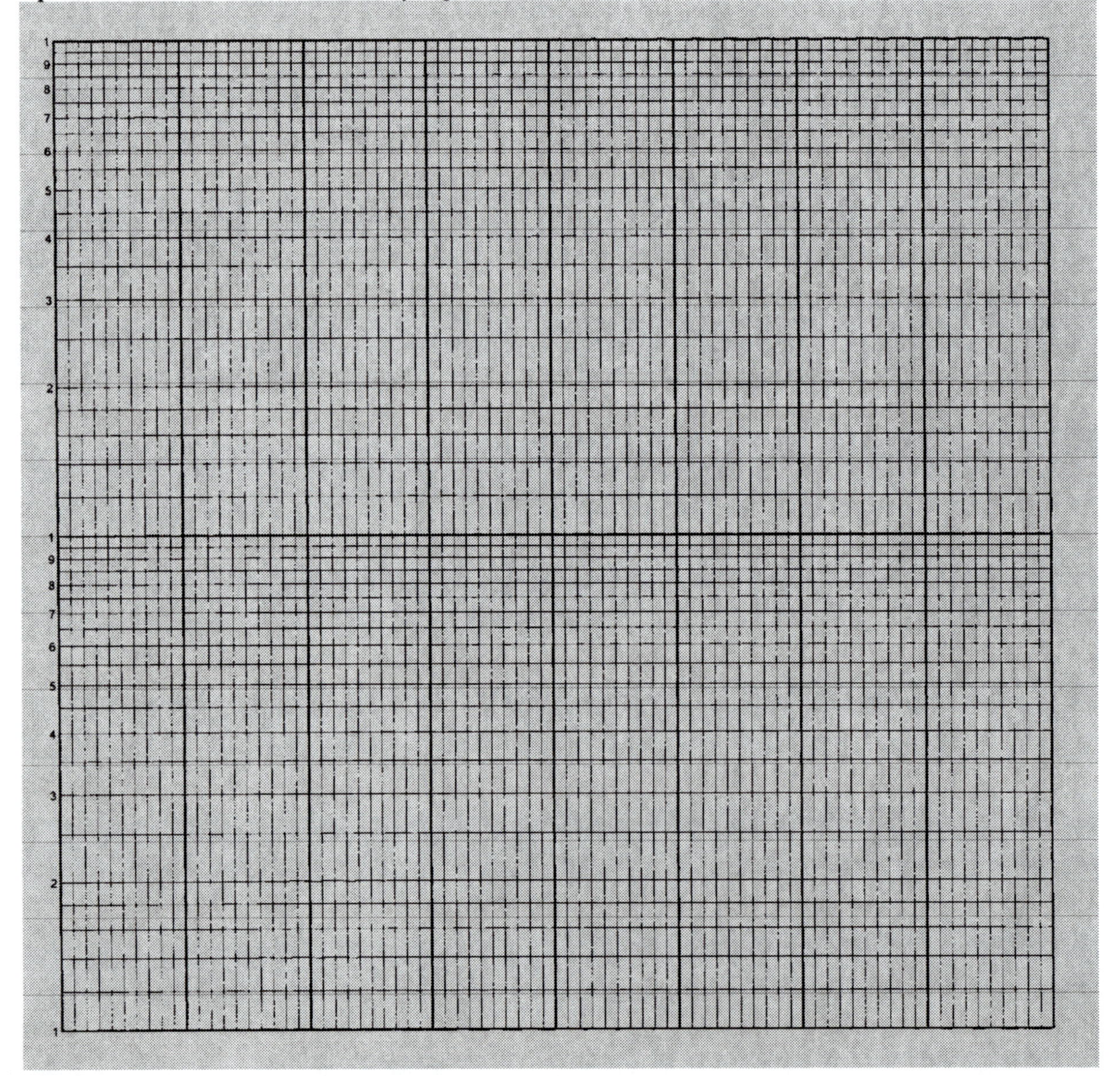

APPENDIX III

Properties of Fossil Fuels

Energy Content of Fossil Fuels

Fuel	**Energy Content**
Ethyl alcohol	3×10^6 J/kg
Coal	$(17 - 29) \times 10^6$ J/kg
Gasoline	48×10^6 J/kg, or
	35×10^6 J/L
Natural gas	38×10^6 J/m^3
Wood	$(17 - 19) \times 10^6$ J/kg
Kerosene (jet fuel)	40×10^6 J/kg
Oil	42×10^6 J/kg
1 barrel of oil	5800×10^6 J

Properties of Coal

	Anthracite	Bituminous	Sub-bituminous	Lignite
Moisture %	4.4	2 - 6	~20	~40
Volatiles %	4.8	20 - 40	~30	~30
Carbon %	82	65 - 45	~40	~30
Ash %	9	12 - 4	10 - 8	~5
Energy J/kg	30×10^6	30×10^6	22×10^6	16×10^6

APPENDIX IV

NUMERICAL CONSTANTS, ETC.

FUNDAMENTAL PHYSICAL CONSTANTS

Name	*Symbol*	*Value*
|Charge of electron|	e	1.602×10^{-19} C
Avogadro's number	N_A	6.022×10^{23} mol^{-1}
Planck's constant	h	6.626×10^{-34} J·s
Atomic mass unit	u	1.661×10^{-27} kg
Speed of light in vacuum	c	2.998×10^{8} m/s
Mass of electron	m_e	9.109×10^{-31} kg
Mass of proton	m_p	1.673×10^{-27} kg
Mass of neutron	m_n	1.675×10^{-27} kg
Boltzmann's constant	k_B	1.381×10^{-23} J/K
Molar gas constant	R	8.315 J·mol^{-1}·K^{-1}

OTHER USEFUL CONSTANTS

Name	*Value*
Density of water	1.00×10^{3} kg/m^3
Specific heat of water	4186 J·kg^{-1}·K^{-1}
Latent heat of vaporization of water	2.26×10^{6} J/kg
Latent heat of fusion of water	3.33×10^{5} J/kg
Average radius of Earth	6.37×10^{6} m
Average Earth-Sun distance	1.50×10^{11} m

MOLAR MASSES (in g/mol)

Element	*Molar Mass*
C	12
O	16
S	32
Ca	40

ENERGY UNIT CONVERSIONS

1 cal = 4.184 J (exact)

1 Cal = 1 food calorie

= 1 kcal

= 4184 J

= 4.184 kJ

1 Btu = 1055 J

1 quad = 1 quadrillion Btu's = 10^{15} Btu

1 eV = 1.60×10^{-19} J

1 barrel of crude oil is approx. equal to 5.8×10^{9} J
1 tonne of crude oil is approx. equal to 4.2×10^{10} J

SI PREFIXES

Factor	*Prefix*	*Symbol*	*Factor*	*Prefix*	*Symbol*
10^{18}	exa	E	10^{-1}	deci	d
10^{15}	peta	P	10^{-2}	centi	c
10^{12}	tera	T	10^{-3}	milli	m
10^{9}	giga	G	10^{-6}	micro	μ
10^{6}	mega	M	10^{-9}	nano	n
10^{3}	kilo	k	10^{-12}	pico	p
10^{2}	hecto	h	10^{-15}	femto	f
10^{1}	deka	da	10^{-18}	atto	a

APPENDIX V
RECOMMENDED FURTHER READING

The publication "Annual Reviews of Energy" offers a selection of in-depth articles by specialists. While the level of technical difficulty in some cases exceeds that of the present book, many articles can be appreciated in full or in part, and they in turn supply many references. Journals such as 'Scientific American", "Endeavour", "New Scientist", "Science" and "American Scientist" offer a wide variety of articles on energy topics. These range from refereed original works by experts to articles that are essentially news stories on current developments. A selection is given below. In addition, some journals on occasion present a special issue with extended coverage of a particular subject. In September 2006, Scientific American published a special issue (volume 295 #3) entitled "Energy's Future Beyond Carbon;" this is excellent supplementary reading material.

CHAPTER 1

N. Smith, *The origins of the water turbine.* Scientific American 242, 138-148 (1980).

C. Starr, M.F. Searl and S. Alpert, *Energy sources: a realistic outlook.* Science 256, 981-987 (1992).

CHAPTERS 2 and 3

A.A. Bartlett, *Forgotten Fundamentals of the Energy Crisis,* (20th Anniversay of this Paper), 1998: www.npg.org/specialreports/bartlett_index.htm

C.J. Campbell and J.H. Laherrère, *The End of Cheap Oil,* Scientific American 278, 78-83 (1998).

R.L. George, *Mining for Oil,* Scientific American 278, 84-85 (1998).

R.N. Anderson, *Oil Production in the 21st Century,* Scientific American 278, 86-91 (1998).

K.S. Deffeyes, *Beyond Oil — The View from Hubbert's Peak,* Hill and Wang Publ., New York, 2005.

CHAPTER 4

H.S. Geller and D.B. Goldstein, *Equipment Efficiency Standards: Mitigating Global Climate Change at a Profit,* Physics and Society 28 (2), 1-5 (1999)

CHAPTER 5

V.A. Mohnen, *The challenge of acid rain.* Scientific American 259, 30-38 (1988).

CHAPTER 6

R.A. Houghton and G.M. Woodwell, *Global climate change.* Scientific American 260, 3644 (1989).

T.E. Graedel and P.J. Crutzen, *The changing atmosphere.* Scientific American 261, 58-69 (1989).

R.M. White, *The great climate debate.* Scientific American 263, 36-45 (1990).

D.M. Etheridge, L.P. Steele, R.L. Langenfelds, R.J. Francey, J.-M. Barnola, V.I. Morgan. *Natural and anthropogenic changes in Atmospheric CO_2 over the last 1000 years from Antarctic ice and firn.* Journal of Geophysical Research, Vol. 101, 4115-4128 (1996).

P.D. Jones, M. New, D.E. Parker, S. Martin, I.G. Rigor, *Surface air temperature and its changes over the past 150 years.* Reviews of Geophysics, Vol. 37, 173-199 (1999).

P.A. Stott, S.F.B. Tett, G.S. Jones, M.R. Allen, J.F.B. Mitchell, G.K. Jenkins, *External contro lof 20th century temperature by natural and anthropogenic forcings.* Science Vol. 290, 2133-2137 (2000).
W.G. Graham and J.L. Hunt, *Radiation in the Environment.* Dept. Of Physics, Univ. Of Guelph (1996)
J. Hansen, *Defusing the Global Warming Time Bomb*, Scientific American 290 (3), Mar. 2004, 68-77.
T. Flannery, *The Weather Makers*, HarperCollins Publ., Toronto, 2005.
S.C. Doney, *The Dangers of Ocean Acidification,* Scientific American 294 (3), Mar. 2006, 58-65.
G. Stix, *A Climate Repair Manual*, Scientific American 295 (3), Sept. 2006, 46-49.

CHAPTER 8

J.W. Coltmann, *The transformer.* Scientific American 258, 86-96 (1988).

CHAPTER 9

R. Stone, *Polarized debate: EMFs and cancer.* Science 258, 1724-1725 (1992).
Health effects of low-frequency electric and magnetic fields. Environmental Science and Technology 27, 42-58 (1993): this feature includes the executive summary of the Oak Ridge Associated Universities Report, followed by commentary by D.A.Savitz and T.S.Tenforde.
A.L. Malozemoff et al, *High-Temperature Cuprate Superconductors Get to Work,* Physics Today 58 (4), April 2005, 41-47.
P.M. Grant et al, *A Power Grid for the Hydrogen Economy,* Scientific American 295 (1), July 2006, 76-83.

CHAPTERS 11 and 12

J.J. Taylor, *Improved and safer nuclear power.* Science 244, 318-325 (1989).
M.W. Golay and N.E. Todreas, *Advanced light-water reactors.* Scientific American 262, 82-89 (1990).
R.S. Yalow, *Radiation and Society.* Interdisciplinary Science Reviews 16, 351-356 (1991).
A.C. Upton, *Health effects of low-level ionizing radiation.* Physics Today 44, 34-39 (1991).
J.F. Ahearne, *The future of nuclear power.* American Scientist 81, 24-35 (1993).
(Various Authors) *Radiation in Medicine.* Physics in Canada (special edition). 51, No.4, (1995)
W.J. Hannum et al, *Smarter Use of Nuclear Waste,* Scientific American 293 (6), Dec. 2005, 84-91.

CHAPTER 13

J.A. Fillo and P Lindenfeld, *Introduction to Nuclear Fusion Power and the Design of Fusion Reactors.* American Association of Physics Teachers (1983).
M.G. Haines, *Tokamak physics.* Contemporary Physics 25, 331-353 (1984).
R.S. Craxton, R.L. McCrory and J.M. Soures, *Progress in laser fusion.* Scientific American 255, 68-79 (1986).
R.W. Conn, V.A. Chuyanov, N. Inoue and D.R. Sweetman, *The international thermonuclear experimental reactor.* Scientific American 266, 103-110 (1992).

J.G. Cordey, R.J. Goldston and R.R. Parker, *Progress Toward a Tokamak Fusion Reactor*. Physics Today Jan. 1992, 22-30.

J.A. Lake et al, *Next-Generation Nuclear Power*, Scientific American 286 (1), Jan. 2002, 73-81.

J.M. Deutch & E.J. Moniz, *The Nuclear Option*, Scientific American 295 (3), Sept. 2006, 76-83.

CHAPTER 14

T.R. Penney and D. Bharathan, *Power from the sea*. Scientific American 256, 86-92 (1987).

D.M. Kammen, *The Rise of Renewable Energy*, Scientific American 295 (3), Sept. 2006, 84-93.

CHAPTER 15

Frank Bason, *Energy and Solar Heating*. American Association of Physics Teachers, (1983)

L. Hodges, Ed., *Solar Energy: Book II, Selected Reprints*. American Association of Physics Teachers, (1986)

R. Winston, *Non-imaging optics*. Scientific American 264, 76-81 (1991).

I. Dostrovsky, *Chemical fuels from the sun*. Scientific American 265, 102-107 (1991).

AJ. McEvoy, *Outlook for solar photovoltaic electricity*. Endeavour 17, 17-20 (1993).

J.G. McGowan, *Tilting towards windmills*. Technology Review, July 1993, 40-46.

C. Smith, *Revisiting Solar Power's Past*. Technology Review, July 1995, 38-47.

CHAPTER 16

R.C. Marlay, *Trends in industrial use of energy*. Science 226, 1277-1283 (1984).

A.H. Rosenfeld and D. Hafemeister. *Energy-efficient buildings*. Scientific American 258, 78-85 (1988).

E.K. Jochem, *An Efficient Solution*, Scientific American 295 (3), Sept. 2006, 64-67.

CHAPTER17

Gene Waring, *Energy and the Automobile*. The Physics Teacher, Oct. 1980, 494-503.

J. Manassen, *The new role of rechargeable batteries*. Endeavour 16, 164-166 (1992).

S.R. Ovshinsky, M.A.Fetcenko and J. Ross. *A nickel metal hydride battery for electric vehicles*. Science 260, 176-181 (1993).

J. DeCicco and M. Ross, *Improving Automobile Efficiency*. Scientific American, Dec. 1994, 52-57

W. Drenchhahn and H-E. Vollmar, *Fuel Cells for Mobile and Stationary Applications*. Siemens Review Fall 1995, 20-23.

D. Sperling, *The Case for Electric Vehicles*. Scientific American, Nov. 1996, 54-59.

S. Ashley, *On the Road to Fuel-Cell Cars*, Scientific American 292 (3), Mar. 2005, 62-69.

J.J. Romm & A.A. Frank, *Hybrid Vehicles Gain Traction*, Scientific American 294 (4), Apr. 2006, 72-79.

J.B.Heywood, *Fueling Our Transportation Future*, Scientific American 295 (3), Sept. 2006, 60-63.

J. Ogden, *High Hopes for Hydrogen*, Scientific American 295 (3), Sept. 2006, 94-101.

CHAPTER 18

R.S. Claassen and L.A. Girifalco, *Materials for energy utilization.* Scientific American 255, 103-118 (1986).

C.L. Gray, Jr. and J.A. Alson, *The case for methanol.* Scientific American 261, 108-114 (1989).

E. Corcoran, *Cleaning up coal.* Scientific American 264, 106-116 (1991).

The future for coal. A multi-author survey published in New Scientist 137, 20-41 (1993).

S.A. Fouda, *Liquid Fuels from Natural Gas*, Scientific American 278, 92-95 (1999).

R. Lal, *Carbon Sequestration, Terrestrial*, in Encyclopedia of Energy, Editor C.J. Cleveland, Elsevier Academic Press, New York, 2004, Vol. 1, 289-298

H. Herzog & D. Golomb, *Carbon Capture and Storage from Fossil Fuel Use*, in Encyclopedia of Energy, Editor C.J. Cleveland, Elsevier Academic Press, New York, 2004, Vol. 1, 277-287

T.L. Johnson & D.W. Keith, *Fossil electricity and CO_2 Sequestration: how natural gas prices, initial conditions and retrofits determine the cost of controlling CO_2 emissions,* Energy Policy 32 (2004), 367-382.

M.L. Wald, *Questions about a Hydrogen Economy,* Scientific American 290 (5), May 2004, 66-73.

R.H. Socolow, *Can We Bury Global Warming?*, Scientific American 293 (1), July 2005, 49-55.

D.G. Hawkins et al, *What to do about Coal,* Scientific American 295 (3), Sept. 2006, 68-75.

R.H. Socolow & S.W. Pacala, *A Plan to Keep Carbon in Check,* Scientific American 295 (3), Sept. 2006, 50-57.

W.W. Wayt Gibbs, *Plan B for Energy*, Scientific American 295 (3), Sept. 2006, 102-114.

EPILOGUE

P.H. Abelson, *Energy futures.* American Scientist 75, 584-593 (1987).

J.H. Gibbons, P.D. Blair and H.L. Gwin, *Strategies for energy use.* Scientific American 261, 136-143 (1989).

H.M. Hubbard, *The real cost of energy.* Scientific American 264, 36-42 (1991).

APPENDIX VI
SYMBOLS

"η"	efficiency (refrigerator)
α	albedo
α	alpha particle
β	beta particle
β	fraction of delayed neutrons
γ	gamma ray
ε	emissivity
η	efficiency
η	efficiency
λ, λ_p	physical decay constant
λ_b	biological decay constant
λ_e	effective decay constant
μ	linear attenuation coefficient
μ_m	mass attenuation coefficient
ρ	electric resistivity
ρ	density
σ	standard deviation
σ	Stefan's constant
σ	cross-section
τ	confinement time
τ	neutron lifetime
φ	magnetic flux
ω	angular frequency
A	area
A	atomic (mass) number
A	activity
A	wave amplitude
B	magnetic field strength
c	specific heat
c	speed of light
C	growth constant
C.O.P.	coefficient of performance
D	absorbed dose
e	fuel consumption
e	electron
E	electric field strength
E	energy
E_T	effective dose
f	frequency
$\mathbf{F}$	force vector
g	acceleration due to gravity
h	Planck's constant
h	conduction-convection parameter
H_T	equivalent dose
I	electric current
I	moment of inertia
I	intensity
J	heat flux
k	radius of gyration
k	thermal conductivity
k	reactor constant
k	growth constant
k_B	Boltzmann's constant
KE	kinetic energy
ℓ	length
L_F	latent heat of fusion
L_V	latent heat of vaporization
$L_{½}$	half thickness
m	mass
n	number density

n	neutron
N	quantity
N	rate of production
N	number of nuclei
N_0	initial quantity
N_a	Avogadro's number
N_M	value at maximum
p	proton
p	pressure
P	power
q	fuel heating value
Q	heat quantity
Q, q	electric charge
Q_D	resource discovered
Q_F	quality factor
Q_P	quantity of resource
Q_R	reserves
Q_∞	total quantity of resource
r	Internal resistance
R, r	radius
R	universal gas constant
R	growth rate
R	electric resistance
R	R-value
RBE	Relative Biological Effectiveness
n	number of moles
S	entropy
t	time
T	temperature
T	wave period
T_2	doubling time
T_M	time of maximum
$T_{½}$, $T_{½,p}$	physical half life
$T_{½,b}$	biological half life
$T_{½,e}$	effective half life
u	atomic mass unit
U	potential energy
v	speed
V	volume
V	electric potential
W	work
w_R	radiation weighting factor
w_T	tissue weighting factor
Y	Young's modulus
z	time interval measured in standard deviations
Z	atomic number

APPENDIX VII
ACRYONYMS

AC	Alternating current
AGR	Advanced Gas cooled Reactor
B(a)P	Benzo(a)pyrene
BWR	Boiling Water Reactor
C.O.P.	Coefficient of performance
CANDU	Canadian Deuterium Uranium
DC	Direct current
DNA	d-oxy ribo nucleic acid
E-M	Electromagnetic
ECCS	Emergency Core Cooling System
HDD	Heating-degree-days
HEU	Highly Enriched Uranium
HTGR	High Temperature Gas cooled Reactor
HVAC	Heating Ventilating and Airconditioning
ICRP	International Commission on Radiation Protection
ITER	International Thermonuclear Energy Reactor
JET	Joint European Torus
LET	Linear Energy Transfer
LEU	Lightly Enriched Uranium
LOCA	Loss of Coolant Accident
MAGNOX	A Be-Mg alloy
MOX	Mixed Oxide
MWe	Megawatts of electricity
NASA	National Aeronautics and Space Administration
NIF	National Ignition Facility
NIMBY	Not in my back yard
NO_X	Nitrogen oxides
NRU	Canadian research reactor
NRX	Canadian research reactor
PM	articulate matter
PWR	Pressurized Water Reactor
RBE	Relative Biological Effectiveness
RME	Rape methyl ester
RMS	Root mean square
TMI	Three Mile Island
URL	Underground Research Laboratory
VOC	Volatile organic compound

APPENDIX VIII

THE THREE MILE ISLAND AND CHERNOBYL NUCLEAR ACCIDENTS

Three Mile Island

The first step leading to this accident in 1979 occurred when the feed water pumps supplying water to the steam generators for conversion to steam shut down (for reasons that are not clear), leading to automatic shutdown of the reactor and the steam-generating system. A loss-of-coolant accident was in progress, and the automatic safety systems came into action. Three auxiliary feed water pumps began to operate, but the water could not reach the reactor because two valves had incorrectly been left closed after maintenance. Heat from fission products in the fuel caused the temperature and pressure in the reactor core to rise, and a pressure relief valve opened, and the reactor was shut down by insertion of the control rods. The relief valve should have closed within a few seconds, but it stuck open, allowing coolant water to drain from it. After a couple of minutes, emergency high-pressure water pumps began operating automatically, but these were cut back by the operators, who were confused by an instrument that reported the relief valve was now closed (it was not). The valve remained open long enough for major loss of water to occur, and so part of the core became uncovered, the fuel cladding failed and a portion of the fuel melted. Radionuclides from the fuel passed into the coolant water, which was draining from the pressure relief valve and filling a drain tank. A seal on this tank ruptured, and water began to collect in the reactor basement. Finally, operators recognized what was happening, the relief valve was closed, and about six hours after the accident began, the core was again covered with water. The core temperature had risen so high that zirconium in the fuel cladding reacted with steam to produce hydrogen. Roughly 24 hours after the accident had begun, a small hydrogen explosion occurred, but produced little damage.

The accident liberated a very small amount of radioactivity into the environment. When the contaminated cooling water flowed into the reactor basement, pumps turned on and sent this fluid into an auxiliary building, from which some radioactivity was released into the atmosphere. The impact on human health in terms of cancer and leukaemia was too small to measure.

Chernobyl

The disaster in April 1986 at the Chernobyl reactor in the Ukraine region of the former USSR was vastly more serious than the TMI accident. An explosion and fire released a great deal of radiation to the environment, and a number of people were killed in the initial blast. The reactor involved was one of four at the plant, each producing about 3200 MW of thermal power, generating about 1000 MWe. Each reactor, a BWR with graphite moderator, had two 500-MW turbine-generators for the production of electricity, and was housed in a separate building having just a standard industrial factory roof instead of concrete and steel containment. (There was better containment on the sides and bottom of the reactor.) The shutdown rods were extremely slow-moving, requiring about 18 s to come into place.

The control of the reactor was much more manual than in a Western plant, and the operators had no specific training for dealing with accident situations. Nor did they have training on simulators, which is common practice in Western countries. There were no independent safety audits, and operators were allowed to work 36-hour shifts. The reactor that exploded had been very reliable and had a high capacity factor.

The accident occurred as operators were doing a test on one of the turbine generators; they wanted to determine its ability to generate electricity while "spinning down" after the steam supply was shut off. The reason for doing this was that some of the pumps in the emergency core-cooling system were normally powered by the turbine, and in the event of a reactor shutdown it was desirable that this system could still be powered by the turbine for a short time (30 s or so) until backup diesel generators could come into operation. The reactor staff were eager to complete the test successfully on this particular occasion — the reactor was to be shut down for scheduled maintenance in a few days, and future tests would be delayed.

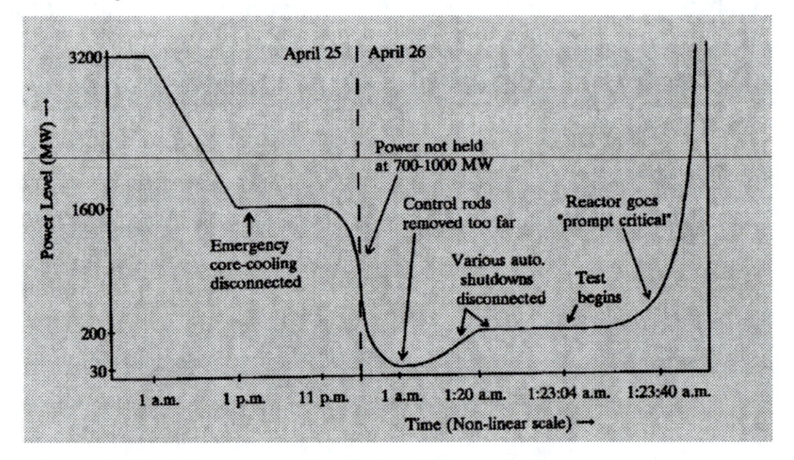

Figure AVIII-1 Sequence of events at Chernobyl

Figure AVIII-1 shows the sequence of events. At 1 a.m. on April 25, the power began to be reduced from its normal level of 3200 MW and at about 4 am, the reactor was at half -power, and one of the two turbo-generators was disconnected.

Instead of continuing the power descent immediately, the operators had to maintain half-power until about 11 p.m. because of unexpected electrical demand. At 2 p.m. they disconnected the emergency core-cooling system, in violation of operating procedures, because they did not want this system coming on during the test and draining energy from the turbine.

At 11 p.m. the power descent resumed, but at 12:28 a.m. on April 26, the operators made an error in entering an instruction into the computer and instead of the power levelling off between 700 MW and 1000 MW, it continued to drop to 30 MW, which was much too low a power level for the test. At such low power, there is a build-up of the fission product ^{135}Xe in the fuel; this isotope strongly absorbs neutrons and that tends to decrease the power level even more. In addition, at the 30-MW level, the water in the core area was not boiling as much as usual, and absorption of neutrons by the liquid water was also tending to decrease the power. Therefore, in order to increase the power, at 1:00 a.m. the operators withdrew most of the control rods from the reactor core beyond the allowable limit and managed to get the power to 200 MW. This control-rod withdrawal was a huge violation of operating procedure; the control rods are also used as emergency shutdown rods, and if they are withdrawn beyond a certain limit, it takes too long for them to return to the core in case of an emergency. (As we discuss later, there was an additional unexpected problem that arose as a result of withdrawing the control rods too far.)

The operators were having other problems, since the reactor was not designed to run at such low power. They had to switch to manual control of the water flow returning from the turbine, and had to make many adjustments to get reasonable steam pressure and water flow. These adjustments were difficult, and the reactor was on the verge of shutting down automatically because of fluctuating flow and pressure. Because such a shutdown would abort the test, the operators disconnected the automatic shutdowns related to steam pressure and water flow (another violation of operating procedure). By 1:22 a.m. the operators felt that the reactor was as stable as it was going to be, and decided to begin the test. But first they turned off the safety device that shuts down the reactor when the turbine is disconnected, again violating proper procedures; they wanted the reactor not to shut down in case they needed to perform the test a second time.

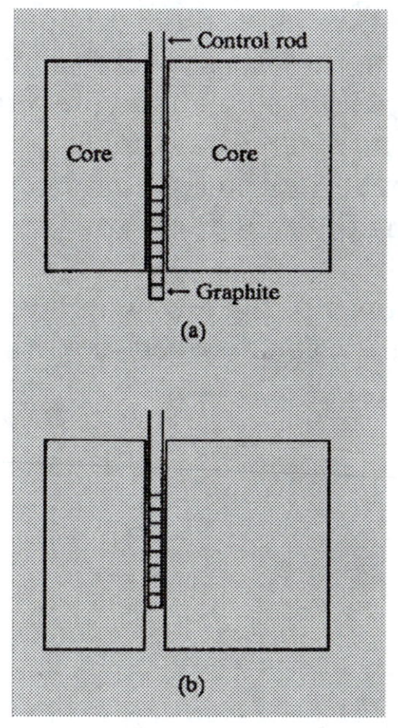

Figure AVIII-2 (a) normal control-rod position: (b) position prior to explosion

The reactor was now in a highly unstable state, requiring much more manual control than normal; it should have been shut down when the power reached 30 MW. Most of the control/shutoff rods had been withdrawn too far, and many of the automatic shutoff signals had been disabled. Recall also from Section 11.6 that a graphite-moderated water-cooled reactor has a positive void coefficient; as more water boils, the bubbles (voids) of vapour absorb fewer neutrons, and the fission rate accelerates. As long as the reactor is critical on the delayed neutrons, this situation is acceptable, but if it becomes "prompt critical" (able to continue a chain reaction using only prompt neutrons), then the fission rate increases uncontrollably. At 1:23:04 the turbine was disconnected to begin the test. As it slowed down, the regular water-coolant pumps that it powered also slowed down, decreasing the flow of cooling water over the core. This produced increased boiling in the water, and the power began to rise slowly, then more quickly. (Remember the positive void coefficient.) At 1:23:40 an operator noticed this increase and pushed the "scram" button to drive the emergency shutoff rods into the core. As will be explained in the next paragraph, because the shutoff rods had been withdrawn so far from the core, their movement back led not to a decrease in power, but to an increase. The reactor had become "prompt critical", and within four seconds, the power output rose to about 100 times normal.

The control/shutoff rods travel in vertical tubes, and in normal operation, part of the rod projects above the core, and part below (Fig. 2a). The rods for this particular type of reactor are rather unusual in design: the top portion is standard, constructed of a neutron-absorbing material, but the bottom section is graphite, a neutron moderator. Raising the rod removes neutron-absorbing material from the top of the core, and introduces graphite at the bottom, thus increasing the rate of fission. When the rod is lowered, graphite moves out of the core area at the bottom, neutron-absorber comes in at the top, and the reaction rate decreases. Just prior to the Chernobyl explosion, the rods had been raised so far that their bottoms were above the bottom of the core and the lower parts of the tubes contained water (Fig 2b). Water is both a neutron absorber and a neutron moderator, but compared to graphite, it is a more effective absorber than moderator. When the "scram" button was pushed and the rods were lowered (very slowly, remember), the effect in the bottom section was to displace water in the tubes with graphite, thus increasing the fission rate.

The sudden burst of heat in the fuel broke it into small pieces and caused the cooling water to boil extremely rapidly; the resulting high steam pressure blew the roof off the reactor building, sending flaming fuel and graphite onto the roofs of adjacent buildings. Local fire-fighters extinguished the exterior fires by 5 a.m., but many of them suffered lethal doses of radiation. However, graphite in the core was still burning and the updraft was carrying radioactive particles high into the atmosphere. Starting on April 28, military helicopters dropped about 5000 tonnes of boron, dolomite, sand, clay, and lead onto the exposed core. This decreased emissions for a few days, but also thermally insulated the core, raising its temperature and increasing emissions again. On May 4 and 5, nitrogen was pumped under pressure into the space beneath the reactor, cooling it effectively. The reactor is now entombed in a concrete sarcophagus; it is cooled by air, which is filtered afterward.

The impact of the massive radioactive release on human and environmental health was very large, as described in section 12.8.

Index